Evolution and Speciation in Fungi and Eukaryotic Biodiversity

T.J. Pandian
Valli Nivas, 9 Old Natham Road
Madurai 625014, TN, India

CRC Press
Taylor & Francis Group
Boca Raton London New York

CRC Press is an imprint of the
Taylor & Francis Group, an **informa** business

A SCIENCE PUBLISHERS BOOK

Cover page: Proposed phylogenetic tree of fungi. For details, see Fig. 12.3.

First edition published 2024
by CRC Press
2385 NW Executive Center Drive, Suite 320, Boca Raton FL 33431

and by CRC Press
4 Park Square, Milton Park, Abingdon, Oxon, OX14 4RN

Library of Congress Cataloging-in-Publication Data (applied for)

ISBN: 978-1-032-42141-4 (hbk)
ISBN: 978-1-032-42143-8 (pbk)
ISBN: 978-1-003-36135-0 (ebk)

DOI: 10.1201/9781003361350

Typeset in Palatino
by Radiant Productions

Preface

Fungi are sessiles as plants are, but heterotrophics as motile animals are. Besides, they have opted for external digestion, which has restricted them at surface dependent tissue-grade organization. The need for faster spore dispersal and avoiding rapid dilution of externally secreted digestive enzymes have almost limited them to terrestrial habitats. This book explores the strategies and adaptations of the fascinating fungi.

"For a long time, the idea of evolution was there among scientists and even with religions like Hinduism. With keywords "Variations, Struggle for existence and Survival of the fittest by Natural Selection", Charles Darwin established the theory of evolution and its by-product speciation. Subsequently, a large number of publications by microbiologists, botanists and zoologists have confirmed the correctness of Darwin's evolutionary theory. Presently, there are more concerns for species diversity than for evolution. The year 2010 marked the International Year of "Species Diversity". This book identifies some life history features of eukaryotes and environmental factors that accelerate species diversity and others that decelerate it. Some of these features and factors are known but are not adequately recognized. That requires quantification of the identified factors and features."

"In Shakespearean language, one may say, 'Oh, variation, thy name is evolution'. Hence, the idea of quantification of the identified factors and features may look odd and not possible at a time, when information on *per se* is not known for many species and when taxonomy of eukaryotes itself is in a fluid but dynamic state. However, I was a little emboldened, as taxonomy itself represents quantification of species, genus and so on, despite variation(s) among individuals within a species. The onerous task of quantification required a great deal of computer search and a few compromises on the number of some taxa". In addition to website citations, the search included 490 publications covering 750 species. "Yet, the quantifications may neither be exhaustive nor precise. But the proportions arrived and inferred generalizations shall remain valid. A separate chapter to highlight new findings are not included, as there are too many (shown in italics) of them. The Holy Bible states: "Let your light so shine that people may see your good work and praise the Lord". Being innovative and informative, I earnestly hope that this book stands up to the Biblical statement."

September 2022   **T.J. Pandian**
Madurai 625 014

Acknowledgements

It is with pleasure that I wish to place on record my grateful appreciation to Drs. P. Murugesan, V. Sarma and E. Vivekanandan for reviewing the manuscript of this book and offering valuable comments. The manuscript was ably prepared by Mr. T.S. Surya, M.Sc. and I wish to thank him for his competence, patience and cooperation.

I sincerely thank many authors/publishers, whose published figures are simplified/modified/compiled/redrawn for an easier understanding. To reproduce original figures from published domain, I welcome and gratefully appreciate the open access policy of Acta Botanica Brasilica, Agronomy, Frontiers in Microbiology, Fungal Diversity, Fungal Biology and Biotechnology, Plant and Soil, PLoS Biology and PLoS Genetics. For advancing our knowledge on this subject by their rich contributions, I thank my fellow scientists, whose publications are cited in this book.

September 2022   **T.J. Pandian**
Madurai 625 014

Contents

Preamble

The last century witnessed two most important discoveries that characteristically unifies all eukaryotes. The first one is the Calvin cycle (Calvin, 1964) in plants, in which simple organic molecules like glucose and others are synthesized from water and carbon-dioxide, using solar energy, i.e., the simple molecules serve as 'batteries' to store solar energy and sustain metabolism. The second one is the discovery of Kreb's cycle (Krebs, 1940), in which simple molecules are decomposed to generate ATPs to sustain metabolism. The last two centuries were devoted to experimentally demonstrate the past geological climate, in which simple inorganic molecules combined to form amino acids and others, from which life emerged. Thanks to the towering biologists Charles Darwin (1809–1882) and Gregor Johann Mendel (1822–1884), the hypothesis of evolution and speciation was developed, which explains how the (Darwinian) variations or preferably, new gene combinations and their (Mendelian) inheritance have sustained evolution and led to species diversity. Subsequent experiments on microbes, plants and animals by several authors have all brought adequate evidences in support of the Darwinian hypothesis. And the hypothesis has now become a universally accepted theory of 'Origin of Species'. Presently, there are more concerns for species diversity than for evolution. In recognition of its importance, the year 2010 was marked as the International Year of Species Diversity. This innovative series is devoted to discover the causes for biodiversity in eukaryotes. From scattered relevant information followed by incisive analysis, it has brought to light several new findings.

Many authors consider that 'asexual' or clonal multiplication is derived from sexual reproduction. This is a wrong notion. The blue green algae namely the cyanobionts arose around 2.5 Billion Years Ago (BYA). Their reproduction is limited to clonal multiplication alone. To tide over unfavorable conditions, they produce cysts called akinetes. Sex was discovered ~ 2 BYA (Butlin, 2002). Subsequently, it was successfully manifested in microbes, plants and animals at different times during the geological past. During the checkered history of evolution, the supplementary role played by clonal multiplication has been progressively reduced. So much so, it occurs in all protozoans (except perhaps *Entamoeba*), 39% of plants and 1% of metazoans (Pandian, 2021b, 2022, 2023).

Animals are heterotrophs and motile. In them, motility facilitates mate searching and outbreeding. The ensuing sexual reproduction involves meiosis (segregation) during gametogenesis and recombination at fertilization. Though costly (Stelzer, 2015), meiosis and recombination generate new gene combinations that have led to rapid evolution and speciation. Consequently, animals are enriched with > 1.5 million species (see Pandian, 2021b). Conversely, plants are autotrophs and sessile. In most of them, the sessility has imposed clonal multiplication. However, ~ 74% of plants achieve cross pollination by engaging symbiotic motile animals and other strategies. A quarter of plant species are unable to generate as much as new gene combinations (see Pandian, 2022), as animals can do it. This inability of plants has limited their diversity to < 374,000 species (Fig. 1). Nevertheless, plants are more amenable to variety diversity than animals. For example, 1,664 variety/crop species have been generated by farmers during the last 10,000 year (y) history of agriculture. But it is limited to 277 variety/animal species during the last 25,000 y history of domestication (see Pandian, 2022). Hence, plants have compensated the limited species diversity by opting and amenable to variety diversity.

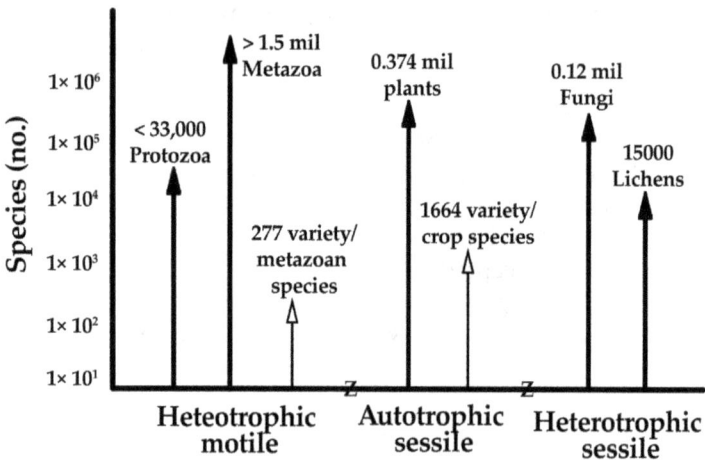

FIGURE 1

Distribution of species number in eukaryotic taxonomic groups. mil = million.

In metazoans, the combination of heterotrophy and motility has demanded the development of different tissues, organs and systems. On the other hand, the heterotrophic and motile protozoans have opted perhaps for the 'wrong' choice of performing almost all metazoan functions by specialization and differentiation of subcellular organelles. Briefly, to sustain heterotrophy and motility, the metazoans have opted for multicellularity but the protozoans for unicellularity. The first consequence of it is the limitation of living species number to 32,950 (see Fig. 1). Another consequence is the secondary loss of sexual reproduction by many protozoans. Approximately, 3,300 species or 10% protozoans can ill-afford sexual reproduction and can multiply clonally alone. Surprisingly, ~ 1,600 species (e.g., Volvocida, Cryptomonadida, Coccolithsophorida and 50% Dinoflagellida) or 5% flagellates are autotrophs but motiles (Pandian, 2022). It is from these ancient mastigophores, protozoans, fungi and metazoans are considered to have emerged. Barring the unusual high proportion of 33.8% parasites, the remaining free-living protozoans internally digest the captured food.

Surprisingly, the 0.12 million (Hawksworth and Lucking, 2018, however see p 204) speciose Fungi are heterotrophs (Fig. 1), as animals are but sessile, as autotrophic plants are. Unlike animals, they digest the food externally and absorb the digested micronutrients through the body surface. That the sessile Fungi are not as speciose as the motile animals indicates that species diversity in heterotrophic organisms have to necessarily depend on motility. The sessile but heterotrophic fungi have retained structural simplicity by opting for external digestion and acquisition of micronutrients. Irrespective of sessility, the multicellular 374,000 autotrophic plant species have developed structurally complex organs (roots, shoots and so on) and systems (vascular system inclusive of xylem and phloem). For example, the number of tissue types increases from < 7 in algae to 60 in flowering plants but from 6–7 in sponges and cnidarians to > 200 in mammals (Pandian, 2022). Remarkably, that the 374,000 speciose sessile plants have also undergone complex structural organization, especially in higher plants clearly indicates that sessility alone does not deter complex structural organization in multicellular plants. Hence, it is rather the unicellularity and structural simplicity that have limited protozoan diversity to < 33,000 species. Similarly, the adoption of heterotrophism demands complex structural organization, when the organisms choose to engulf larger food and digest it internally, as in animals. However, the other alternative chosen by fungi is to retain structural simplicity and acquire micronutrients after digesting the food externally. In other words, heterotrophism may demand motility and consequent complex structural organization, when the choice of the animals is internal digestion, as in animals or opt for external digestion that may not demand motility and associated complex structural organization, as in fungi.

Lichens are fascinating eukaryotes and represent the symbiotic association between sessile heterotrophic fungi and autotrophic algae (Fig. 1). Their diversity is limited to ~ 15,000 species (*Wikipedia*). In a way, they are the terrestrial equivalents of the aquatic combination of sessile heterotrophic cnidarians and autotrophic zooxanthellae. The coral – zooxanthellae are widely distributed in tropical and subtropical seas around the Earth up to the depth of 200 m. But that of lichens are limited between certain levels of altitude in montane ecosystem only (e.g., 3,100 and 3,400 m in Nepal, Baniya et al., 2010, Fig. 2B). In the context of trophic dynamics, the autotrophic plants are the producers; they occur and thrive in marine, freshwater and terrestrial habitats. So are the metazoans, which serve as consumers. The distribution of protozoans is, however, limited mostly to marine and freshwater habitats (Fig. 2A). Being decomposers, the fungi are more common on land than in water. The reverse is true for bacteria (Fig. 2B). It remains to be seen whether the Fungi are indeed less common in the aquatic system, as we know less of them from aquatic ecosystem. On the whole, the objective theme of this innovative book is to explore the causes for the limited species diversity in Fungi.

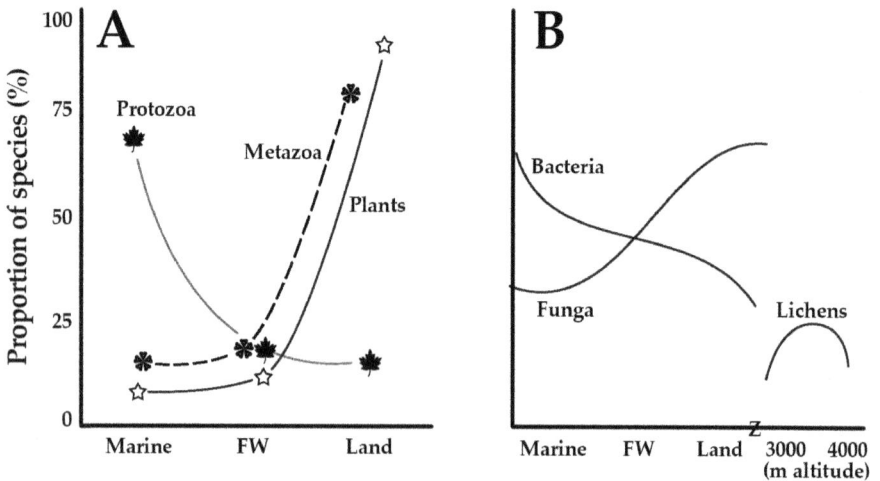

FIGURE 2

(A) Proportion of species distribution in marine, freshwater (FW) and terrestrial habitats. (B) Proportion of bacteria and fungi distribution in marine, freshwater and terrestrial habitats. The right end shows the distribution of lichens at different altitudes in montane systems.

Summary: 1. In heterotrophic motile Protozoa, unicellularity has limited diversity to < 33,000 species. 2a. Sessility has imposed clonal multiplication in > 39% autotrophic plants. b. It has reduced diversity to 374,000 species, c. which is partially compensated by their amenability to variety diversity. d. In them, multicellularity, has, however, facilitated structural complexity. 3a. In motile multicellular Metazoa, the combination of heterotrophy and motility has led to diversity of > 1.5 million species. b. Further, the need for food capture and internal digestion have demanded complex structural organization. 4. In contrast, that of heterotrophy and sessility has limited diversity to > 120,000 species in Fungi, which have opted for external digestion and absorption of micronutrients; in its turn, this has led to the retention of relatively simpler structural organization than plants and metazoans.

1

General Introduction

Introduction

Fungi were traditionally aligned with plants. The old classification schemes considered them as degenerated plants lacking chlorophyll (e.g., Evert and Eichhorn, 2013). As heterotrophs, fungi are more closely related to animals than plants. Unlike plants with a cellulose cell wall, their cell wall is chitinous like that of animals (e.g., arthropods). However, fungi differ from animals and plants in the life cycle, mode of nutrition, pattern of development and so on (see Cole and Baron, 1996). They are heterotrophic (saptorophic/osmotrophic, e.g., *Candida*), mostly sessile, tissue grade eukaryotes (McConnaughey, 2014, Voigt et al., 2021). They comprise agarics, chanterellus, molds, morels, mushrooms, penicillia, polypores, puffballs, rusts, smuts, stinkhorns, yeasts and so on (Fig. 1.1). They exist as immotile, unicellular yeasts and/or filamentous hyphae. Neocallimastigomycotes are motile and possess a whip-like flagellum. Of ~ 550 Gigaton Carbon (GtC) biomass on the Earth, the fungi constitute only 0.4% (Bar-On et al., 2018) with their diversity spanning over six phyla. They grow in almost all habitats on Earth, surpassed only by bacteria in their ability to withstand thermal extremes and availability of carbon sources (e.g., yeast, Raspor and Zupan, 2006). Water is an essential resource to facilitate (i) nutrient uptake, (ii) germination, (iii) penetration into the host tissue and (iv) dispersal in aquatic fungi. Most fungi thrive between 10° and 40°C with optimum of 25°–30°C. At higher than 40°C, they are thermophilics (cf Sharma, 1989) and at lower than 10°C, they are psychrophics (Gould, 2009). Some strains of *Cladosporium herbarum* that infect meat in cold storage can grow at a temperature as low as –6°C. In contrast, one species of *Chaetomium* grow optimally at 50°C and survive even at 60°C

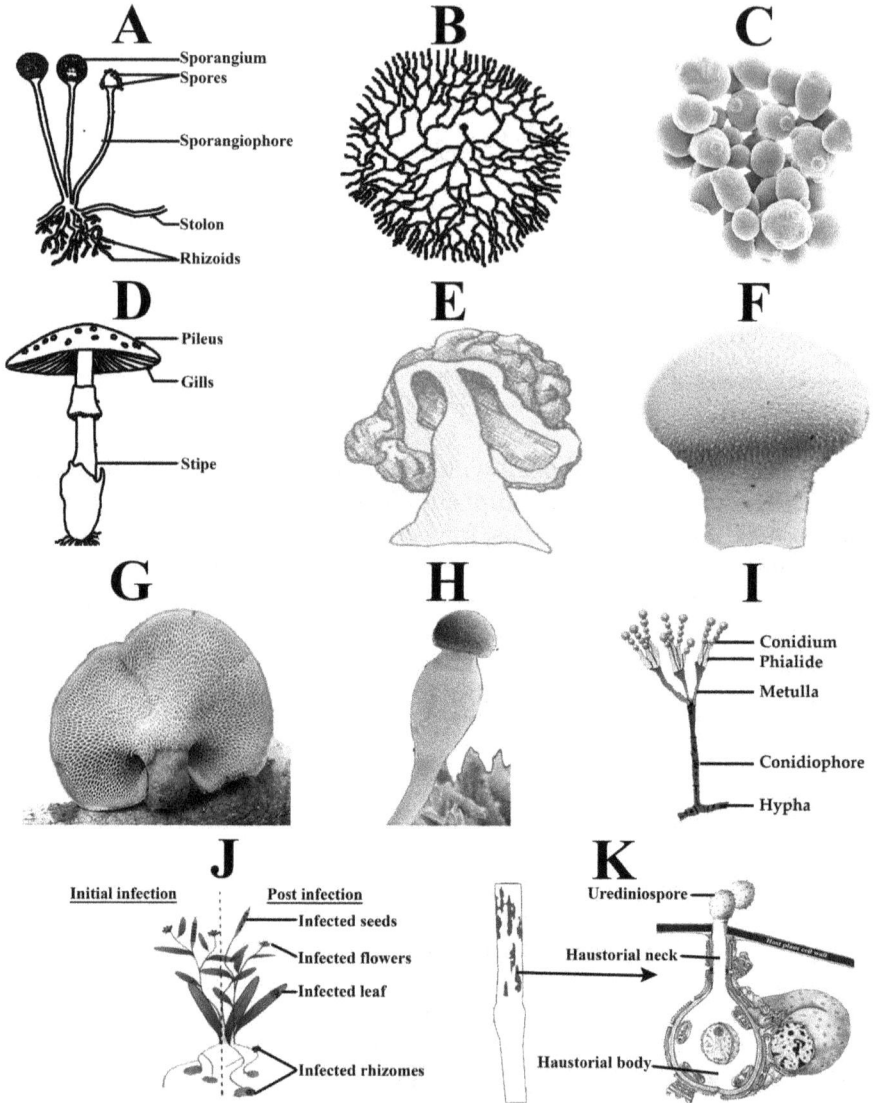

FIGURE 1.1

Representative examples for fungi: (A) *Rhizopus stolinifera* (*backyardnature.net*), (B) *Coprinus sterquilinus* (free hand drawing based on Buller, 1931), (C) *Saccharomyces cerevisiae* (*publons.com*), (D) *Agaricus bisporus* (free hand drawing), (E) *Gromita esculenta* (*pngegg.com*), (F) *Lycoperdon perlatum* (*free-nature.com*), (G) *Polyporus brumalis* (*wikiwand.com*), (H) *Pilobolus* (free hand drawing), (I) *Penicillium notatum* (*biologyreader.com*), (J) the smut fungus *Ustilago esculenta* showing initial infection and post-infection on rhizome, leaf, flower and seed of *Thecaphora thlaspeos* (modified from Linde and Gohre, 2021), (K) stem rust caused by *Puccinia graminis* on wheat (modified from Kolmer et al., 2009).

(see Evert and Eichhorn, 2013). Nevertheless, due to their size, mushrooms were the only known fungi up to 17th century. The invention of the microscope and its improved versions up to electron microscope aided the recognition, identification and description of a large number of fungal species living on dead (decomposers) or live (symbionts or parasites) animals and plants, bacteria and other fungi, as well.

Being simple in structural organization, the fungi consist of mostly (i) the spherical slender filamentous hypha, a collection of which constitutes the **mycelium** (Fig. 1.1B), (ii) the reproductive body named as the **sporophore**, which arises vertically from the horizontal hypha and terminates mostly in an open or closed umbrella-shaped cap or **pileus** and (iii) rarely rhizoids, as in *Rhizopus stolonifera* (Fig. 1.1A). The visible sporophores vary in size, shape, color and longevity. Some of them are microscopic (e.g., yeasts, Fig. 1.1C), others are no larger than a pinhead (e.g., *R. stolonifera*, Fig. 1.1A) and a few are truly gigantic. The largest sporophore of the bracket fungus *Formitiporia ellipsoidea* reaches a size of 10.8 meter (m) in length and 85 cm in width and weighs up to 500 kg. It holds 450 million spores (*brittanica. com*). The following examples may suggest the amazing enormity of their spore-producing ability. A single **basidiome** of *Agaricus campestris* discharge 2.7 billion spore/day (d) or 31,000 spore/second (s). On rotting wood, the 2 million tubes of *Ganoderma applanatum* shed 30 billion spore/d or > 5 trillion during the 6 months (mo), when the perennial bracket is active each year (see Watkinson et al., 2015). Most of them appear like an umbrella with a short (e.g., *Russula brevipes*) or long (e.g., up to 8 cm in enoki *Flammulina velutipes*) stem/stalk or **stipe**. The umbrella may be opened (e.g., *Pleurotus ostreatus*) or closed (e.g., button mushroom *Agaricus bisporus*). Due to the structural simplification, the parasitic fungi like rusts are found only as dots, spots and patches on the surface of the host plants (Fig. 1.1K). The pileus or cap may be colored; mostly milky white (e.g., *Calocybe indica*), gray (e.g., *P. ostreatus*), pink (e.g., *P. djamor*), dark (e.g., *Lentinula edodes*) or red (e.g., *Amanita muscaria*) and so on. The lifespan of most fungi is short and lasts for a season or year. However, there are others, which live for long years; for example, *Armillaria gallica* is estimated to live for 2,500 y (*reviewjournal.com*).

For ages, fungi have served man in many ways. (i) Many mushrooms are tasty meals in our menu. With a rubbery texture and relishable taste, some like the Mexican *Ustilago* spp are delicacies. (ii) The yeast *Saccharomyces cerevisiae* thrive on the sugar-rich diet and metabolize glucose to produce Carbon-di-Oxide (CO_2) and alcohol (under low oxygenic condition). This metabolic process facilitates making bread, brew beer, ferment wine and other alcoholic beverages. (iii) Using traditional knowledge of the fermenting technique, the conversion of perishable milk to less perishable curd, butter and ghee is known since the days of Mahabharadha (400 BC, *wikipedia*). For thousands of years, the South and East Asians have known how to ferment boiled rice mixed with water to make nutritious *kanji*. (iv) Thanks to modern technical development, commercially the most successful fungal drugs have become available, for example, the antibiotic penicillin from *Penicillium chrysogenum* and *P. notatum*, and the immune suppressant cyclosporine from *Tolypocladium inflatum*. These discoveries and techniques have indeed revolutionized medicinal treatment and organ transplantation in surgery (see Blackwell, 2011).

On the other hand, (v) parasitic fungi have a devastating effect on human health and wealth. Some 125 million people chronically suffer physically and psychologically from psoriasis caused by the fungi *Candida* spp (*psoriasis.org*). A large number of parasitic fungi (~ 39,000 species, see Tables 3.8, 3.9) drastically reduce the yield of commercial crops like wheat, maize and rye. Two historically important events may also be mentioned. The potato famine (1845–1848) caused by *Phytophthora infestans* drove many Europeans to migrate to America (Bourke, 1991). The incidence of downy mildew of grapes caused by *Plasmopara viticola* destabilized the economy of wine manufacturers (Large, 1940).

Of academic interest, 89% of terrestrial plants depend on the arbuscular mycorrhizae for direct intake of water and phosphate, and indirect acquisition of inert nitrogen through bacteria (Table 11.1). The neocallimastigomycote fungi lack mitochondria and are anaerobic. In association with ciliates and bacteria, they symbiotically facilitate cellulose digestion in herbivorous mammals. This process generates and injects methane, a greenhouse gas, into the atmosphere leading to climate change.

Not surprisingly, thousands of publications are available on fungal biology. They are reviewed from time-to-time by many authors in some dedicated journals and books. Almost all them view fungi from the point of medical biology (e.g., Arora and Arora, 2020), agricultural parasitology (e.g., Vidhyasekaran, 1997) or biotechnology–nanotechnology (e.g., Prasad, 2018). **This book approaches fungi from the angle of species diversity, and environmental factors and biological attributes that accelerate or decelerate the diversity.**

1.1 Form and Function

Fungi exist as unicellular motile chytrids/non-motile yeasts or multicellular hyphae. The unicellular fungi comprise the motile (1a) 1,840 speciose Chytridiomycota-Myxomycota and 260 speciose Neocallimastigomycota, i.e., 2,100 species or 2.0% of fungi as well as the non-motile Microbotryomycetes

(200 species), Taphrinomycota (140 species) and Saccharomycotina (1,309 species). In all, they constitute 3,749 species or 3.5% fungi. Besides, yeasts serve as a stage during the ontogeny of ~ 225 species or < 0.3% fungi. Yeasts are regarded as reverted from multicellular hyphae. *Yet, the reversion has not accelerated species diversity as much as the chytrids. Nevertheless, the symbiosis in the clonal Neocallimastigomycota (260 species or 0.25% fungi) has also not accelerated species diversity.*

In the remaining 96.3% fungi, the filamentous body consists of multicellular hyphae. In turn, the somatic (thallus) hyphae of the Zygomycota, for example, extend horizontally as **stolons**. On maturation, the clonal hypha develops the vertical **sporophore** terminally bearing the **sporangium**, as in *Rhizopus stolonifera* (Fig. 1.1A). The basic unit of a filamentous fungi is the hypha, which is a tough, cylindrical, thread-like syncytial or multicellular tube of 2–10 μm diameter. It contains the streaming cytoplasm strewn with organelles like nuclei, mitochondria, Golgi Apparatus (GA), ribosomes, liposomes and Endoplasmic Reticulum (ER) and other inclusions like (i) vacuoles, which serve to store water, nutrients, wastes or enzymes like nucleases, phosphatases, proteases and (ii) plastids. The latter hold pigments or enzymes (see Gould, 2009). A hypha may be divided into three regions: (i) the apical **Spitzenkorper** zone of ~ 5–10 μm in length, (ii) the subapical zone of 40 μm rich in cytoplasmic organelles but devoid of vacuoles and (iii) the vacuolar zone characterized by the abundant vacuoles and accumulated lipids. The individual fungus is potentially immortal, as long as conditions are favorable. However, the older parts like the vacuolar zone may die sooner or later and are discarded (*brittanica.com*).

The tubular hypha may be either a continuous structure, as in the primitive lower fungi chytridiomycotes and zygomycotes (see Ramesh et al., 2021), or divided into compartments or cells by perforated or non-perforated septa. In the non-septate continuous **coenocytic** (syncytial) hyphae, the nuclei are scattered throughout the streaming cytoplasm, which facilitates the flow of nutrients to flow from one end to other. The perforated septate hypha permits the cytoplasmic streaming along with the rapid movement of various molecules, and organelles like mitochondria but not the larger ones like nuclei. In some fungi, the small pores in the sieve-like pseudosepta prevent the movement of nuclei to adjacent cells, while the larger ones permit the passage of smaller molecules and organelles. Most basidiomycotes have **dolipore** septa that permit the flow of flowing cytoplasm but not nucleus. In the non-perforated septate filamentous basidiomycotes, the nuclei of one parent may often invade the hypha of the other eligible parent, as the septa are degraded prior to the nuclear invasion allowing their passage but reforming after the establishment of the passaged nuclei.

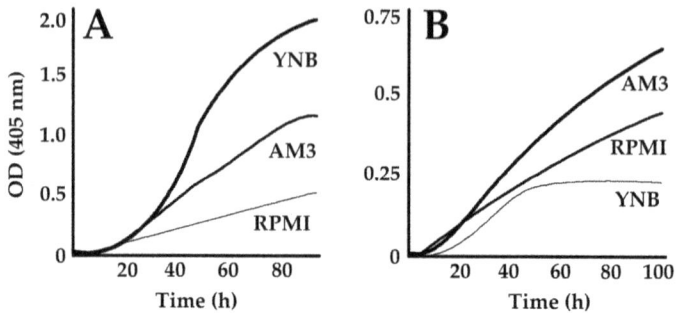

FIGURE 1.2

Effect of nutrient media on growth (as measured by OD) in (A) *Aspergillus fumigatus* and (B) *Scedosporium prolificans* (modified and redrawn from Meletiadis et al., 2001).

In most filamentous fungi, the assimilative hyphae grow **monopodially** on the main axis and are potentially capable of unlimited growth up to 1 km/d (*csus.edu*). However, their growth is controlled by species-specific and nutrient-specific factors. On the YNB medium, for example, it is fastest for *Aspergillus fumigatus* but slowest for *Scedosporium prolificans* (Fig. 1.2). In most fungi, the hyphal growth occurs exclusively in the apical zone, where the cell wall extends continuously to produce a long hyphal tube. As an internal factor, the Spitzenkorper plays a key role in deciding the directionality and rate of growth. According to the molecular mechanism described by Steinberg (2007), Golgi apparatus, concentrated in the apical 5–10 μm of the hypha, releases secretory vesicles (for example, 38,000 vesicle/min [min] in *Neurospora crassa*, Collinge and Trinci, 1974) in a controlled manner and thereby generates an exocytosic gradient and carries the newly synthesized proteins to the apex. Studies with *Ustilago maydis*, *A. nidulans* and *Candida albicans* have demonstrated that three classes of kinesins (1, 3 and 7) and three classes of myosins (1, 5 and 7) are involved in polarized (directional) growth. Endocytosis is necessary to retrieve membrane material in excess of that which is required for the extension of the growing apex (Webster and Weber, 2007). Kinesin-3 supports not only the vesicular secretion but also mediates the tip toward the traffic of early endosomes; these endosomes take up endocystosed material from the primary endocytic transport vesicle and return their content back to the surface for reuse or to the vacuole for degradation (Steinberg, 2007). In this process, the Spitzenkorper itself moves forward.

In a growing hypha, branches do appear suggesting that the apical dominance is inhibited by an unknown factor present at some distance behind the apex. They usually arise singly, as in vegetative hyphae; however, whorls of branching occur in reproductive structures. Branching may thus be under genetic or external control (Burnett, 1976). According to Webster and Weber (2007), an even spacing between vegetative branching results from a combination of chemotropic growth toward a source of diffusible nutrients; their growth and branching on a substratum by staling products secreted by co-occurrent hyphae. The repeated laterally branching hyphae result in the formation of mycelium (Fig. 1.1B).

Some variants of mycelium may be mentioned. In filamentous fungi, dichotomous branching is rare. But it does occur in *Allomyces* and *Galactomyces geotrichium* (see Webster and Weber, 2007). Some mycelial network called **sclerotium** – measuring 20–25 cm in *Wolfiporea extensa* – develops into an extremely hard structure, and serves to carry the fungus over periods of adverse conditions of extreme temperatures and/or non-availability of water (e.g., *Claviceps purpurea*). In the black colored honey mushroom *Armillaria mellea*, the hyphae are intricately constructed and differentiated to conduct water and nutrients from one part of the fungus to other parts (*brittanica.com*). On soil, inside leaf tissue or within the skin of the animal, the hyphae grow not through the extension of the hyphal tip but by intercalary growth and spread in a sort of loose network. Due to parasitic structural simplification, the hypha grows as slowly as the host plant cell(s) (e.g., rust). Among rusts, the parasitic body consists of only the externally visible patches containing the **urediniosporangium** on the surface of the host plant and internal **haustorium** (Fig. 1.1K).

Reproductive structures: Except for budding or fragmenting unicellular yeasts, the filamentous fungi are propagated mostly through spores. Usually being unicellular, the spores are produced by fragmenting mycelium or within specialized structures. On maturation, the germinal thalli are differentiated and developed into an array of reproductive structures, which may be considered under two headings: 1. Clonal structures: (i) sporangium, (ii) zoosporangium, (iii) conidium, (iv) sporocarp and 2. Sexual structures: (i) gametangium, (ii) basidium and (iii) ascidium. The sporangium is a four tissue-typed structure (see Fig. 1.9A). From it, non-motile spores are released (e.g., zygomycotous *Rhizopus stolonifera*, pezizomycotous *Junewangia globulosa*, Fig. 1.3A, B). From the somatic coenocytic hyphae, zoosporangium is directly developed; it releases biflagellated motile zoospores (e.g., Oomycota: *Saprolegnia*, Fig. 1.3C). In *Allomyces macrogynus*, the 2n sporophytics develop into mitosporangium and meiosporangium. The latter generates on the tip of their unicellular sporangiophores, the lower female and upper male gametangia (Lee et al., 2010, Fig. 1.3D, E).

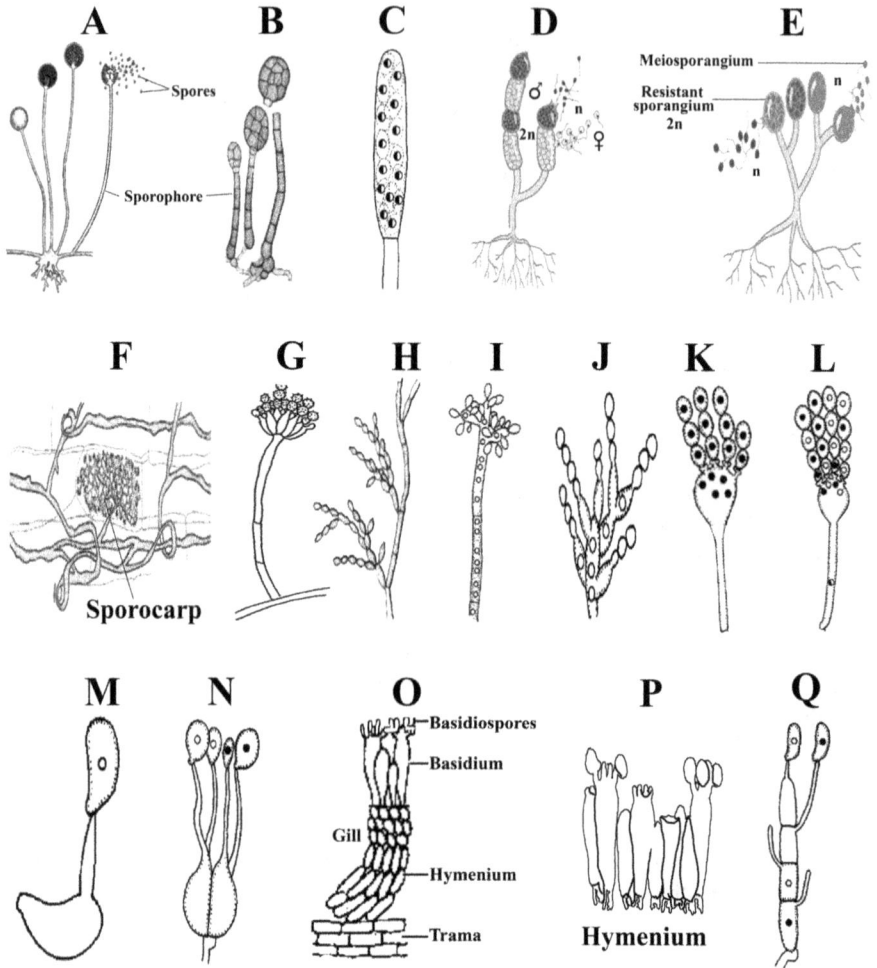

FIGURE 1.3

Sporangium. (A) Zygomycota: *Rhizopus stolonifera*, (B) Ascomycota: *Junewangia globulosa*. (C) **Zoosporangium**: Oomycota: *Saprolegnia*. (D-E) **Gametangium**: Oomycota: *Allomyces macrogynus*. (F) **Sporocarp**: Glomeromycota. (G) **Conidium**: *Penicillium*. (H) *Tyrannosorus pinicola*. Leotiomycetes: *Botryotinia fuckeliana*: (I) macroconidium, (J) micrononidial phialides. Tremellomycetes: *Filobasidiella neoformans*: (K) monokaryotic mycelial **conidium**, (L) dikaryotic mycelial **basidium**. Tremellomycetes: *Tremella mesenterica*: (M) ballistoconidium, (N) tremelloid basidium. (O) A cross section through a gill to show the tissue types as ~ four only. (P) A bunch of basidia called Hymenium. (Q) Auriculariales: *Auricularia auricula-judae*: Phragmobasidium (free hand drawings from Webster and Weber, 2007, Margulis et al., 2009, Watkinson et al. 2015).

Sporocarp is produced in all Glomeromycota (Fig. 1.3F). Conidium is also simple in structure, as its conidiophore directly generates conidial spores in a vertical row (Fig. 1.3G). The conidiophore may arise terminally or laterally

(Fig. 1.3H), as fused synnema or over a flat saucer-shaped acerculum. In some heterogonic life cycle involving alternation of generation between sexual reproduction and clonal multiplication, as in *Botryotinia fuckeliana* (Fig. 1.39), the

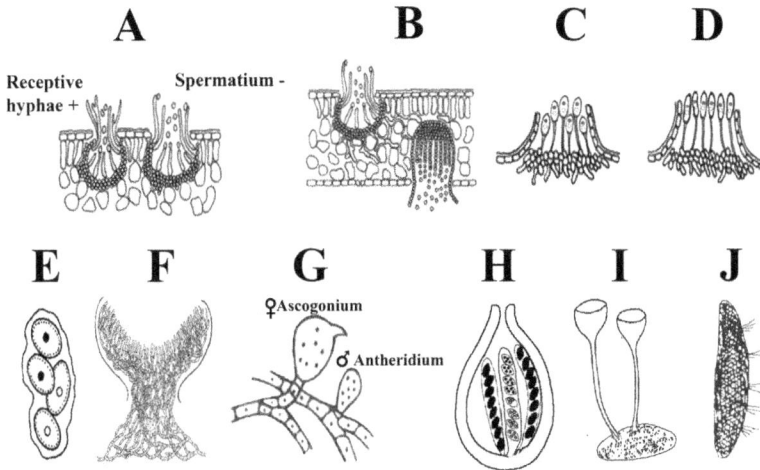

FIGURE 1.4

Reproductive structures of heteroecious *Melampsora larici-epitea*: (A) spermagonia, (B) aecium, (C) uredinium and (D) telium. (E) Ascus of *Schizosaccharomyces octosporus*. (F) Ascomycota and (G) Ascogonium and antheridium of Ascomycetes. (H) Perithecium of *Neurospora crassa* showing the development of mating types A and a. (I) Carpogonium of *Botryotina fuckeliana*, (J) Sclerotium with receptive hyphae (free hand drawings from Webster and Weber, 2007 Margulis et al., 2009, Aime et al., 2014).

multinucleate hypha may directly develop into the macroconidiophore and bear the macroconidium or microconidialphialides (Fig. 1.3I, J). In another heterogonic fungi *Filobasidiella neoformans*, the monokaryotic mycelium produce α-mating type haploid conidium on conidiophore (Fig. 1.3K) and both A- and α-mating type basidium (Fig. 1.3L). Typically, the basidiomycotes develop individual two tissue-typed basidium (Fig. 1.3G) or a group of them called **hymenium** (Fig. 1.3P). A number of variants of the basidium are known (e.g., three tissue-typed phragmobasidium, tremelloid basidium, Fig. 1.3Q, N). In heteroecious (alternating life cycle involving two hosts) smuts like *Melampsora larici-epitea*, spermagonium, aecium, uredinium and telium are produced each with three tissue-types (basal tissue, hypha, spores) (Fig. 1.4A–D). The two tissue-typed ascus is the simplest reproductive structure in *Schizosaccharomyces octosporus* (Fig. 1.4E). In the gametangium of ascomycotes, a two tissue type structure, the female ascogonium and the male antheridium are developed (Fig. 1.4G). In perithecium of *Neurospora crassa*, vertical rows of conidial A- and α-mating type spores are generated

(Fig. 1.4H). Sclerotium is a hardened complex of somatic hyphae, which ensure survival of the fungi over unfavorable conditions (Fig. 1.4J). On the whole, the number of tissue types in the reproductive fungal structure ranges from none, as in zoosporangium to two in ballistoconidium and ascus, three in spermagonium, aecium, uredinium and telium, or at the maximum of four tissue types in a typical Zygomycota namely the peridium, sporangiospore, columella and sporangiophore (see Fig. 1.9A).

1.2 Classification – Systematics – Taxonomy

According to Hyman (1940), Webster's New Collegiate Dictionary and Chambers Dictionary of Science and Technology (2002), the terms classification, systematics and taxonomy are indistinguishably interrelated. They are explained as a branch of biology that systematically arrange the different category of organisms based on some similarities and dissimilarities. This account, however, distinguishes and defines them. Accordingly, *Taxonomy is a branch of biology that identifies a taxon, which includes a group of individuals, which are similar in structure and/or molecular sequence. Systematics is another branch of biology that deals with placement of a group of similar taxa into a type of hierarchical arrangement. Classification is yet another branch of biology that is concerned with comprehensive arrangement of organisms at different systematic hierarchial levels.*

Due to their small size and structural simplicity, collection and cataloging of fungi had to wait for the invention of instruments, and technical developments in culturing and molecular methods for identification. Hence, the invention of (i) microscope (1609, *microscopemaster.com*) and (ii) electron microscope (1931, *wikipedia*), as well as (iii) culturing techniques during 1980s and (iv) molecular analysis using sequences like the Sr RNA are milestone events in recognition, identification and description of fungal species. The use of molecular tools has resulted in recognition of more numbers of fungal species than by morphological traits. For example, by engaging molecular sequences, it has been possible to recognize 15 species within the genus *Neurospora*, in which only five morphological species have been recognized earlier (see Blackwell, 2011).

The electron microscopic techniques have facilitated the description of ultra structures like the spindle polar bodies (cf Pucciniomycotina), cell wall, septa and septal pores that are used as keys in fungal taxonomy. With the development of culture techniques, secondary metabolites serve as a component of integrated taxonomy. For example, chemotaxonomic characterization has been used in systematics of *Penicillium* and Xylariaceae. With the advent of PCR, many gene sequences are used in fungal taxonomy. Some of them are (i) ribosomal RNA genes inclusive of internal subunits

(ITS), which are used to identify fungi at species level (see Branco, 2011), (ii) intergeneric spacers, the large subunit (LSU) and small subunit (SSU), (iii) house-keeping genes (a) actin (ACT), (b) calmodulin (CAL), histones (H_3, H_4) and others. Based on 192 protein-coding loci, for example, the issues at phylum level have been resolved in 46 taxa including 25 zygomycotes. Since the first genome for *Saccharomyces cerevisiae* was published in 1996, ~ 800 fungal genomes were sequenced, of which 50% were published (Zhang et al., 2017). Besides cellular morphology, assays on fermentation and growth (assimilation) in different media are used to identify yeast taxa. These physiological assays include (i) fermentation of eight carbohydrates, (ii) growth rates on carbon and nitrogen sources and (iii) vitamins and others (Dujon and Louis, 2017).

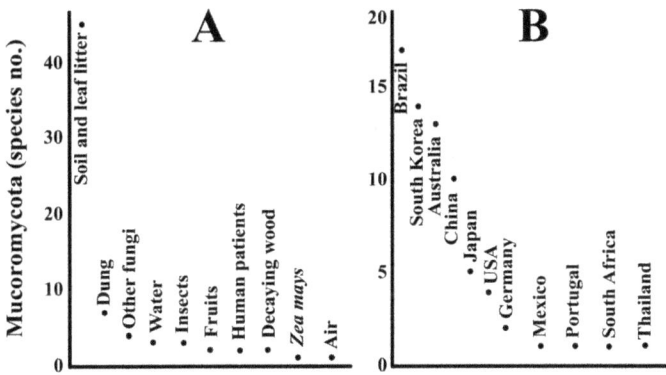

FIGURE 1.5

Number of Mucoromycota species described during the period from 1950 to 2010 from (A) different sources and (B) countries (modified and redrawn from Voigt et al., 2021).

According to Blackwell (2011), 12,000–20,000 fungal species were collected from (i) The Arctic and Antarctic zones, (ii) freshwater (> 2,863 species, see Grossart et al., 2021), (iii) sea (1,112, see Ramesh et al., 2021), (iv) soil (3,150) and (v) desert (1,971 lichens from the Sonoran Deserts, see Margulis et al., 2009). Another 20,000–50,000 species were also collected as endophytics (vi) from plant roots and stems, leaves, flowers and lichens, and (vii) within insects and vertebrates. For example, the number of species collected for the Mucoromycota (recognized as a phylum by Voigt et al. [2021] but as an order within Zygomycota by Blackwell [2011]) decreases in the following descending order: soil/leaf litter > dung > other fungi > water > insects > fruits and so on (Fig. 1.5A). Their number also decreases from those collected from Brazil > South Korea > Australia > China > Japan > USA > Germany and others (Fig. 1.5B). Clearly, there are many other resources (e.g., parasitic species) and countries (e.g., India), from which fungi remain to be collected. Notably, many pharmacuetically important fungi like *Penicillium chrysogenum* (from a northern temperate city) and *Tolypocladium inflatum* (from Norwegian soil) were collected from temperate countries. The tropical countries remain goldmines for collection, identification and development of beneficial fungi.

Species number: Notably, as taxonomy is based more on solid structural characteristics, the systematics and classification of plants and animals remain relatively more definitive. Being based on structural and molecular sequence, the taxonomy, systematics and classification of fungi remain in a fluid but dynamic status; hence, it is indeed a task to assess or estimate the number of fungal species. However, the publication series on the Dictionary of Fungi is a reliable source to count the number of fungal species. Accordingly, the number increased from 38,000 in 1943 to 97,330 in 2008 (Kirk et al, 2008, Fig. 1.6). To this, the 1,300 speciose Microsporidea, which are considered as Protozoa (see Pandian, 2023), may have to be added (see Blackwell, 2011). Similarly, Webster and Weber (2007) considered the amoeba *Dictyostelium discoideum* and the 175 speciose parasitic piroplasmean rhizopods as fungi (see Pandian, 2023). As the number increased at the rate of 1,400 species/y as of now (see p 204), there may be ~ 1,12,000 species. Providing valuable information on species number for each genus and its related publications, *speciesfungoram.org* does not report data for species number for each taxonomic group. Hence, repeated computer searches must be made to know the number of species for the selected taxonomic groups. Mostly due to the explosive increase in Pezizomycotina, the assembled counts

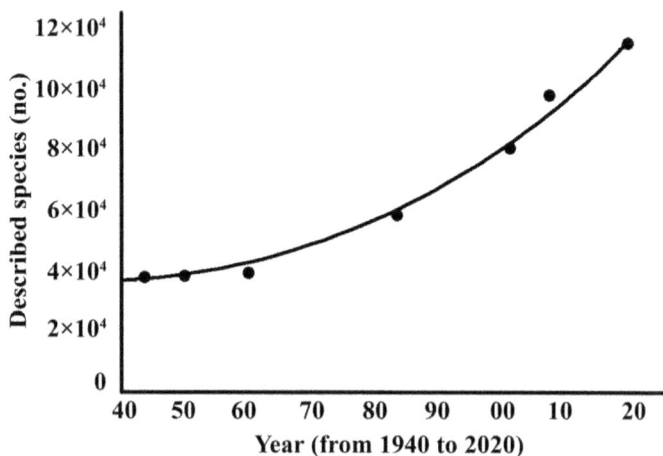

FIGURE 1.6

(A) Number of described fungal species known from the *Dictionary of Fungi* during the period from 1943 to 2008 (modified and redrawn from Blackwell, 2011).

TABLE 1.1

Systematics of fungi. Phylum = bold, Subphylum = italicized. Classes and orders are located at different distances *including miscellaneous 359 species (*onezoom.org*).

Phylum/others	Species (no.)		Example	Reference
Chytridiomycota/ Oomycota	1340 500	**1840**	*Batrachochytrium dendrobatidis*	*ucmp.ber-keley.edu*
Chytriomycetes	900			
Blastocadiales?	100			Grossart et al. (2021)
Monoblepharidomycetes	340			
Neocallimastigomycota		**260**		
Zygomycota/ Blastocladiomycota		**1065**	*Allomyces macrogynus*	*Wikipedia*
Mucoromycotina	325		*Rhizopus stolonifer*	*Wikipedia*
Kickxellomycotina	348		*Harpella melusinae*	*onezoom.org*
Zoopagomycotina	65		*Cochlonema*	
Entomophothoromycotina	322		*Basidiobolus*	
Glomeromycota		**230**	*Glomus, Geosiphon*	*Wikipedia*
Myxomycota		**500**		*botany.-hawai.edu*
Basidiomycota		**23975**		
Ustilaginomycotina	1000		*Ustilago maydis*	
Agaricomycotina	14559			
Dacrymycetes	329		*Dacrymyces stillatus*	
Tremellomycetes	1117		*Tremella fuciformis*	
Agaricomycetes	13113			Watkinson et al. (2015)
Agaricales	8500		*Agaricus bisporus*	
Boletales	300		*Boletus edulis*	
Russulales	1767		*Russula lepida*	
Polyporales	1800		*Gonoderma lucidum*	
Auriculariales	402		*Auricularia auricula – judae*	
Phallales	172		*Phallus impudicus*	*eol.org*
Pucciniomycotina	8416			

Table 1.1 contd. ...

...Table 1.1 contd.

Phylum/others	Species (no.)		Example	Reference
Agaricostilbomycetes	330	2 orders	*Agaricostilbum*	Aime et al. (2014)
Atractiellomycetes	50	1 order, 9 genera	*Atractiella*	
Classiculomycetes	2		*Classicula fluitans*	
Cryptomycocolacomycetes	2		*Cryptomycocolax abnorme*	
Cystobasidiomycetes	25	9 genera	*Bannoa*	
Microbotryomycetes	200	50 orders, 7 families, 9 genera	*Microbotryum violaceum*	
Mixiomycetes	1		*Mixia osmundae*	
Pucciniomycetes	7800	150 genera	*Herpobasidium*	
Tritirachiomycetes	6		*Tritirachium*	
Subtotal	**27870**			
Ascomycota			*Neurospora*	
Taphrinomycotina		140	*Schizosaccharomyces pombe*	*Wikipedia*
Pneumocystidomycetes	14		*Pneumocystis jirovecii*	*onezoom.org*
Archaeorhizomycetes	2			*onezoom.org*
Neolectomycetes	5		*Neolecta*	*eol.org*
Saccharomycotina		1,309 13 families	*Saccharomyces cerevisiae*	*Wikipedia,* Kurtzman and Sugiyama (2015)
Pezizomycotina		77,083		
Pezizomycetes	1,684	15 families	*Cladosporium*	Kirk et al. (2008), Pfister (2015)
Orbiliomycetes	288		*Orbilia*	Pfister (2015)
Dothideomycetes	19,000		*Mycosphaerella*	Pem et al. (2021)
Dothidiales		3 families		Schoch and Grube (2015)
Myriangiales		3 families		
Botrysphaeriales		8 families		

Table 1.1 contd. ...

...Table 1.1 contd.

Phylum/others	Species (no.)		Example	Reference
Arthoniomycetes	1,500		*Arthonia*	Schoch and Grube (2015)
Eurotiomycetes	3,000		*Penicillium*	Jaklitsch et al. (2016)
Sordariomycetes	10,000		*Neurospora*	Zhang and Wang (2015)
Leotiomycetes	14,714		*Monilia*	*onezoom.org*
Lecanoromycetes	14,199		*Cladonia*	Wang et al. (2010)
Laboulbeniomycetes	2486		*Rhizomyces*	*onezoom.org*
Coelomycetes	10,000		*Phoma*	Margulis et al. (2009)
Hyphomycetes	160	23 genera	*Pyrigemmula*	Seifert and Gams (2011)
Mycelia sterilia	52**		*Ozonium*	
Subtotal for Ascomycota	**78,532**			
Subtotal for others	**27870**			
Total	**106,761***			

in this account – updated as far as possible – total to ~ 106,761 species (Table 1.1). This value is lower than those (1,20,000 and 1,50,266 species) recognized by Watkinson et al. (2015) and *speciesfungorum.org*. Hence, there is a need to consider some hypothetic values estimated by different authors. They are based on fungus : plant ratios. Considering 10.6 : 1 ratio, O'Brien et al. (2005) put the number in the range of 3.5–5.1 million. For the parasitic Mucoromycota alone, the ratio ranges from 230 : 1 during the 1970s to 720 : 1 during the 1980s, as if to support that of O'Brien et al. Lo, it may require 2,840–4,170 y to describe the 3.5–5.1 million species, at the rate of describing even 2,000 fungal species/y. Based on fungus : plant ratio of 6 : 1, the 'Estimate G' of 1.5 million fungal species is acceptable to Blackwell (2011) as a reasonable working hypothesis.

1.3 Systematics and Characteristics

For description of systematics and characteristics, this account is based on mostly Webster and Weber (2007), Watkinson et al. (2015), Voigt et al.

(2021) and others. The Kingdom Fungi is divided into two subkingdoms: Dikarya and Non-Dikarya (Voigt et al., 2021). Dikarya compromise the phyla Basidiomycota, Ascomycota and possibly Entomophthoromycota (see Voigt et al., 2021). They are characterized by the formation of cells containing two nuclei during sexual reproduction. Like animals and higher plants, the Dikarya are diplontic/diplophasic, i.e., with relatively longer dominant somatic phase and are characterized by formation of cells containing n + n nuclei during sexual reproduction. The non-Dikarya include Chytridiomycota, Zygomycota, Neocallimastigomycota and Glomeromycota. In contrast, the non-Dikarya are haplontics/haplophasics like the algae, i.e., with relatively longer, dominant gametic phase. Interestingly, the former includes phyla that are more diverse and speciose than those of non-Dikarya (Fig. 1.7). Clearly, *the diplophasic Dikarya with reproductive cells containing two nuclei have accelerated species diversity and haplophasic non-Dikarya with longer dominant gametic phase have decelerated the diversity.*

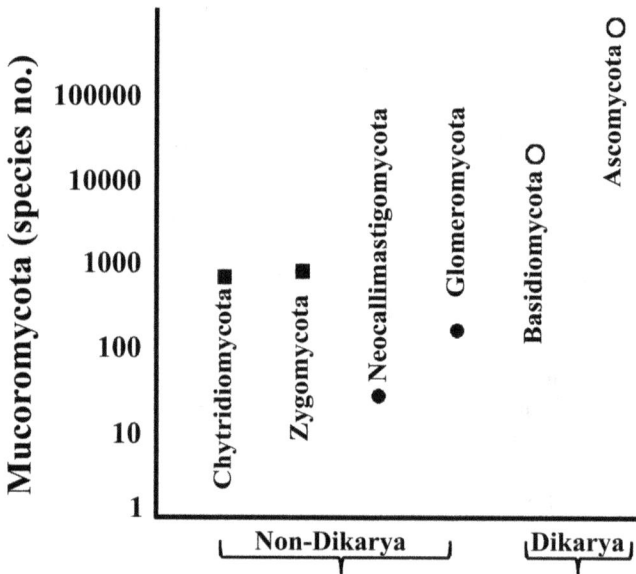

FIGURE 1.7

Number of fungal species in phyla and subkingdoms (drawn from data reported in Table 1.1).

TABLE 1.2

Classification of Fungi by selected authors. Species number for one classification alone, as given by the respective authors and collected from computer search, is also provided. Phyla that are common to the classification are indicated in bold letters.

Blackwell (2011)		Webster and Weber (2007)	Voigt et al. (2021)	Watkinson et al. (2015)
Chytridiomycota	206	Myxomycota	**Ascomycota**	**Chytridiomycota**
Monoblepharidiomycota	26	Plasmodiophoromycota	**Basidiomycota**	**Zygomycota**
Neocallimastigomycota	20	Straminipila	Entorhizomycota	**Glomeromycota**
Blastocladiomycota	179	**Chytridiomycota**		**Neocallimastigomycota**
Microsporidea	1300	**Zygomycota**	**Glomeromycota**	**Basidiomycota**
(Zoosporic fungi)		**Ascomycota**	Mucoromycota	**Ascomycota**
Zygomycota	1071	Archiascomycetes	Calcarisporiellomycota	
Entomophthorales	277	Hemiascomycetes	Mortierellomycota	
Glomeromycota	169	Plectomycetes	Basidiobolomycota	
Ascomycota	64163	Hymenoascomycetes	Olpidiomycota	
Basidiomycota	31515	Lichenized fungi	Zoopagomycota	
Orders of Zygomycota		Loculoascomycetes	Kickxellomycota	
Mucorales		**Basidiomycota**	Entomophthoromycota	
Morterilliales		Homobasidiomycetes	**Blastocladiomycota**	
Entogonales		Heterobasidiomycetes	**Chytridiomycota**	
Kickxellales		Urediniomycetes	Monoblepharomycota	
Harpellales		Ustilaginomycetes	**Neocallimastigomycota**	
Asellariales		Basidiomycetes	Aphelidiomycota	
Zoopagales		Anamorphic fungus	Rozellomycota	
Species number total	99326			

Vertical grouping labels in the Blackwell column: "Zoosporic fungi" and "Zygosporic fungi".

Vertical grouping labels spanning the Voigt et al. (2021) and Watkinson et al. (2015) columns: "Dikarya" and "Non-dikarya".

Traditional taxonomy splits Fungi into two basal lineages consisting of the phyla Chytridiomycota including the flagellate motile zoosporic fungi and Zygomycota comprising the sessile conjugating filamentous fungi (Voigt et al., 2021). The others like Basidiomycota and Ascomycota are derived lineages (Table 1.2). Barr (2001) considered that the former basal phyla are aquatic and ancestral, from which the more terrestrial Basidiomycota and Ascomycota were derived lineages. The Neocallimastigomycota and Glomeromycota may represent the directly derived lineages from Chytridiomycota and Zygomycota, respectively. As the life cycle involves both zoosporic yeast-like and filamentous phases, the Basidiomycota and Ascomycota may have been derived from both Chytridiomycota and Zygomycota, and may represent a polyphylogenic origin. Incidentally, the generic term yeasts include those fungi that clonally multiply by budding or fission. Due to reversion or parasitic structural simplification, they occur as a lifelong stage among the Basidiomycota (e.g., Cystobasidiomycetes) and Ascomycota (e.g., Saccharomycotina) and as a stage during ontogeny of some fungi (e.g., Microbotrymycetes).

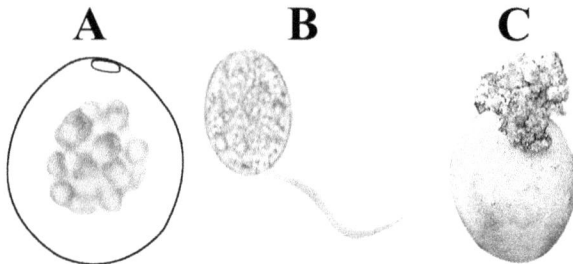

FIGURE 1.8

Representative examples for chytrids: (A) zoosporangium of *Cladochytrium replicatum* on onion skin, (B) zoosporangial stage of *C. tenue* (drawn and modified from Voigt et al., 2021) and (C) wart of *Synchytrium endobioticum* on potato (*Wikipedia*).

1. The Chytridiomycota are aerobic mostly aquatic zoosporic fungi. They are haplophasic with dominant haploid phase. They produce multinuclear (syncytic) spheroidal bodies that are called **thalli** (Fig. 1.8A). When nutrients are limited, the nutritive thalli are transformed into **sporangia** that release haploid uniflagellated motile zoospores (Fig. 1.8B). A diagnostic feature is the presence of (operculate) or absence (inoperculate) of a lid at the tip of the discharging tube. Anchored by finely branched rhizoids, the thalli may produce a colony like a chain. Thus, most chytrids multiply clonally. When sexual reproduction rarely occurs, it is usually followed by the development of a resting spore; meiosis generates the recombinant

haploid gametophytes (Fig. 1.23). The chytrids are monophylogenics. Pathogenic species include *Synchytrium endobioticum*, causing wart disease in potatoes (Fig. 1.8C), and *Batrachochytrium dendrobatidis* and *B. salamandrivorans*, causing the amphibian chytridiomycosis (see Watkinson et al., 2015). The anaerobic chytrid *Orbinomyces joyonii* plays a key role in the digestive process of ruminants (see *microbewiki.kenyon.edu*). According to Voigt et al. (2021), the morphologically identified chytrids include seven orders: Chytridiales, Gromochytriales, Lobulomycetales, Mesochytriales, Polyphagales, Rhizophydiales and Synchytriales that are parasitic on planktonic algae. But Grossart et al. (2021) reported that the chytrids comprise 1,340 freshwater species and includes them into Chytridiomycetes (900 species), Monoblepharidomycetes (340 species) and Chytridiomycota (*incertae sedis*, 100 species). Considering brackish water and marine chytrids (see Watkinson et al., 2015), there may be many more chytrid species.

2. The 1,065 speciose Zygomycota are also haplophasics. Their structurally simpler body consists of dispersed (not condensed stipe as in Agaricomycotina) (i) horizontal hypha, the **stolon**, (ii) rhizoids and (iii) the vertical erection, the **sporangiophore** with (iv) terminal sporangium consisting of mitotically generated haploid non-motile spores (Fig. 1.9). On the whole, Zygomycota may contain no more than three tissue types, as ulval algae do (see Pandian, 2022). Two distinctive modes of reproduction characterize zygomycotes: (i) clonal multiplication through mitotic spores, which directly germinate into hyphae and (ii) sexual reproduction through conjugation leading to nuclear fusion to produce zygospore followed by meiosis during germination. Hence, *the zygomycotes may be the first terrestrial fungi to discover and establish sexual reproduction via conjugation.*

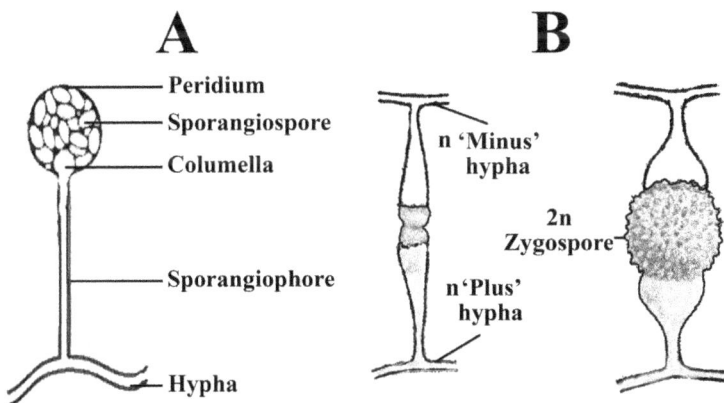

FIGURE 1.9

Structure of a typical clonal Zygomycota: (A) hyphal structure and (B) sexual reproduction by conjugation (free hand drawings).

Taxonomists have recently revised the phylum Zygomycota to include four subphyla: (i) Mucoromycotina, (ii) Kickxellomycotina, (iii) Zoopagomycotina (see Table 1.1, Voigt et al., 2021) and (iv) Entomophthoromycotina. Mucorales are also saprotrophs, drawing nutrients from terrestrial dead organic matter and sexually reproduce via conjugation. Two species among them merit mention. The photosensitivity of *Phycomyces* is astounding. Only a few photons are adequate to initiate the development of sporangium. Growing on horse dung, *Pilobolus* shoot the black mitosporangium to a distance of 2 m away from the dung. The zoopagomycotines comprise ~ 65 species in 10 genera. They are symbionts and parasites on amoebae and nematodes. The hyphae of kickxellales are septated and develop some of the most bizarre and complex clonal reproductive structures of any fungi (see Margulis et al., 2009).

3. The phylum Neocallimastigomycota consists of 260 species (see Grossart et al., 2021) in 6 genera (listed below) and are all accommodated in one family, one order and one class (Gruninger et al., 2014). It represents an earlier diverging lineage of the zoosporic chytrids. It consists of mostly unicellular fibrolytic

Genus	Thallus	Rhizoids	Flagella/zoospore
Caecomyces	Monocentric	Bulbous	Uniflagellate
Neocallimastix	Monocentric	Filamentous	Polyflagellate
Piromyces	Monocentric	Filamentous	Uniflagellate
Anaeromyces	Polycentric	Filamentous	Uniflagellate
Cyllamyces	Polycentric	Bulbous	Uniflagellate
Orpinomyces	Polycentric	Filamentous	Polyflagellate

highly specialized anaerobic symbionts in the digestive tract of herbivorous mammals and reptiles (e.g., Iguanas). Inhabiting the anoxic tract, the neocallimastigomycotes catalyze pyruvate oxidation and ferredoxin oxidoreductase in hydrogenosomes, and donate electron to hydrogen catalyzed by (Fe)-hydrogenase (Watkinson et al., 2015). Unlike the other eukaryotic co-symbionts, the ciliates, they exhibit little host specificity – a trait that may have accelerated species diversity except for *Ontomyces anksri* in camels, *Piromyces finnis* in horses and *Ghazallomyces constrictus* in deer. Incidentally, the combination of symbiotic ciliates and Glomeromycota seems to complete digestion of cellulose and other recalcitrant substances. According to Voigt et al. (2021), of 14 monocentric + three polycentric and two bulbous thalli, 12 genera are characterized by 16 flagella (Fig. 1.10A), while the remaining 7 genera are reverted to four flagella (Fig. 1.10B, C).

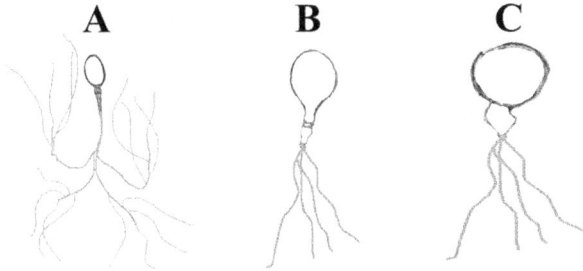

FIGURE 1.10

Representative examples for Neocallimastigomycota: (A) monocentric 16 flagellated *Aestipascuomyces*, (B) monocentric four flagellated *Buwchfawromyces* and (C) bulbous four flagellated *Cyllamyces* (free hand drawing from Voigt et al., 2021 and others).

4. The phylum Glomeromycota is characterized by the non-septate hyphae bearing large multinucleated (up to hundreds and thousands) spores (800 μm in diameter) with multilayered walls. Mixed populations of genetically distinct nuclei coexist within an individual spore. They have never evolved meiotic sexuality (see Watkinson et al., 2015) or secondarily lost it (Voigt et al., 2021). As they are derived from Zygomycota, it is likely that they have secondarily lost sex. The 230 speciose Glomeromycota comprise all the mycorrhizal symbionts in the root system (Fig. 1.11A, B) of 89% terrestrial plants (Table 11.1). Their hyphae are short, stubby, dwarf-like and facilitate rapid uptake of minerals especially phosphate and water (Dighton, 2009). Some like *Arthrobotrys oligospora* snare nematodes in a manner reminiscent of predatory constrictor snakes (Fig. 1.11C).

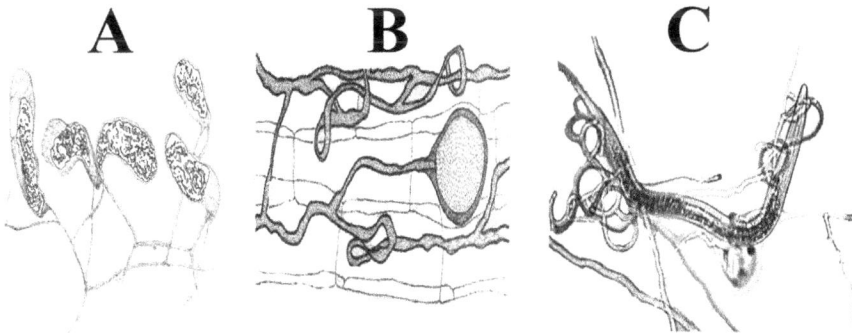

FIGURE 1.11

Representative examples for Glomeromycota: (A) *Geosiphon* hyphae with the terminal sporangia (*wikipedia*), (B) typical mycorrhizal hyphae bearing sporocarp surrounded by host's rectangular root cells (modified and redrawn from Watkinson et al., 2015) and (C) *Arthrobotrys oligospora* snaring a nematode (redrawn from *diark.org*).

5. The Basidiomycota are (i) doliporic septated **monokaryotic** or dikaryotic filamentous fungi and (ii) characterized by basidiospores borne on the club-shaped terminal (McConnaughey, 2014), extracellular structure, the **basidium**, in which the haploid nuclear fusion is followed by meiosis and (iii) formation of clamp connections after the nuclear fusion (Watkinson et al., 2015). Consisting of two nuclei derived from two parents, their **heterokaryotic hyphae**, following germination, develop hyphal compartments, stipe, pileus and gills. The fertile tissue of the fruit body is referred to as the **basidiome**. The Basidiomycota include mushrooms, jelly fungi, yeasts, rusts and smuts, and play a multitude of ecological roles as decomposers, mycorrhizal symbionts, partners with termites and leaf-cutter ants and parasites of animals (e.g., dandruff causing yeast *Malassezia globosa*, a ustilaginomycotine, psoriasis causing *Candida* spp) and plants (e.g., *Puccinia sessilis* on *Arum maculatum*).

This phylum is divided into three subphyla: A. Agaricomycotina (14,559 species), B. Ustilaginomycotina (smuts, ~1,000 species) and C. Puccinomycotina (rusts, 8,416 species, Table 1.1). In its turn, the Agaricomycotina is divided into three classes. (i) Agaricomycetes, (ii) Dacrymycetes and (iii) Tremellomycetes. Dacrymycetes include 329 species (Table 1.1) and hold tuning fork-like basidiome and form one spore at the tip of each of their two branches (e.g., *Dacrymyces stillatus* Fig. 1.12A, 1.29). The class tremellomycetes are characterized by basidiome, which is longitudinally divided into four chambers by septa and involves heterogonic life cycle switching between hyphal (e.g., *Tremella fuciformis*, Fig. 1.12C, 1.30) and yeast phases (e.g., *Cryptococcus neoformans*, Fig. 1.12B). It consists of 1,117 species (Table 1.1) in 50 genera and 11 families (*wikipedia*).

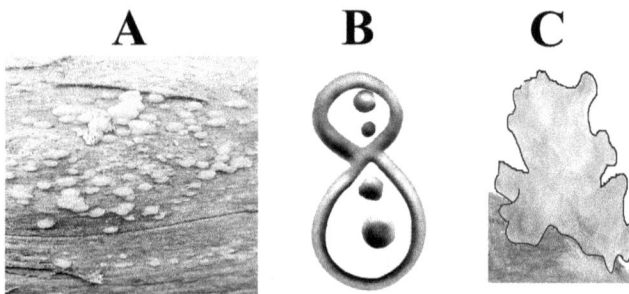

FIGURE 1.12

(A) Dacrymycetes: the sprouting sporangia of *Dacrymyces stillatus* on a log (from *fungalpunknature.co.uk*) and (B) yeast phase of *Cryptococcus neoformans* (free hand drawing), (C) Snow fungus *Tremella fuciformis* (from *inaturalist.org*).

The agaricomycetes are stalked umbrella-shaped mushrooms and brackets or shelf fungi (e.g., *Ganoderma*, Fig. 1.13D), which produce basidiospores on the surface of gills or teeth, and over the inner tubular surface. They are free-living decomposing fungi on dead organisms. They are split into six orders. 1. **Agaricales** are 8,500 speciose, the largest and more described. They are saprotrophs on plant debris (e.g., *Armillaria gallica*, Fig. 1.13A). 2. The spores of **Boletales** are discharged from the tubular surfaces beneath the pileus. In them, *Serpula lacrymans* (Fig. 1.13B) is saprotrophic and tremendously important as wood decomposer and wall-paint corrosive fungus in buildings. 3. **Russulales** comprise mycorrhizal, saprotrophs and a few parasites. Their fruit bodies include mushrooms with gills (e.g., *Russula lepida*), teeth (e.g.,

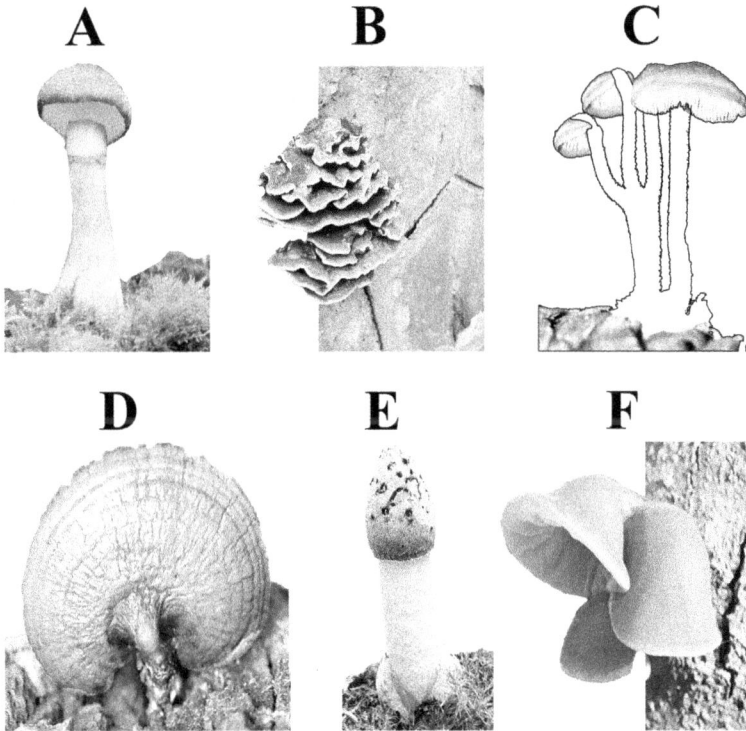

![A B C D E F mushroom figures]

FIGURE 1.13

Representative examples for Agaricomycetes: (A) Agaricales: *Armillaria gallica* growing on plant debris (*first-nature.com*), (B) Boletales: *Serpula lacrymans* growing on walls (*alchetron.com*), (C) Russulales: *Auriscalpium vulgare* (*wikipedia*), (D) Polyporales: *Ganoderma* growing on a log, (E) Phallales: *Phallus impudicus* growing among plant debris (*wikimedia*) and (F) Auriculariales: *Auricularia auricula-judae* on a log (*bnss.org.uk*).

Auriscalpium vulgare (Fig. 1.13C), tubes (*Bondarzewia*) or flattened ones (e.g., *Sterna ostrea*). 4. The shelf fungi **polyporales** are another speciose (1,800) order. They cause rot in live standing trees and fallen logs, and are the key player in recycling resources in the forest ecosystem. The fruit bodies of corticoid polyporales are flattened crusts on log surfaces. *Ganoderma lucidum* is used as natural medicine in China. 5. **Phallales** have lost the **ballistosporic** mechanism of spore discharge. They engage animals for dispersal. The fruit body of the common stinkhorn *Phallus impudicus* (Fig. 1.13E) is covered by slime with embedded spores attracting vectors like carrion flies and other invertebrates for spore dispersal. 6. In **auriculariales**, basidiospores are held on basidiome that are transversely partitioned into four chambers. The edible Judas ear mushroom *Auricularia auricula – judae* (Fig. 1.13F) is the best-known fungus in this order.

The 1,000 speciose smuts of Ustilaginomycotina also lack **ballistospore** discharge mechanism – the emblematic hallmark of the phylum Basidiomycota. They comprise plant pathogens. The family Septobasiodiaceae, symbiotic with scale insects, is an exception. It also includes the dandruff causing parasitic yeast *Malassezia globosa* (Fig. 1.14C). As plant pathogens, the smuts cause tumors or galls. Within the tumor, their transformed heterokaryotic hyphae produce the blackened teliospores (e.g., *Ustilago maydis* in the corn *Zea mays*, Fig. 1.14A$_1$, A$_2$). Following fusion, the diploid spores are dispersed by winds. The spore germinates into a filamentous **promycelium**, in which meiosis occurs. On infection of the female flower, formation of the ovary and seed production are suppressed. Others like *Microbotryum violaceum* display simple mating systems (Fig. 1.14B).

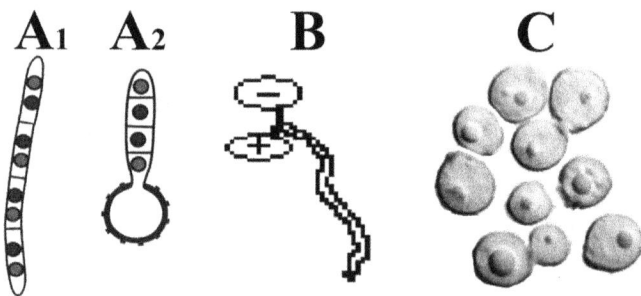

FIGURE 1.14

Representative examples for Ustilaginomycotina: (A$_1$) *Ustilago maydis* hypha (A$_2$) promycelium emerging from its spore, (B) hyphae with terminal sporangia of plus (+) and minus (–) mating types in *Microbotryum violaceum* (free hand drawing) and (C) *Malassezia globosa* (modified from *Wikimedia*).

According to Aime et al. (2014), the subphylum Pucciniomycotina is characterized by the following features: (i) their members range from yeasts to macrobasidiocarp forming filamentous hyphae, (ii) they include predominantly phytopathogens, a few phylloplanes and others, (iii) they display an array of life cycles ranging from simple teleosporic yeasts to the elaborate five stages in the **biotrophic** (involves penetration of haustoria surrounding the host cells for sustained nutrient absorption) rust fungi, (iv) their clonal life cycle is predominant in most lineages and (v) in contrast to the Agaricomycotina and Ustilagiomycotina, their cell wall consists of mostly mannose and lacks xylose and (vi) they possess Spindle Pole Bodies (SPBs), the organelles that organize microtubules during nuclear division. The SPBs are more or less internalized within the nucleus.

Pucciniomycotina comprise nine classes divided into 20 orders and 37 families (Pem et al., 2021). (i). Agaricostilbomycetes, exemplified by *Agaricostilbum pulcherrimum* (Fig. 1.15A), are either saprotrophs or mycoparasites. As some groups are polyphyletics, this class is in a more fluid status. They include **anamorphic** (lacking the sex tissue) genera *Bensingtonia*, *Kurtsmanomyces*, *Sterigmatomyces* and *Sporobolomyces* and form a yeast-like stage except in *Cystobasidiopsis nirenbergiae*. Most **teleomorphic** species produce **phragmobasidia**, while others like *Kurtzmanomyces* and *Streigmatomyces* bear ballistoconidia on conidiophores. *Spiculogloea* and *Kondoa* are ballistosporic. In studied members of the families Agaricostilbaceae and Chionosphaeraceae, an unusual pattern of mitosis occurs in the yeast phase. In them, the nucleus is divided in the parent cell itself rather than migrating into the bud prior to division. (ii). Atractiellomycetes contain a single order Atractiellales with < 50 species in the anamorphic genera *Hobsonia*, *Infundibura*, *Proceropycnis* and others like *Atractiella* (Fig. 1.15B). They possess a special organelle, the microscala. (iii). With only two species, Classiculomycetes are hyphal with hyaline cells (e.g., anamorphic *Classicula fluitans* = *Naiadella fluitans*, Fig. 1.15C). (iv). Cryptomycocolacomycetes also consist of two species (e.g., *Cryptomycocolax abnorme*, Fig. 1.15D) characterized by the extremely elongate holobasidia. (v). The Cystobasidiomycotes are a small group and consist of predominantly anamorphic yeast fungi. They are mostly mycoparasites and display a multiallelic mating system (e.g., *Bannoa hahajimensis*). (vi). With 200 species, Microbotryomycetes (e.g., *Kriegeria eriophori*, Fig. 1.15F) are the third largest class in Pucciniomycotina. They are dimorphic with haploid yeast morph and phragmobasidiate teleomorph. They comprise five orders in seven families and 21 genera including the anamorphic *Zymoxenogloea*. (vii). The single speciose (*Mixia osmundae*) mixiomycete is a parasite of the fern genus *Osmunda*. (viii). Pucciniomycetes are a diverse class with ~ 7,800 species in 150 genera. They

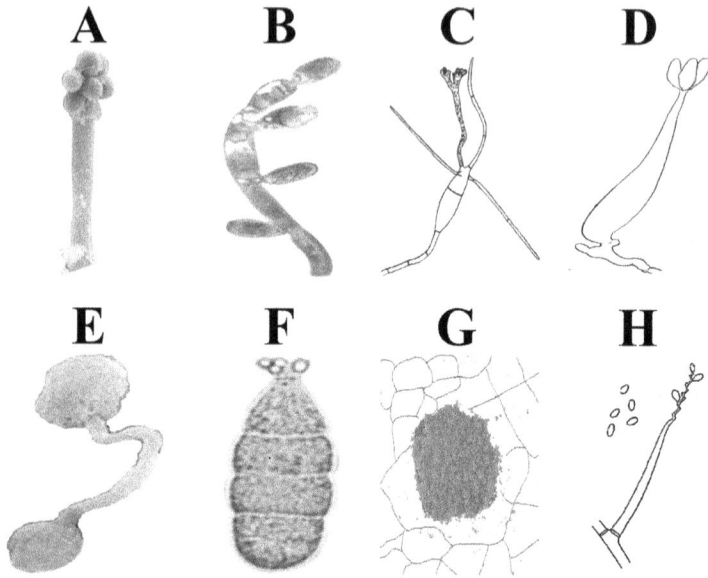

FIGURE 1.15

Pucciniomycotina: (A) Agaricostilbomycetes: *Agaricostilbum pulcherrimum*, (B) Atractiellomycetes: *Atractiella*, (C) Classiculomycetes: *Classicula fluitans*, (D) Cryptomycocolacomycetes: *Cryptomycocolax abnorme*, (E) Cystobasidiomycetes: unicellular yeast like fungi *Bannoa*, (F) Microbotryomycetes: *Kriegeria eriophori*, (G) haustorial cells of Pucciniomycetes on a leaf, (H) Tritirachiomycetes: parasitic *Tritirachium oryzae* (some are modified and redrawn from Aime et al., 2014; all others are free hand drawings).

are plant pathogens and cause some of the most devastating diseases. In them, the dikaryon is the dominant phase. (ix). The class Tritirachiomycetes contain a single order Tritirachiales with six species. They are all anamorphic molds (e.g., *Tritirachium oryzae*, Fig. 1.15H).

Ascomycota is the largest phylum and consists of 78,532 described species (Table 1.1) and a vast number of yet to be described species. According to Wang et al. (2010), > 98% of terrestrial fungi belong to this phylum. With selected DNA sequence, Wang et al. have also shown that the ascomycotes have a faster evolutionary rate and display richer biodiversity than the basidiomycotes. Within the former, Sordariomycetes have undergone the fastest evolution, while Leotiomycetes the slowest (however, see p 181). The septa of its members have a single pore. In them, mobile organelles with dense protein cores called **Woronian Bodies** (WBs) plug the pore and isolate damaged cells from the rest of a hypha. However, the WBs are present in 90% of them but are missing among the non-pezizomycotines. The Ascomycota can be free living saprotrophic decomposers or parasitic on plants and animals or constitute the mycobionts in lichens. The phylum includes members ranging from non-motile unicellular yeasts to filamentous fungi. They can be symbiotic partners with cyanophytes or other algae and terrestrial plants with mycorrhiza. Some of them are known only as clonals bearing no sexual tissue (anamorphic) and produce **conidial spores** on the stipe **conidiophore**. In the life cycle of most ascomycotes, the teleomorphic sexual phase has been identified. Their sexual structures may be an open cup-shaped (with a single vent for spore release) **perithecia** or cleistothecia, open in a variety of ways to release spores. The sexual spores of Ascomycota are called **ascospores**, which arise internally from the cells named **ascus**. Whereas basidiospores are formed outside the basidium, those of ascospores are developed internally. Incidentally, the hyphae of Ascomycota lack the doliporous septa and clamp connections, the characteristics of Basidiomycota.

In Ascomycota, two basic kinds of haploid spore producing asci exist. (a) uninucleate with a homogenous wall and pressure-sensitive apical spore discharging mechanism and operculum or ring-like sphincter and (b) binucleate with a double wall – a thin inelastic outer wall and a thick inner wall that on absorption of water ruptures the walls and expands upward carrying the spores with it. Ascomycotes have a much more restricted dikaryotic phase, which is initiated by compatible monokaryotic (haploid) hyphae. An ascus is produced, when two hyphae of complementary mating types conjugate (cytogamy) and subsequent fusion of conjugant nuclei (karyogamy), with the same number but in different allelic combinations of chromosomes. Then, the transient diploid nucleus undergoes meiosis followed by mitosis to produce eight haploid ascospores, each with a haploid nucleus, some cytoplasms and a protective spore wall. On maturation, the spores are released to be dispersed by wind, water or animals. Clonal multiplication, through conidiospore, is very widespread in ascomycotes. Specialized hyphae develop a succession of conidial spores from conidiogenous cells in the modified terminal hypha called **conidium** borne on **conidiophore**.

This account on the phylum Ascomycota is based on the description by Margulis et al. (2009) and others. Accordingly, the phylum is split into three subphyla: (i) the most speciose Pezizomycotina, (ii) Saccharoycotina and (iii) Taphrinomycotina. In its turn, the first one is divided into 10–12 classes, of which the last three are entirely anamorphic and depend solely on clonal multiplication. In these three classes, two are placed as ascomycotes, although their affinity to Basidiomycota or Ascomycota remains to be known. The reproductive structure like perithecia or cleistothecia and the nuclear number such as the uninucleate or binucleate asci are some key features considered in classification of the Ascomycota.

Having lost many fungal characteristics due to a parasitic mode of life, the small group consisting of unicellular yeasts of Taphrinomycotina do not morphologically look like fungi at all (Fig. 1.16A) but have a strong affinity with fungi at a molecular level. In it, the only class Taphrinomycetes are structurally the simplest among Ascomycota. They may have a short hyphae and develop restricted mycelia on their host and lack ascomata.

Their unitunicate asci are formed directly, instead of on the ascus-forming ascogenous hyphae, which grow from the conjugated cells. Their ascospores bud, while still in the ascus. *Penumocystis* is a representative example for this class.

The 1,309 speciose Saccharomycotina comprise the unicellular yeasts that have reverted to a single-celled saptrotrophic mode of life (Fig. 1.16B$_1$). They are included in one class: Saccharomycetes. Within it, the only order Saccharomycetales includes 84 genera in 13 families. Of the latter, some like Ascoidea and Saccharomycopsidaceae are constituted by a single genus each, but Dabarymycetaceae (12 genera) and Saccharomycetaceae (13 genera) are rich in genera (Kurtzman and Sugiyama, 2015). They multiply by mitotic budding (Fig. 1.16B$_2$). Following karyokinesis, one daughter nucleus is injected into the daughter bud by microtubules through spindle elongation. The bud enlarges to the parent's size, following which cytokinesis produces two approximately equal offspring. The haploid yeast may sexually reproduce by conjugation to form a diploid zygote, which forms a meiosporangium and is followed by meiosis. The yeast ascospore arrangement is often tetrahedrics (Fig. 1.16C). The ascospore germinates by budding following its release from the ascus.

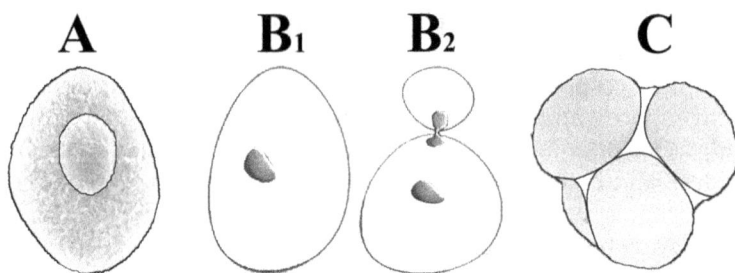

FIGURE 1.16

(A) Taphrinomycotina: *Pneumocystis*, (B$_1$, B$_2$) Saccharomycotina: *Saccharomyces cerevisiae* (note nuclear injection into the bud) and (C) tetrad of *S. cerevisiae* (free hand drawings).

Developing from the hyphae, Pezizomycotina are filamentous fungi. Their dikaryotic ascogenous hyphae grow through the monokaryotic sterile stage to the hymenium. Their asci may be (i) rigid uninucleates, in which the walls do not separate during ejection of spores, (ii) binucleates, in which the inner elastic wall expands beyond the outer wall, when spores are released and (iii) pronucleates, in which the ascal walls are dissolved at maturity to disperse the ascospores. The Pezizomycotina are the largest and best-known subphylum and include almost all ascomycotes, except for yeasts and the leaf-curled fungi.

The class Pezizomycetes comprise the operculate discomycetes that produce apothecial asci with an apical (operculate) lid. They comprise a single Order Pezizales with 16 families, of which Pyronemataceae is the most speciose (see Pfeister, 2015). Most of them are saprotrophs. But the others are ectomycorrhizal symbionts with plants. They may be epigeous (above ground) or hypogeous (underground). The former is characterized by apothecial or cleistothecial or highly reduced with a few asci in clusters; their spores are forcibly discharged. Hypogeous members are mostly mycorrhizals. *Aleuria aurantia* and *Tarzetta spurcata* (Fig. 1.17A, B) are some examples. *Phytomatothrichopsis ominivora* is parasitic on cotton.

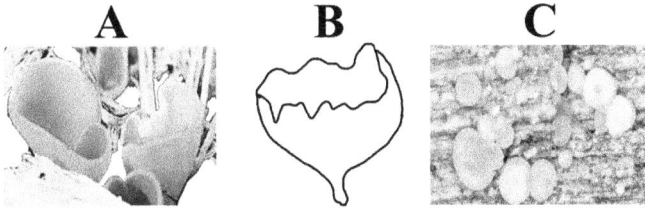

FIGURE 1.17

(A) Pezizomycetes: *Aleuria aurantia* on rotting debris (*Wikipedia*), (B) *Tarzetta spurcata* and (C) Orbiliomycetes: *Orbilia* (*fungi.myspecies.info*).

Orbiliomycetes, a sister group of Pezizomycetes, consist of a single family Orbiliaceae with 288 species (Kirk et al., 2008). As a class, it was proposed by Nannfeldt (1932) and confirmed by Svrcek (1954) by his monograph on the 16 speciose *Orbilia* (Fig. 1.17C). It consists of a single order and only three teleomorphic genera (see Pfeister, 2015).

Dothideomycetes are characterized by binucleate asci. Most of them depend on soil organics (e.g., *Megalohypha aqua-dulces*, *Venturia inaequalis*, Fig. 1.18A, B) including rotting corn and others. Members of the genus *Elsinoe* are parasites on citrus, raspberry and avocado. *Cercospora apii* is a parasite on lemon (Fig. 1.18C). The family Arthropyreniaceae consists of entirely of lichen-forming fungi. According to Pem et al. (2021), Dothideomycetes comprise 19,000 species in 1,495 genera, 191 families and 32 orders.

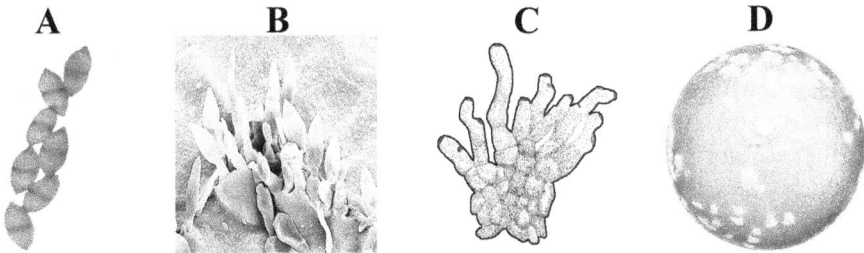

FIGURE 1.18

Dothideomycetes: (A) Ascospores of *Megalohypha aqua-dulces* on decomposing plant debris, (B) emerging conidia of *Venturia inaequalis* on a leaf, (C) *Cercospora apii* parasitic on leaf and (D) parasitic *Elsinoe fawcetti* on lemon (drawn from different sources).

The 1,500 speciose Arthoniomycetes, a sister group of Dothiomycetes, was erected by Eriksson and Winka (1997). Five families including the 500 speciose Arthoniaceae constitute this group. Whereas only a few lineages in Dothideomycetes contain lichenicolous fungi, lichenization appears to be a primary ecological trait of Arthoniomycetes. Some non-lichenized fungi are adapted to extreme conditions on rock surfaces. *Arthonia* (Fig. 1.19A), *Chrysothrix* and the sterile lichen *Lecanactis latebrarum* are some examples (Schoch and Grube, 2015).

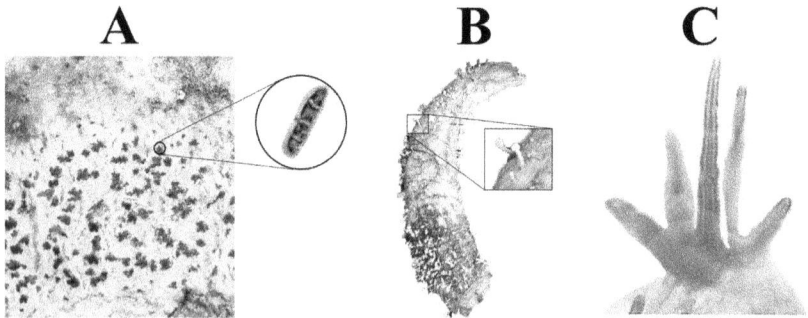

FIGURE 1.19

Representative examples of Arthoniomycetes: (A) *Arthonia* growing on rock. Eurotiomycetes: (B) *Onygena equina* with a window showing the fungus, (C) *Capronia muriformis* (drawn from different sources including *Wikipedia*).

The class Eurotiomycetes comprise fungi that are pronucleate asci borne on small closed ascomata. According to Geiser et al. (2015), they include three subclasses: (i) Eurotiomycetidae (including four orders: Eurotiales, Onygenales, Arachnomycetales, Coryneliales), (ii) Chaetothyriomycetidae (with three orders: Chaetothyriales, Verrucariales, Pyrenulales) and (iii) Mycocaliciomycetidae (with one order: Mycocaliciales). Within Eurotiales, the family Aspergillaceae includes the well-known two genera *Aspergillus* and *Penicillium*. Among Mycocalinomycetes, the myocalician genus *Dactylospora* includes marine species and *Sclereococcum sphaerale* is an anamorph. The order Onygenales includes fungi that specialize in decomposing keratin; their clonals grow on hooves , horns (e.g., *Onygena equina*, Fig. 1.19B), hairs, nails and on the skin also. The 40 speciose dermatophytic clonal molds are placed in three genera *Trichophyton* (24 species), *Microsporum* (17 species) and *Epidermophyton* (2 species). Some of them have many varieties. The class includes *Verucaria* spp too, which grow on coastal rocks, above the high tidal level. *Capronia muriformis* is a newly added species (Fig. 1.19C).

Sordariomycetes are mostly the non-lichenized ascomycetes that produce perithecial ascomata. They are uniquely characterized by a true perithecium and inoperculate, uninucleate asci. They are a large class in Ascomycota with 10,000 species accommodated in three subclasses: (i) Hypocreomycetidae, (ii) Sordariomycetidae and (iii) Xylariomycetidae with 18 orders and 1,119 genera (Zhang and Wang, 2015). Their ascospores are expelled through a narrow opening, the **ostiole**. In each ascus, meiosis is followed by a single mitosis to generate eight asci, which remain fixed in a row in the order, in which they have been generated. More importantly, each ascospore from an ascus can be picked up in that order and cultured to determine its genetic constitution. This has led to the understanding of chromosomal behavior during a single meiosis and the position of the genes on chromosomes. Not surprisingly, researches on the sordariomycete *Neurospora* (Fig. 1.20A) awarded the Nobel Prize to Edward Tatum and George Wells Beadle in 1958. *Fusarium, Magnaporthe oryzae, Xylaria* and *Leotia* are some representative examples (Fig. 1.20B–E).

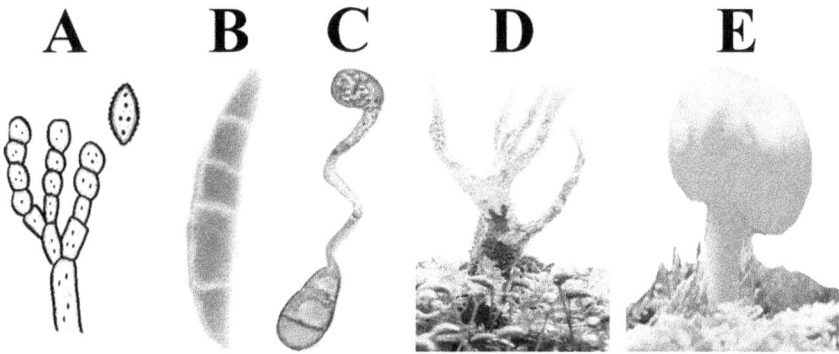

FIGURE 1.20

Sordariomycetes: (A) *Neurospora* (*fgcs.net*). Note spores being released., (B) *Fusarium*, (C) *Magnaporthe oryzae*, (D) *Xylaria hypoxylon* and (E) *Leotia* (drawn from different sources).

Lecanoromycetes (e.g., *Peltigera*, Fig 1.21A) comprise the lichenized ascomycetes, most of which produce apothecial ascomata. In these ascomata, the binucleate asci atypically dehisce apically. A variety of thalloidal structures like foliose, crustose, fruticose lichens belong to this class. *Cladonia* (Fig. 1.21B), *Laconora, Ramalina, Usnea, Lobaria* and *Physcia* are all examples for these groups. Lecanoromycetes, Acarosporomycetidae comprise 13 orders; of them, Lecanorales are rich with 17 families but Trapeliales include a single family (Gueidan et al., 2015). *Parmelia sulcata* is associated with lichens (Fig. 1.21C).

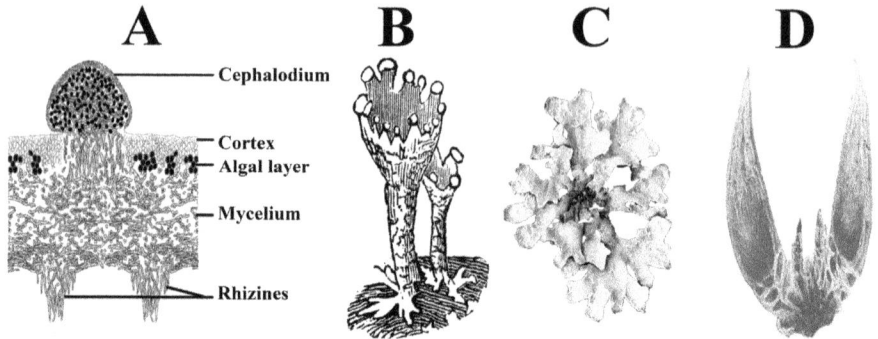

FIGURE 1.21

Lecanoromycetes: (A) *Peltigera* (cross section to show the sites of fungus and alga, (B) *Cladonia*, (C) *Parmelia sulcata* (from *Wikipedia*). Laboulbenomycetes: (D) *Herpomyces* (drawn from different sources).

Laboulbenomycetes are characterized by binucleate asci, which directly germinate and eventually develop the reproductive structures of 0.1–1.0 mm in diameter. Their ascospores produce septate, species-specific number of cells, the upper ones differentiate into spermatia and the lower one becomes the ascogonium. These ascomycetes are all minute ectosymbiotrophs on the legs or wings of insects. They are strictly host specific, organ specific and sex specific. *Herpomyces* (Fig. 1.21D), *Rhizomyces* and *Amorphomyces* are some examples.

Being anamorphs, the nearly 10,000 speciose polyphyletic Coelomycetes (e.g., *Phoma glomerata*, Fig. 1.22A) clonally multiply from conidia borne on short, closely packed conidiophores. These spores grow subcuticularly, intraepidermally, subepidermally or deeper beneath several layers of host cells. *Cryptosporium lunasporum* develop crescent-shaped conidia.

The another 160 speciose Hypomycetes are also anamorphs. They consist of unicellular yeasts that do not form asci and clonally multiply by terminal

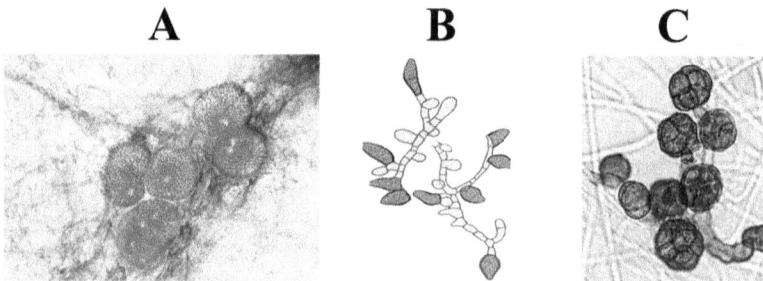

FIGURE 1.22

(A) Coelomycetes: *Phoma glomerata* (from *inspq.qc.ca*), (B) Hypomycetes: *Pyrigemmula* (free hand drawing) and (C) Mycelia sterilia: *Epicoccum nigrum* (*adelaide.edu.au*).

conidia that develop on the tip in various ways. *Pyrigemmula* is an example for Hypomycetes (Fig. 1.22B).

The class Mycelia sterilia (e.g., *Epicoccum nigrum*, Fig. 1.22C) are anamorphs and lack any specialized reproductive structures. Within their ~ two dozen genera, *Rhizoctonia* is the best-known soil fungus.

From the foregone description, the following may be noted: 1. The number of species ranges (i) from one/genus in *Cryptomycocolax abnorme* and *Colacosiphon filiformis* (Cryptomycocolacomycetes) to 500/genus in *Mycosphaerella* (Dothideomycetes), (ii) from one/family in *Mixia osmundae* (Myxiomycetaceae) to 288/family in Orbiliaceae (Orbiliomycets), (iii) from 172/order in Phallales to 8,500/order in Agaricales, (iv) from 2/class in Classulomycetes to 19,000/class in Dothideomycetes, (v) from 65/subphylum in Zoopagonomycotina to 77,083/subphylum in Pezizomycotina and (vi) 230/phylum in Glomeromycotina to 78,532/phylum in Ascomycota. This overview on the number of species accommodated at different hierarchical levels of classification may provide an idea to enthusiastic taxonomists and systematics in erecting new species, genus and so forth. In fact, even after erecting 19 phyla (Table 1.2), groups such as Oomycota could not be placed in any classification. 2. Voigt et al. (2021) consider that Ascomycota and Basidiomycota are derived lineages from two basal lineages, the zoosporic (Chytridiomycota) and zygosporic (Zygomycota) fungi. Firstly, it should be noted that there have been frequent reversions (see also Margulis et al., 2009) from one lineage to the other for unknown reasons (e.g., Saccharomycetes) or due to parasitic structural simplification. Secondly, Barr (2001) defined that the two ancestral lineages, i.e., Chytridiomycota and Zygomycota. In them, there are many free-living saprotrophic (e.g., *Rhizopus stolonifera*) and parasitic (e.g., *Synchytrium endobioticum*) fungi that are terrestrials. 3. Like algae, sponges and cnidarians, *fungi are also tissue grade eukaryotes. All of them are constituted by a maximum of six or seven tissue types. They are all clonals and clonality is known to limit species diversity.* Accordingly, the species count ranges from 8,853 for sponges to 10,856 for cnidarians among metazoans and to 44,000 for algae (Pandian, 2021b, 2022). However, the count for fungi ranges from 106,761 (Table 1.1) to 150,000 (*speciesfungorum.org*).

There are many predictions to indicate that the fungi may consist of 3.5 to 5.1 million species (see p 15). An explanation in favor of these predictions, claiming the potential existence of more than a million species, is that > 96% fungi are terrestrial, whose lability facilitates greater species diversity. In fact, among angiosperms, for example, the aquatic Nymphiaceae is diversified into 70 species over a period of 90 Million Years (MYs), whereas the terrestrial Poaceae into 1,200 species during the last 63 MYs (see Pandian, 2022). Nevertheless, the existence of clonality in 89% fungi (Table 15.10) has imposed a superseding impact on species diversity over the labile terrestrial habitats. *At the current rate of description, i.e., 1,400 species/y* (see Fig. 15.4), *the number of fungal species may not exceed 260,000 over the next hundred years.* However, there is a need to explain why the 12,277 speciose agaricale complex (comprising agaricales, boletales, russulales, phallales), which have resorted to sexual reproduction alone, have not diversified as much as ascomycotes. In these basidiomycotes, homothallism limits species diversity over the propagation by sexual reproduction alone. The complex regulation of species diversity by selected environment factors and biological attributes are elaborated in Chapter 15.

1.4 Life Cycles

Life cycles provide the base to build all other aspects of diversity and speciation. Unlike most authors, this account views the cycle more from the angle of ploidal status during different ontogenic stages. In their description of life cycles, some authors do not indicate the parentage of the two nuclei in dikaryotes and their meiotic reproduction. Others do not distinguish the n microsporidium from n + n plasmogamic microsporidium and so on. They are improved as far as possible. However, prior to the description of life cycles, a couple of aspects need to be explained: (i) the uniqueness of fungi and (ii) definitions of selected frequently used terms (see Petersen, 1974, Gould, 2009).

Unlike the other eukaryotic protozoans, metazoans and plants, the process of fungal sexual reproduction is unique in many respects: (i) Whereas nuclear division in other eukaryotes involves the dissolution and reformation of the nuclear membrane, that of fungi remains intact. (ii) Their nuclei are pinched at midpoints and the diploid chromosomes are pulled apart by spindle fibers formed within each intact cores. (iii) Their sexual reproduction includes three sequential stages: (a) **Plasmogamy** involves the fusion of two cytoplasms of the complementary cells/hyphae and brings them together the two compatible haploid nuclei. (b) **Karyogamy** results from fusion of these nuclei to form diploid zygote. The dikaryotic nuclear status, resulting from the plasmogamy, may be prolonged over several generations. In lower fungi, the karyogamy follows plasmogamy almost immediately (e.g., Fig. 1.23). But in more advanced fungi, they are temporally separated. Eventually, meiosis is followed to restore haploidy. The resulting haploid nuclei are called **meiospores**. (c) Fungi employ an array of methods to bring together the two haploid nuclei arising from plasmogamy. Some fungi generate specialized sex cells, the **gametes** that are released from differentiated reproductive structures, the **gametangia**. In others (e.g., Zygomycota), the gametangia come in contact with each other and their nuclei are exchanged and thereby they assume the function of gametes. In still other fungi (e.g., Fig. 1.38), the gametangia themselves may fuse to bring their nuclei together. Consequently, some of the advanced fungi produce no gametes at all. Their somatic hyphae take over the sexual function, come in contact, fuse and form zygote (see *brittanica.org*).

Anamorph	-	The fungus in a clonal multiplying state, e.g., Hyphomycetes (Figs. 1.43, 1.44)
Teleomorph	-	The fungus in a sexually reproducing state, e.g., *Agaricus bisporus*, Fig. 1.31
Holomorph	-	The fungus, which exhibits both anamorphic and teleomorphic states, e.g., *Melamspora larici-epitea* (Fig. 1.34)
Rhizomorph	-	The root-like strand of somatic hyphae that facilitate survival and dispersal in some basidiomycotes
Saprotroph	-	The fungus that derives nutrition from dead or dying organic matter
Biotroph	-	The (parasitic) fungus that obligately derives nutrition from a living source
Holocarp	-	The thallus, whose entire body is converted into a fruiting body
Sporocarp	-	Fruiting structure bearing spores in a fungus (e.g., Fig. 1.11B)
Teliospore	-	Spore produced from telium (e.g., Fig. 1.34)
Sclerotium	-	Compact masses of mycelium consisting of a central core of hyphae protected by a thick-walled rind; survival structure (e.g., Fig. 1.39)
Plasmodium	-	Multinucleate mass of protoplasm lacking the cell wall
Heterogony	-	Life cycle involving alternate reproductive phases
Macrocycle	-	Life cycle, in which the rust fungi have all the five reproductive structures: spermagonium, aecium, uredinium, telium and basidium (e.g., *Puccinia graminis*)

Demiocycle	-	Life cycle, in which the rust fungi has lost uredinium (e.g., *Uromyces striatus*)
Microcycle	-	Life cycles, in which the rust fungi have lost aecium, uredinium, and rarely spermagonium (e.g., *P. helianthi*)
Autoecy	-	Life cycles of the rust fungi involving a single host, e.g., *P. helianthi*
Heteroecy	-	Life cycles of the rust fungi involving two hosts, e.g., *Melampsora larici-epitea*. Basidiospore infects larch leaf and aeciospore willow leaf (e.g., Fig. 1.34)
Dikaryon	-	Fungal cells with two genetically different but compatible haploid nuclei
Homothallism	-	The monoecious fungi that produce both female and male gametangia and may be self-fertile (e.g., *Saprolegnia*). These gametangia may remain undifferentiated or differentiated into oogonium and antheridium (e.g., *Polyphytophthora debaryanum*)
Heterothallism	-	The diecious fungi that generate female and male gametangia from separate hyphae or mating types and cannot be self-sterile
Zoospore	-	Uni- and multi-flagellated motile vegetative gamete
Oospore	-	A product of sexual reproduction resulting from the oogamic transfer of male into female gamete or gametangium
Holo- or entero-blast	-	A product of clonal division, in which buds arise through the rupture of the parental cell walls and daughter buds separate, leaving a scar, over which no further budding occurs, e.g., yeasts (Fig. 1.14C)
Endospores	-	A clonal cell that is formed within a single cell or from a hyphal cell (e.g., *Candida*)
Chlamydospore	-	A thick-walled non-deciduous intercalary terminal clonal spore formed by rounded cell (*Candida albicans*)

| Fission | - | A clonal process, in which septum grows inward from the cell wall to bisect the long axis of the clonal cell |
| Pseudohypha | - | It represents a filament consisting of a chain of budded cells |

Chytridiomycota consist of mostly sexualized thalli. Their heterogonic life cycle includes two phases namely clonal multiplication and sexual reproduction. Usually, the clonal diplophase lasts longer than that of the short sexual haplophase. Only stress like undernutrition switches them to undertake sexual reproduction. In most chytrids (inclusive of parasitic species, e.g., *Olpidium viciae*) represented by *Polyphagus euglenae* (Fig. 1.23A) and Oomycota (e.g., *Allomyces arbusculus*, Fig. 1.23B), the clonal phase is diphasic and lasts longer. However, it is haplophasic only in the terrestrial parasitic *Synchytrium endobioticum*.

For the life cycle of 12 chytrid species, descriptions are available. From them, the following can be recognized: (i) Understandably, 60% of the cycles are known for parasitic species. (ii) Within the sexual phase, copulation occurs between unicellular complementary planogametes, as in *Polyphagus euglenae* or conjugation between hyphae of complementary mating types, as in *Chytriomyces hyalinus*. (iii) The copulation can be isogamic (e.g., *A. arbusculus*) or anisogamic/oogamic (e.g., *Monoblepharis polymorpha*). (iv) Resting stage or encystment occurs during the sexual phase (e.g., *Rhizophidium couchii*) or clonal phase (e.g., *C. hyalinus*) or during both clonal and sexual phases (e.g., *P. euglenae*). In parasitic *P. euglenae*, the clonal phase is repeated twice to enhance the chances of infecting new hosts. (v) Sexual mycelia are homothallic in *Saprolegnia* but heterothallic in *M. polymorpha*. In *Phytophthora*, the sex ratio remains one oogonium to one antheridium in *P. ultimatum* but one oogonium to several antheridia in *P. debaryanum*. Their life cycles are so diverse that the members of the Phylum Chytridiomycota have every possible combination of them. The sexual phase is secondarily lost in the anamorphic *Nowakowskiella ramosa*. With the presence of gametangium and planogametes in *P. euglenae* and the more advanced somatic hyphae taking over the sexual function in *C. hyalinus*, the phylum has representations to ancient as well as advanced fungi and proves to be the ancestors of all others.

The life cycle of Myxomycota represented by *Physarum polycephalum* (Fig. 1.24) is perhaps the most complicated one for fungi. It includes the 2n clonal cycle between the sporophyte and sclerotium within the sexual phase as well as n mitotic cycle within the clonal phase. The sexual cycle is further complicated by interconversion between myxoamoeba and swarmer.

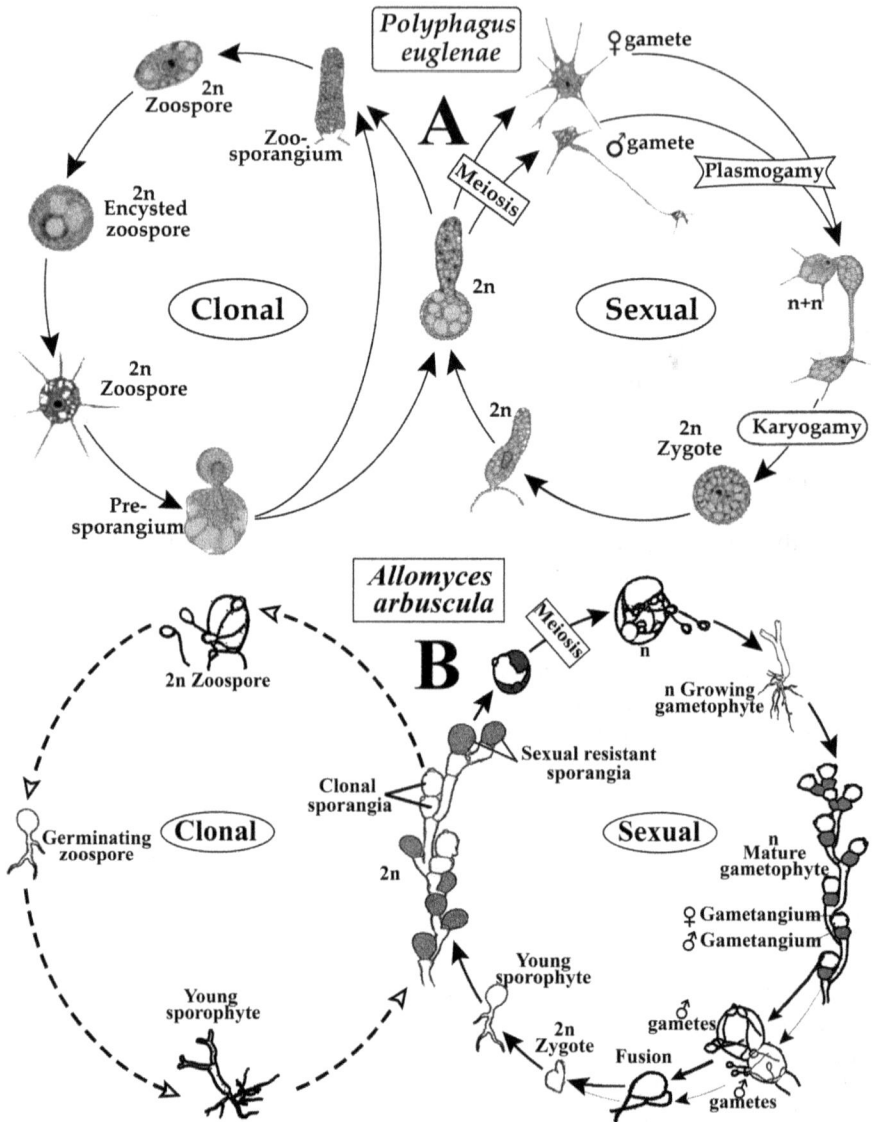

FIGURE 1.23

Life cycles in Chytridiomycota: (A) *Polyphagus euglenae* (modified and redrawn from Alexopoulos and Mims, 1979). Oomycota: (B) *Allomyces arbusculus* (redrawn from Lee et al., 2010). Note the diploid clonal and haploid sexual phases in them.

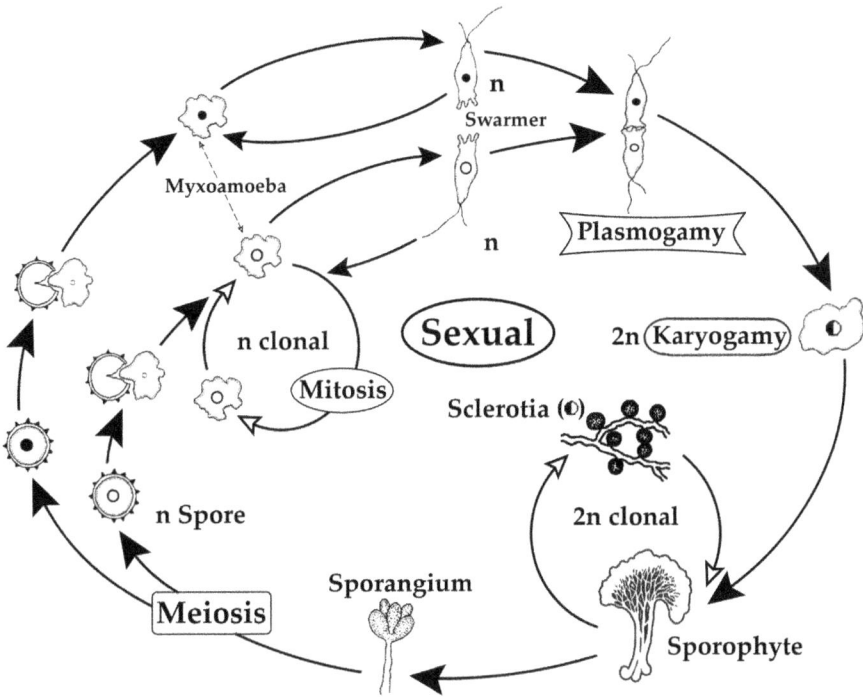

FIGURE 1.24

Myxomycota: Life cycle of *Physarum polycephalum* (modified and redrawn from Webster and Weber, 2007).

Life cycles of the 1,065 speciose Zygomycota are reported for *Rhizopus stolonifera*, *Phycomyces blakeslecanus* (Webster and Weber, 2007) and *Mucor* sp. Commonality characterizes their life cycle, which involves haploid clonal and diploid sexual phases. At conjugation, their complementary hyphae are pheromonally attracted and grow toward each other until they meet. The tips of these hyphae swell, the walls between them dissolve and their cytoplasm merge (cytogamy). Subsequently, their haploid nuclei fuse to form a diploid zygote (Fig. 1.25). During germination, meiosis occurs to restore the haploid status in hypha (see Margulis et al., 2009). Two points may be noted: (a) *unlike in Metazoa and higher plants, in which meiosis precedes fertilization, the meiosis is followed by fusion in Zygomycota.* (b) *In the former, karyogamy, i.e., nuclear fusion precedes cytogamy but cytogamy precedes karyogamy not only in Zygomycota but also in other fungi.*

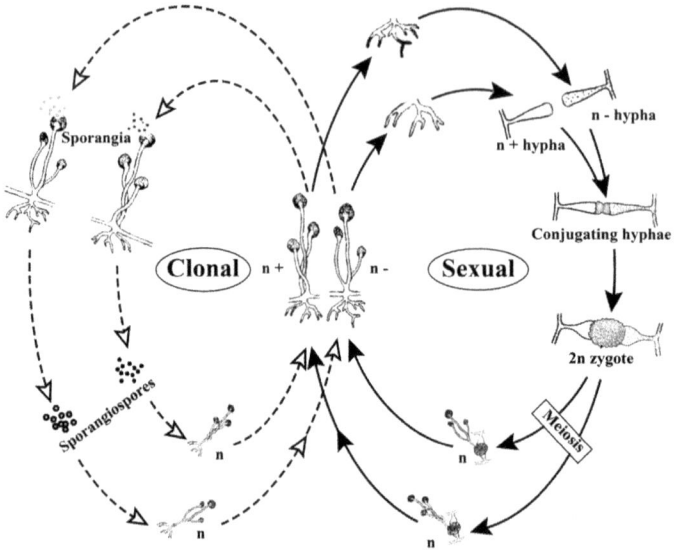

FIGURE 1.25

Life cycle of Zygomycota: *Rhizopus stolonifer* (modified and redrawn from Margulis et al., 2009, Evert and Eichhorn, 2013).

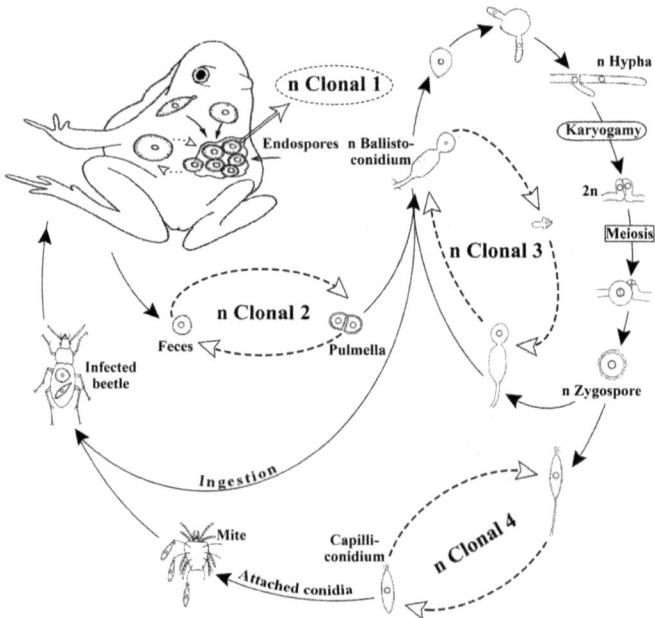

FIGURE 1.26

Entomophthoromycotina: The most complicated life cycle of *Basidiobolus ranarum*. Note the involvement of three hosts intervened by four clonal cycles (modified and redrawn from Webster and Weber, 2007).

The life cycle for the Entomorphthoramycotina, *Basidiobolus ranarum* is illustrated in Fig. 1.26. In it, the cycle is also the most complicated for any known fungi and involves three unusual hosts. To compensate the riskiest cycle involving the infection of the three hosts, four haploid clonal phases are included. The sexual cycle includes conjugation of complementary hyphae and production of 2n zygospore, which germinates into the ballistoconidium.

Derived from the zoosporic chytrids, the Neocallimastigomycota diverged from the iguanid reptiles and spread among ruminants and herbivorous rodents (Gruninger et al., 2014). They are the key players in degradation of recalcitrant lignocellulosic plant fiber in the rumen by physical penetration of the rhizoids and enzymatic digestion. Following attachment to the plant material (Fig. 1.27), the zoospore sheds flagella and encysts. The cysted zoospore, on reaching the rumen along the attached plant, is germinated forming a thickened cell wall and develops anucleated rhizoids in monocentrics (e.g., *Caecomyces, Neocallimastix, Piromyces*) or nucleated rhizoids in polycentrics (e.g., *Anaeromyces, Cyllamyces, Orpinomyces*). On receiving an induction signal from the digesting plants, the rhizomycelium differentiates into the external sporangium, which is filled with zoospores. The fact that it is very difficult to maintain the ruminant free of these anaerobic fungi attests the efficient dispersal of neocallimastigomycotes.

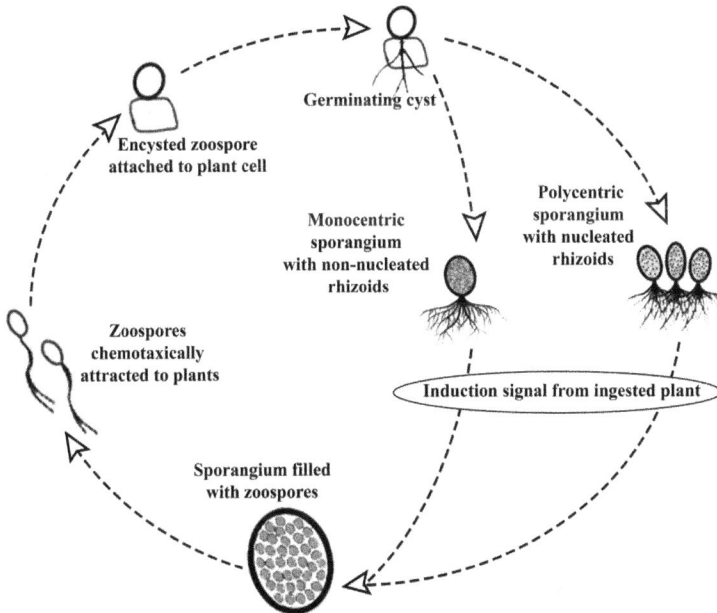

FIGURE 1.27

Schematic life cycle of a typical Neocallimastigomycota (modified and redrawn from Gruninger et al. 2014).

Repeated computer search has not yielded any report on the description of the life cycle for the 230 speciose Glomeromycota.

The phylum Basidiomycota consists of three subphyla namely Ustilaginomycotina, Agaricomycotina and Pucciniomycotina. The 1,000 speciose smuts of Ustilaginomycotina are all plant parasites. Most of them infect angiosperms but a few are parasites of ferns and lycopodiales. Their heterogonic life cycle is described in detail for *Ustilago maydis* (Fig. 1.28). The pheromonally attracted complementary haploid yeasts conjugate to form the plasmogamic dikaryotic hyphae (see Watkinson et al., 2015). On infection, their heterokaryotic hypha penetrates the host tissues and causes the development of tumor or gall. The tumor development within the flower of *Zea mays* leads to the formation of masses of swollen kernels. The mycelium within the tumor is transformed into blackened teliospore; following their nuclear fusion, the diploid teliospores are dispersed by wind. The promycelium germinates from a teliospore after meiosis. The ustilaginomycotine cycle is characterized by uniformity and involves the budding clonal phase at the external saprotrophic yeast and internal sexual phase within the biotrophic mycelial phase. *Tilletia tritici* is another ustilaginomycotine, for which the life cycle is described.

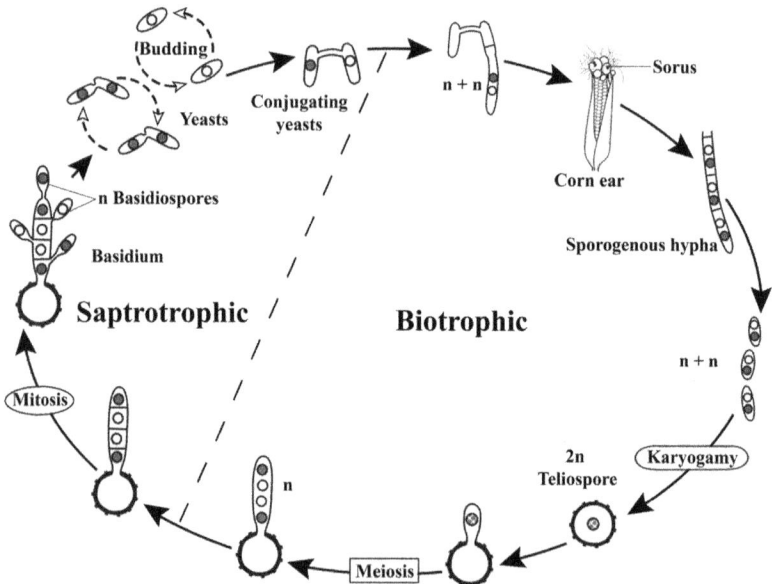

FIGURE 1.28

Life cycle of a Ustilaginomycotina: *Ustilago maydis*. Note the shorter budding clonal cycles of yeast stage (modified and redrawn from Webster and Weber, 2007, Begerow et al. 2014).

The subphylum Agaricomycotina consists of three classes namely (i) Dacrymycetes, (ii) Tremellomycetes and (iii) Agaricomycetes. For the 329 speciose Dacrymycetes, a proposed life cycle for *Dacrymyces stillatus* (Fig. 1.29) is described by Oberwinkler (2014). Their heterogonic cycle involves clonal multiplication through diploid conidiospores or fragmentation of diploid hyphae. Their sexual cycle comprises the haploid basidiospores; their conjugation results in the formation of diploid sporophyte.

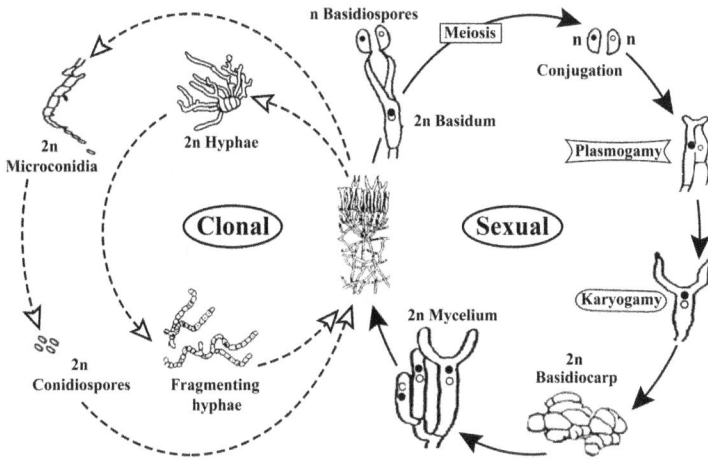

FIGURE 1.29

Dacrymycetes: The tentative life cycle of *Dacrymyces stillatus* (based on Oberwinkler, 2014).

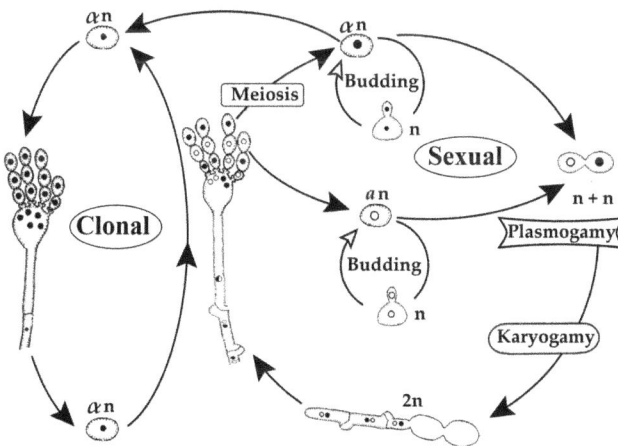

FIGURE 1.30

Tremellomycetes: Life cycle of *Filobasidiella neoformans*. Note that only α (but not the *a*) mating type undergo haploid clonal phase (modified and redrawn from Webster and Weber, 2007).

Only two reports are available for the description of the life cycle in Tremellomycetes. Figure 1.30 illustrates the life cycle of *Filobasidiella neoformans*. In it, the sexual phase is also complicated by the inclusion of clonal budding of α and *a* mating types. Notably, only the α mating type is capable of undertaking the haploid clonal cycle.

The 13,113 speciose Agaricomycetes comprises six orders. For them, description for the cycle is available for the 8,500 speciose Agaricales, 402 speciose Auriculariales and 1,800 speciose Polyporales (Mukhin and Votintseva, 2022). Typically, the cycle is characterized by uniformity and unusual restriction to the sexual phase alone in the Agaricales like *Agaricus bisporus*, *Schizophyllum commune*, *Coprinus* (see Webster and Weber, 2007), *Flammulina velutipes* and *Lentinula edodes* as well as Boletales, Russulales and Phallales. Their cycle begins with the dikaryotic n + n mushroom stage; following meiosis, the complementary haploid basidiospores in the basidium plasmogamically fuse to produce the dikaryotic spores, which are released from the mushroom or the spores are fused externally to form the plasmogamic hyphae. In these spores, the external karyogamy takes place in air. On germination of the spores, their complementary hyphae conjugate to generate the plasmogamic n + n dikaryotic fruiting structure, the mushroom (Fig. 1.31).

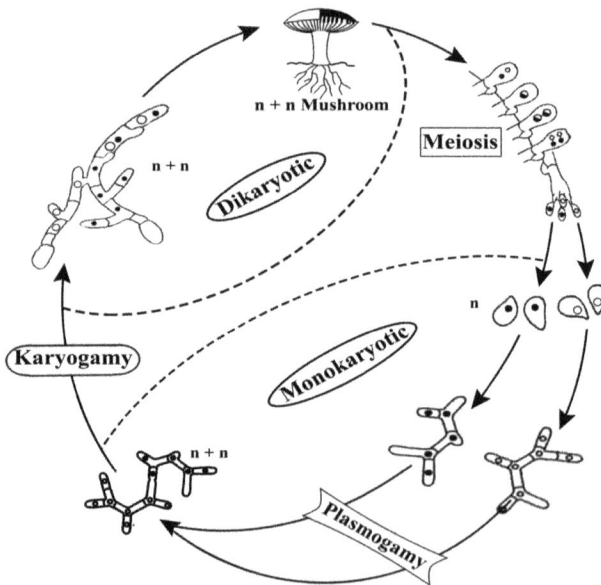

FIGURE 1.31

A generalized sexual life cycle of Agaricales represented by *Agaricus bisporus* and *Schizophyllum commune* (compiled from Palmer and Horton, 2006, Nieuwenhuis, 2012, *Wikimedia*,). Nieuwenhuis indicates hyphal mating but others basidiosporal mating.

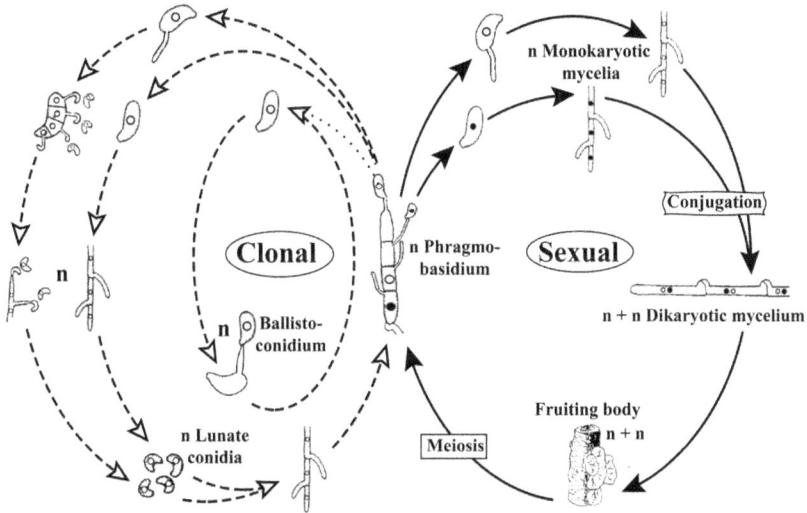

FIGURE 1.32

Auriculariales: Life cycle of *Auricularia auricula-judae*. Depending on the nutrient status, basidiospore may germinate into a monokaryotic mycelium or lunate microconidia (based on Webster and Weber, 2007).

Information is available on the life cycle for only the members auriculariales and polyporales. Understandably, the cycle is not yet described, as the Boletales fungi are ectomycorrhizas in the roots of forest trees (see Watkinson et al., 2015). Description for the cycle of the parasitic, but commercially important delicacy of *Auricularia auricula-judae* is reported. It includes both clonal and sexual phases. The sexual phase comprises conjugation between complementary monokaryotic hyphae resulting in n + n fruiting body. Following meiosis, the phragmatic basidium is germinated (Fig. 1.32). A similar life cycle is also described for the 1,800 speciose Polyporales (Mukhin and Vontinseva, 2002). Hence, the auriculariale complex consists of 2,202 species, but that for Agaricales is 10,911 species.

The subphylum Pucciniomycotina consists of one of the most diverse fungal taxa; but it members are united in possessing simple septal pores that lack dolipores and pore caps. They are ecologically and economically important fungi. Most of them, but not all produce phragmobasidia. Within nine classes, the 25 speciose Cystobasidiomycetes exist as yeasts throughout their lifespan, while yeast is a stage in clonal phase of 314 species in five classes (Table 1.3). The remaining 8,127 species, especially Pucciniomycetes (7,800 species) involve either heteroecious macrocycle or autoecious microcycle. The macrocycle includes five sporous stages namely (i) spermagonium, (ii) aecium, (iii) uredinium, (iv) telium and (v) basidium. To minimize the risks involved in the two-hosted macrocycles,

TABLE 1.3

Modes of life and reproductive characteristics of Pucciniomycotina (from Aime et al., 2014). Myco = mycoparasites, Sapro = saprotroph, Phyto = phytoparasites, Phragmo- = Phragmobasidium.

Class	Mode of life	Clonal		Sexual	
		Yeast	Conidium	Fruiting body	Clamp
Cryptomycocolamycetes	Mycoparasites	2	0	Holo	+
Cystobasidiomycetes	Mycoparasites	23	2	Holo, phragmo	+
Agaricostilbomycetes	Myco + sapro	155	155	Phragmo, holo	+/−
Classiculomycetes	Myco, aquatics	0	2	Phragmo	+
Atractiellomycetes	Sapro + mycorrhizal	0	50	Phragmo	+
Microbotryomycetes	Aquatics, phyto + myco	150	50	Phragmo	+
Mixiomycetes	Phytoparasites	1	0	?	−
Pucciniomycetes	Phyto or myco	−	7800	Phragmo/holo	−
Tritirachiomycetes	Sapro, parasitic, myco	6	0	?	−

Aquatics = 202 species, mycorrhizas = 50 species, plant parasites = ~ 3,984 species, mycoparasites = ~ 4,462 species; Clonal multiplication: yeast = 337 species, conidium – 8,059 species.

one or more stages are deleted. It is a common strategy among organisms; for example, the digenean Platyhelminthes delete one or more life stage(s) and abbreviate the cycle (see Pandian, 2020). Table 1.3 lists counts on life style of these fungi. For Pucciniales, Petersen (1974) classified these deletions into 10 groups. In all of them, the diploid telial and basidial stages are obligately retained. But in 7 of the 10 groups, the spermagonium is deleted. In 6 of the 10 groups, aecium and uredinium are also deleted, in addition to spermagonium, and thereby minimize the cycle to a single host. However, in the bean rust *Uromyces appendiculatus* (also *U. striatus*), all stages are retained, but in a single

Heterocious macrocycle	Host 1	Host 2
Puccinia graminis		Cereals, oat, rye
a) *P. graminis tritici*		*Triticum aestivum*
b) *P. graminis stiiformis*	*Berberis vulgaris*	Stripe rust
c) *P. graminis avenae*		*Avena sativa*
d) *P. graminis secalis*		*Secale cereale*
P. coronata	*Avena* spp	*Rhamnus* spp
Melampsora larici-epitea	Larch leaf	Willow leaf
Gymnosporangium juniperi virginiana	Apple	Cedar
Cronartium ribicola	Pines	*Ribes*
Gymnoconia peckiana	Trotter	*Rubus*
Autoecious microcycle		
Uromyces striatus	Alfalfa (*Medicago sativa*)	

host (*wiki.budwood.org*, see also Figure 7.4). The undermentioned are some examples for the heteroecious macrocycles. The autoecious microcycle takes place in *P. helianthi*, *P. malvacearum*, *Hemileia vastatrix*, which parasitize *Helianthus*, hollyhock and coffee, respectively.

For the non-pucciniales, descriptions are available for the cycles of Cystobasidiomycetes and Microbotryomycetes each for a single species (see Webster and Weber, 2007). In *Sporidiobolus salmonicolor*, the meiosis-failed basidiospores are transformed into diploid yeast cells. These 2n yeast cells and 2n teliospores undergo clonal multiplication by unicellular conjugation (Fig. 1.33B). Following meiosis, the teliospores are transformed into saprophytic basidium. The n basidiospores undertake plasmogamy and karyogamy to generate diploid teliospores.

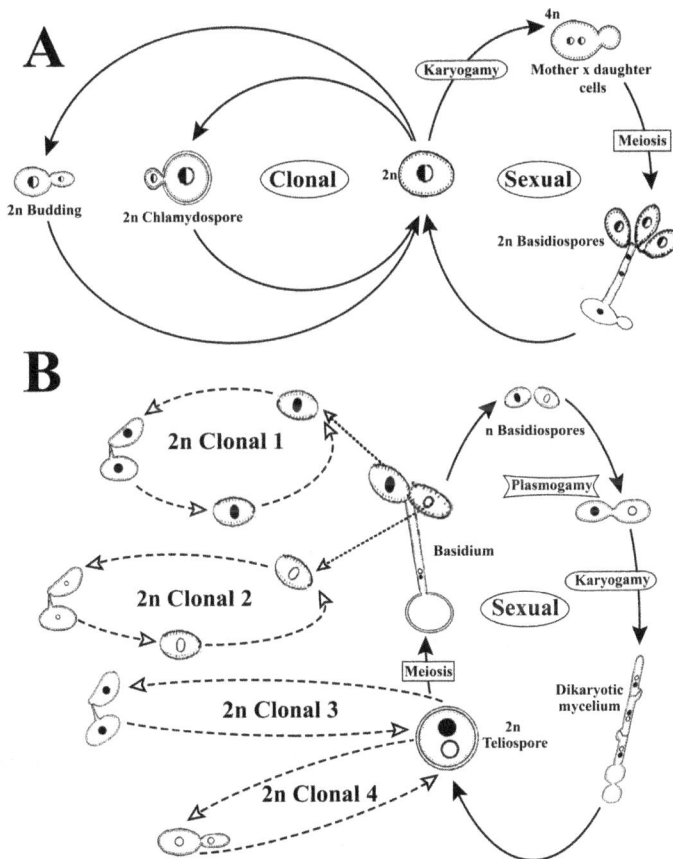

FIGURE 1.33

Pucciniomycotina: (A) Life cycle of cystobasidiomycetic homothallic *Xanthophyllomyces dendrorhous* and (B) microbotryomycetic heterothallic *Sporidiobolus salmonicolor*. Note the inclusion of four clonal phases, of which the first three involves yeast conjugation and the fourth one yeast budding (modified and redrawn from Webster and Weber, 2007).

Xanthophyllomyces dendrorhous is a homothallic cystobasidiomycete. Its life cycle includes the clonal phase through chlamydospores and budding yeasts. The sexual phase includes the fusion between the bud and mother cell of a yeast resulting in the production of basidium carrying basidiospores, which harbor diploid dikaryotic nuclei (Fig. 1.33A).

For pucciniales, the cycle is described in detail for ~ 10 species. It could be known for more numbers of species (see Aime et al., 2018). The representative cycle is illustrated in Fig. 1.34 in greater details for *Melampsora larici-epitea*. It includes the monokaryotic sexual phase consisting of spermagonial and aecial stages in larch leaves and the dikaryotic clonal uredinial stage in willow leaves followed by saprotrophic telial and basidial stages.

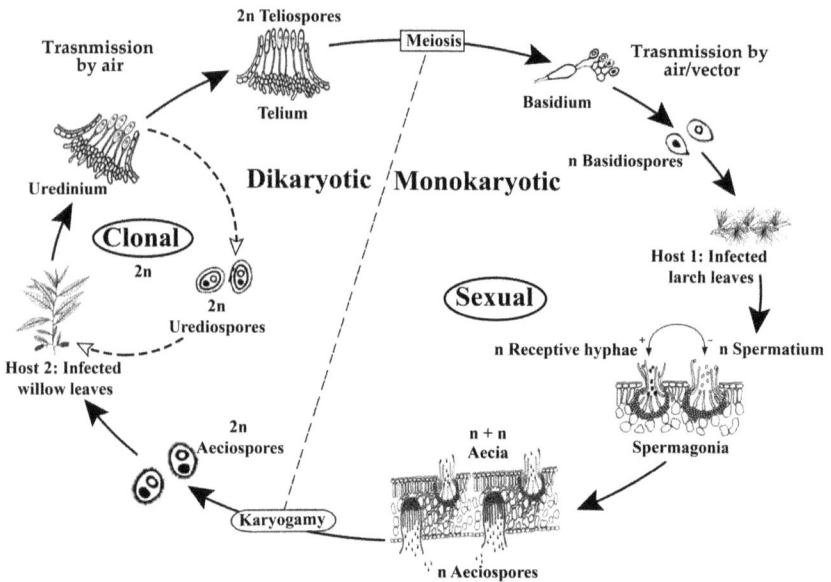

FIGURE 1.34

Puccioniomycetes: Life cycle of the heteroecious macrocyclic rust *Melampsora larici-epitea* (modified and redrawn from Evert and Eichhorn, 2013, Aime et al., 2014).

The life cycles within Saccharomycotina may be considered under three groups: 1. The *Saccharomyces* is characterized by clonal budding and production of four ascospores per ascus (Fig. 1.35A). 2. In *Schizosaccharomyces*, the heterogonic cycle includes the clonal phase by fission and the sexual phase by production of eight ascospores per ascus (Fig. 1.35B). 3. In *Saccharomycodes* spp, the entire sexual cycle is completed within an envelope to produce only four ascospores per ascus (Fig. 1.35C) and they do not clonally multiply. Rarely, the lecanoromycetic Acarosporaceae is highly polysporic. In them, meiosis is followed by several mitoses to generate > 100 ascospore/ascus (Gueidan et al., 2015).

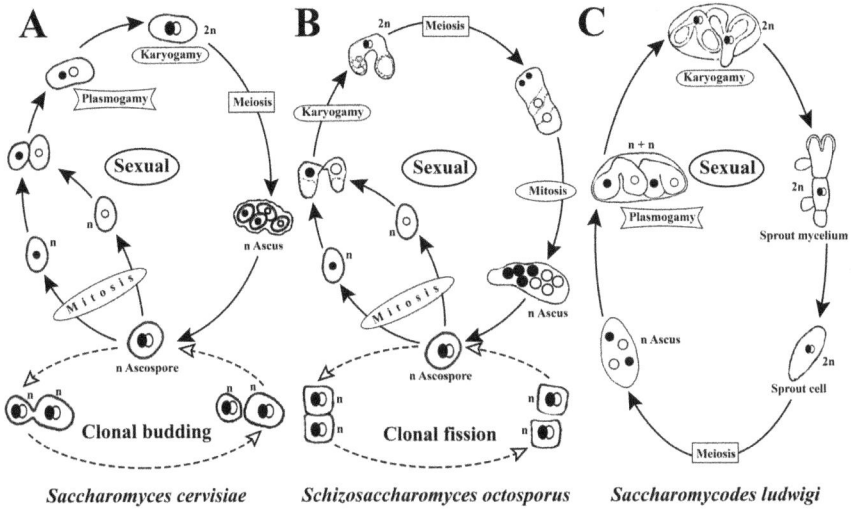

FIGURE 1.35

Representative model life cycles in Saccharomycotina (modified and drawn from Alexopoulos and Mims, 1979).

The life cycle of Taphrinomycotina, as exemplified by *Taphrina deformans*, a peach leaf curl causing parasitic fungi, clonal phase by budding includes yeast cells. In some others, fission in yeast or pseudohyphae may occur. The sexual phase includes karyogamy between homothallic cells followed by formation of eight ascospores per ascus (Fig. 1.36).

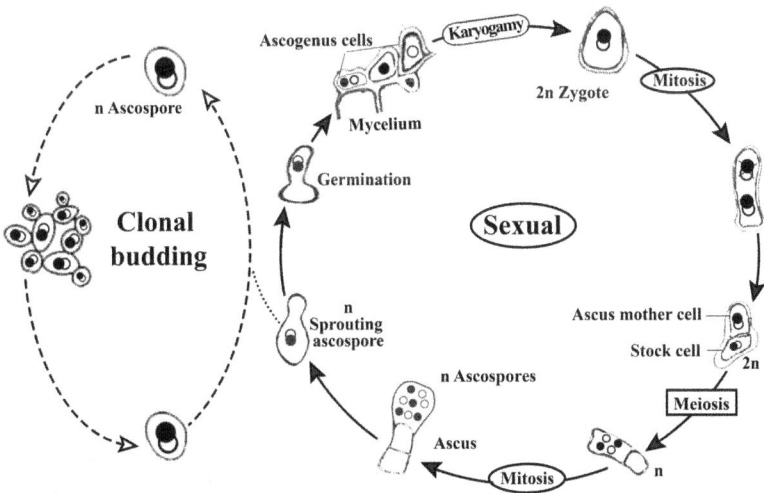

FIGURE 1.36

Taphrimycotina: Life cycle of *Taphrina deformans* (modified from Alexopoulos and Mims, 1979).

Many publications on Lecanoromycetes deal mainly with its taxonomy (e.g., Miadlikowska et al., 2006). They consist of 90% of all known lichen forming fungi. No publication is yet available on reproduction of Lecanoromycetes, although information is available for ascoma development, ascus and ascospores (Gueidan et al., 2015). Gueidan et al. have also provided valuable information on their clonal reproductive phase alone. It is not indicated whether these fungi have a sexual phase also. Accordingly, "Lecanoromycetes disperse clonally by thallus fragmentation aided by undifferentiated or specialized dispersal structures". The structures are corticated and cylindrical thalloidal detachable outgrowths called **isidia** in *Parmelina tiliacea* or ecorticated granules known as **soredia** released through spherical openings, the **soralia** in *Pertusaria flavicans*.

Typically, Dothideomycetes produce a flask-like structure called pseudothecium. In it, the ascus development begins with the accumulation of ascostomum, which eventually develops into germinal hyphae (not shown in Fig. 1.37). Their heterogonic life cycle includes the clonal phase, in which multiplication occurs through haploid conidial spores and the sexual phase, which produces eight ascospores per ascus. Incidentally, lichenization is a primary trait of 266 Arthoniomycetes, a sister group of Dothideomycetes. Their life cycle may be limited to the clonal phase alone (Schoch and Grube, 2015).

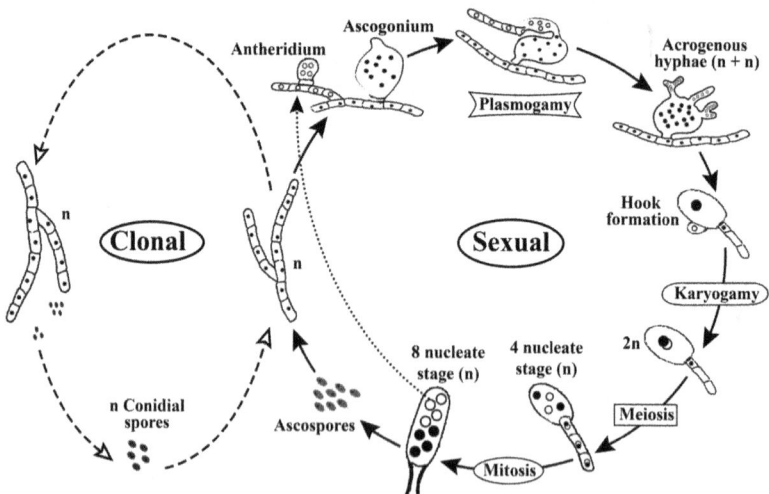

FIGURE 1.37

A generalized scheme of life cycle of Dothideomycetes (based on Pem et al., 2021).

For Eurotiomycetes, the heterogonic life cycle in *Penicillium vermiculatum* includes the clonal phase involving multiplication through conidiospores and the sexual phase including heterothallic mating types and production of eight ascospore/ascus accommodated in cleistothecium (Fig. 1.38). Many authors consider the 250 speciose *Aspergillus* and 225 speciose *Penicillium* as anamorphs (e.g., Geiser et al., 2015). However, Dyer and O'Gorman (2011) have shown the presence and expression of 75 sex-related genes, distribution of mating type genes and detection of recombination from population genetic analyses.

The heterogonic life cycle of Leotiomycetes is complicated with inclusion of three clonal cycles (Fig. 1.39). The macroconidial cycle passes via the dormant n sclerotium stage in the infected host plant → conidiogenic germination → n macroconidiophore → macroconidium → mating type. The more complicated microconidial cycle passes through microconidiophore → n microconidial spores, which add the spore number from microconidial phialides → entry into the sexual phase via plasmogamy between microconidiospore MAT 1 n and n of MAT 2 hyphal nucleus. The latter passes through n + n sclerotium → carpogenic karyogamy → meiosis → ascus → ascospores.

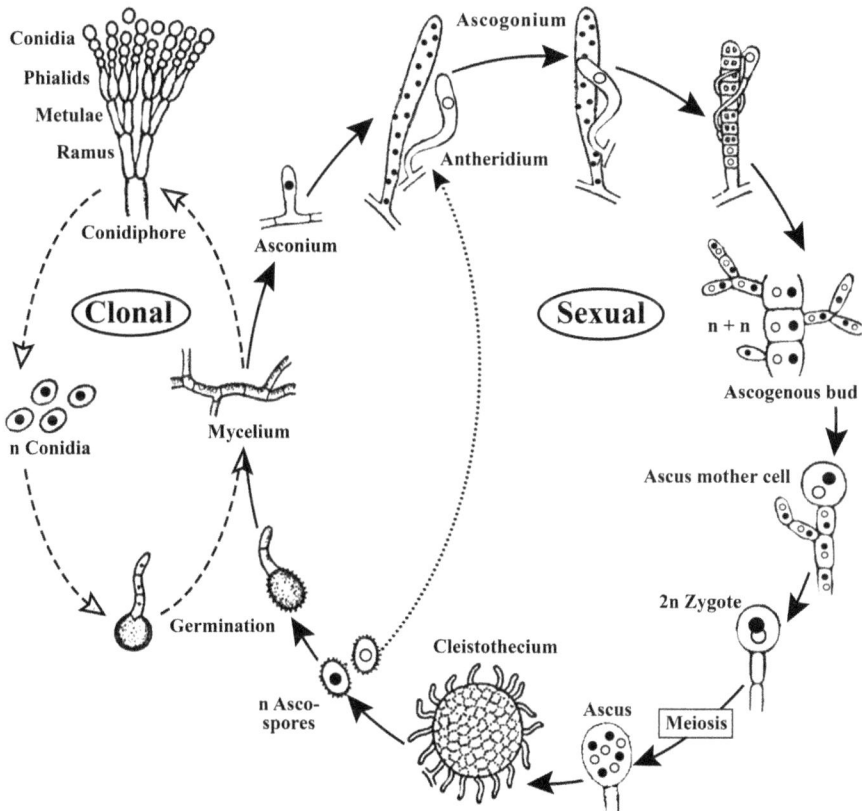

FIGURE 1.38

Eurotiomycetes: Life cycle of *Penicillium vermiculatum* (based on *istudy.pk*).

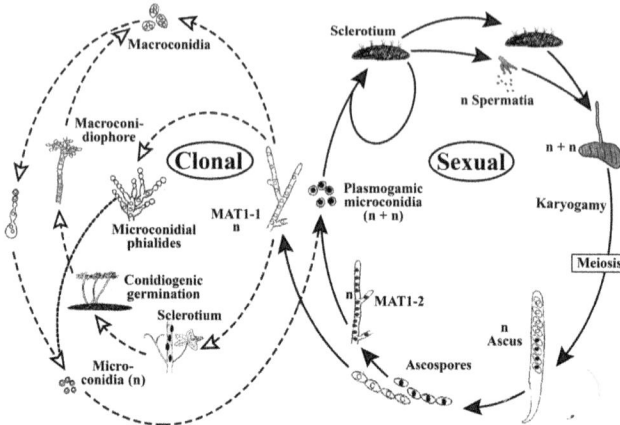

FIGURE 1.39

Leotiomycetes: Compiled life cycle of *Botryotinia fuckeliana* (modified from Webster and Weber, 2007) and *Botrytis cinera* (modified from Amselem et al., 2011).

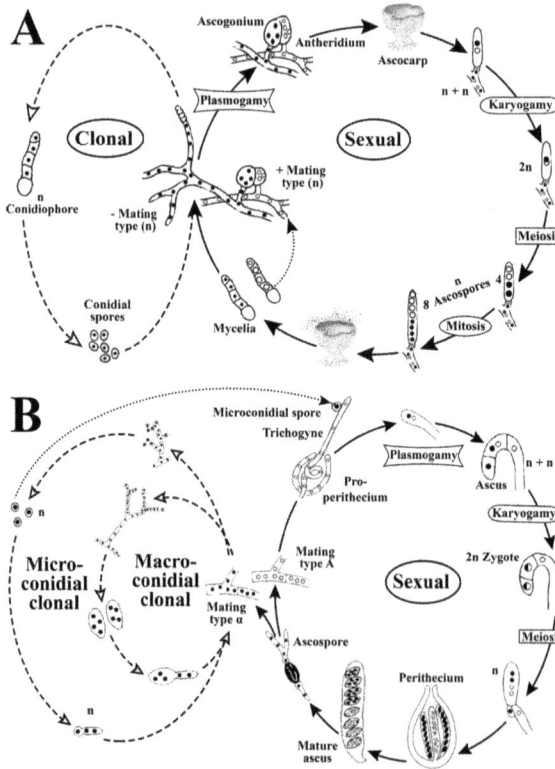

FIGURE 1.40

Sordariomycetes: Life cycle of (A) *Sordaria fimicola* and (B) *Neurospora crassa* (modified from Webster and Weber, 2007).

The cycles for Sordariomycetes are even more complicated. Firstly, they are all heterothallics. Secondly, their clonal phase involves conidial multiplication alone, as in *Sordaria fimicola* (Fig. 1.40A) and both micro- and macro-conidiospores, as in *Neurospora crassa* (Fig. 1.40B). The sexual phase passes through an ascocarp stage in the former but perithecium in the latter. In *N. crassa*, entry into sexual phase commences with plasmogamy between n pro-perithecium and n microconidial spore arising from clonal cycle. This may imply the dependency of the sexual phase on the clonal phase in the leotiomycete *Botryotina fuckeliana* (Fig. 1.39) and sordariomycete *N. crassa* (Fig. 1.40B). Incidentally, the clonal phase of *S. fimicola* occurs on the infected plants but the sexual phase in dung voided by herbivores (see Newcombe et al., 2016).

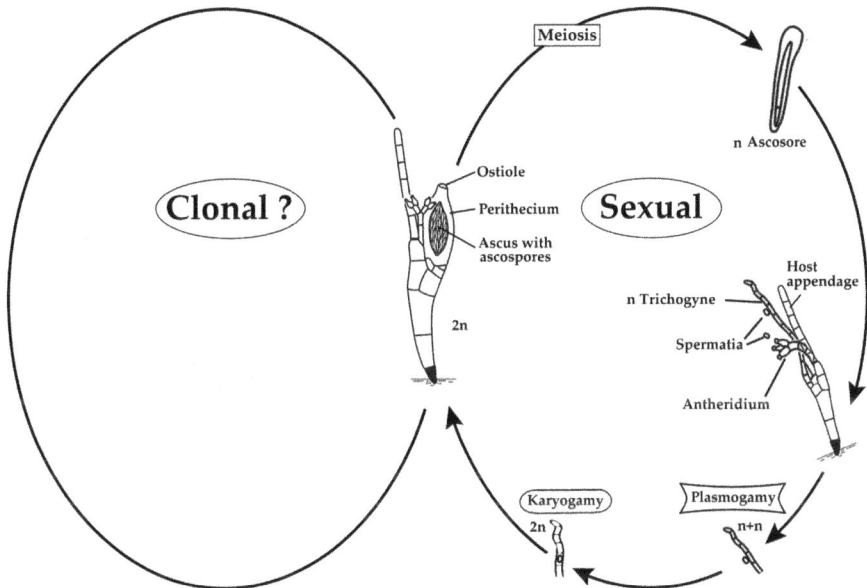

FIGURE 1.41

A generalized reproductive cycle of Laboulbeniomycetes limited to sexual phase alone (modified and redrawn from Pipenbring, 2015b).

Repeated computer search yielded only one description for the generalized life cycle of Laboulbeniomycetes and that too limited to only the sexual phase (Fig. 1.41). The phase is typical of that for ascomycotes. Further searches revealed only passing remarks on the structure of conidium by Humphrey (1891). It remains to be seen whether the conidial spore is a haploid and if so, how diploidy is restored.

For Coelomycetes, descriptions for the life cycle of two species are available. The life cycle of the coelomycete *Erynia conica* is known but it is limited to the clonal phase alone on its host *Simulium* (Hywel-Jones and Webster, 1986, see also Wurzbacher et al., 2011). Whisler et al. (1975) described the complete cycle consisting of both clonal and sexual phases for *Coelomomyces psorophorae*. Being a parasite, its thick-walled sporangia are released from the hyphae of the infected mosquito larva of *Culiseta inornata*. Following meiosis, the haploid motile zoospores of complementary mating types emerge from each sporangium and infect the copepod *Cyclops vernalis*. From it, haploid zoospores are released, which, on mating in water, produce diploid zygospore. In its turn, the zygospore infects a mosquito larva (Fig. 1.42).

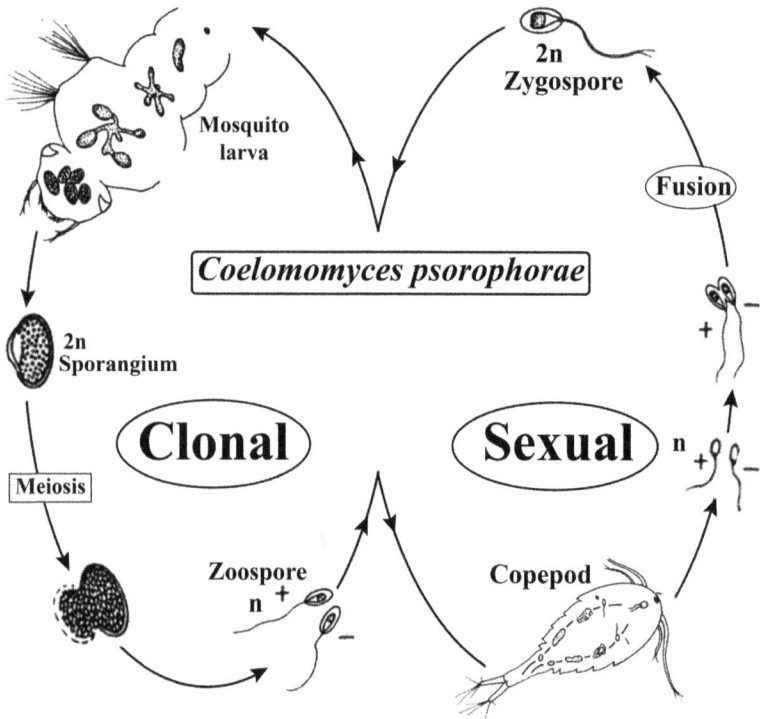

FIGURE 1.42

Life cycle of *Coelomomyces psorophorae* involving mosquito larva and copepod as hosts. Note the occurrence of sexual and clonal phases and external aquatic fertilization (modified and redrawn from Whisler et al., 1975).

For Hyphomycetes, the life cycle is described for only two terrestrial parasitic species. Of the three clonal cycles described for the downy mildew *Plasmopara viticola* parasitic on grapes (Fig. 1.43), the simpler two cycles involve infection of twigs or grapes and eventual production of sporangium, from which zoospores arise. The third cycle commences with infection of the leaf, which includes the retained vestigial relic reproductive structures the antheridium and oogonium. However, no gametic fusion occurs. The oospore germinates into a sporangium, from which zoospores are released, which encyst prior to infecting new host plants. Sporangium may also directly be produced from the infected twigs bearing grapes or leaf.

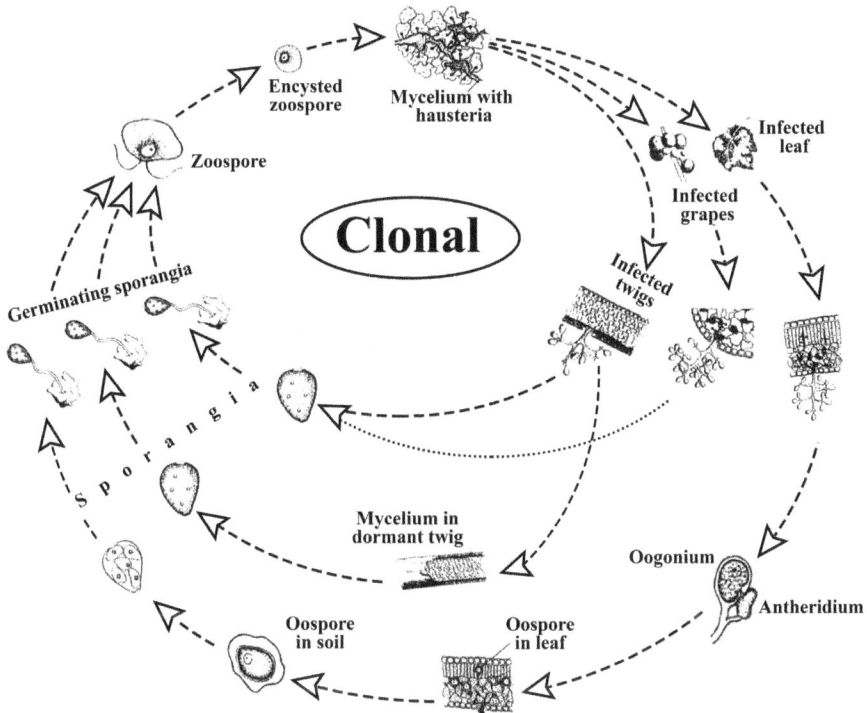

FIGURE 1.43

Hyphomycetes: Life cycle of downy mildew *Plasmopara viticola* on grapes (modified and redrawn from Jackson, 2008).

For the other anamoprhic hyphomycete, the cycle described is also limited to the clonal phase alone in *Phoma ligulicola* (Fig. 1.44). The fungus infects pyrethrum and chrysanthemum. The infection on chrysanthemum is limited to pycnidium, which infects through the conidia arising from (i) the infected pyrethrum or (ii) the infected seeds. New host plants of pyrethrum are infected by pycnidia held in the dormant pseudosclerotium arising from the infected chrysanthemum or ascospores ejected from the perithecium. The source of haploid sexual ascospore is not indicated by Pethybridge et al. (2008). It may be a vestigial relic retained by the fungus, when it has secondarily lost the sexual cycle. Magyar et al. (2011) reported conidial clonal multiplication in anamorphic *Pyrigemmula aurantiaca* and phylogenetic relation for 41 hyphomycete species but have not described the life cycle. Obviously, their cycle is limited only to the clonal phase.

Approximately 30% hyphomycetes (153 species) and 22% coelomycetes (~ 2,200 species) occur and flourish in aquatic habitats. Wurzbacher et al. (2011) listed the life cycles for aquatic species. However, the life cycles for only four species are described for the anamorphic Coelomycetes and Hyphomycetes. Within them, there is an interesting series among clonal species, in which one or more sexual stages are retained as relic(s). *Erynia conica* (see Wurzbacher et al. 2011) has retained none of the sexual stages. However, *Phoma ligulicola* has retained the ascospore stage. *Coelomomyces psorophorae* has retained the entire sequence of sexual phase (Fig. 1.42).

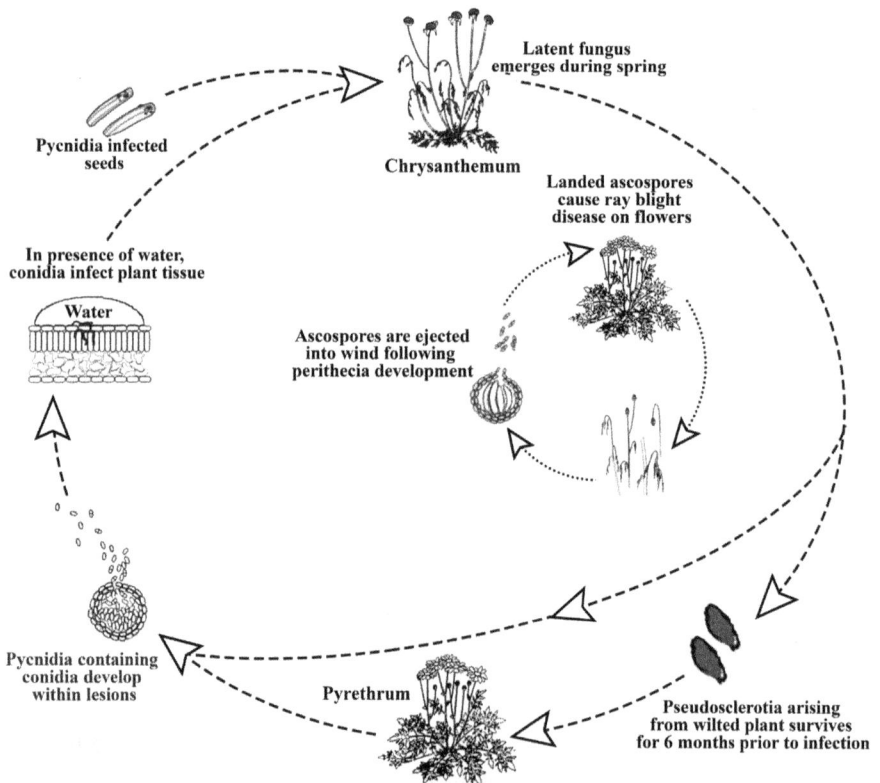

FIGURE 1.44

Hyphomycetes: Clonal life cycle of *Phoma ligulicola* infecting pyrethrum and chrysanthemum (modified from Pethybridge et al., 2008).

The number of life cycles described for fungi is evaluated in Table 1.4. The microscopic size of most fungi and/or their non-amenability to laboratory culture have impeded the description of the life cycle of many fungi, especially the parasitic ones involving two (Fig. 1.34) or three (Fig. 1.26) hosts. Like in plants, the life cycle of fungi is direct and does not involve trophic or non-trophic larval stage(s). A survey revealed the availability of reliable life cycles only for ~ 77 species, i.e., 0.07% of fungi. This value may be compared with < 0.07% for protozoans (Pandian, 2023) and 3.0% for polychaetes (Pandian, 2019). Repeated computer searches have not yielded relevant information for the life cycle of Boletales, Russulales, Phallales among Agaricomycetes, Agaricostilbomycetes, Atractiellomycetes, Classiculomycetes, Mixiomycetes, Cryptomycocolamycetes and anamorphic Tritirachiomycetes among Pucciniomycotina, and Lecanoromycetes among Pezizomycotina (see Table 1.1). Notably, the cycle is known only for a representative species for many clades. The description for *Penicillium vermiculatum* (Fig. 1.38) may hold good for the 354 species belonging to the genus *Penicillium*. This may also hold good for the 180 speciose *Aspergillus* and ~ 20 speciose *Fusarium* among Sordariomycetes. *Descriptions are required for the cycles, especially for Ascomycota and Basidiomycota, for which the life cycles are known only for 0.04% and 0.10%, as compared to 0.65% for Chytridiomycota and 0.38% for Zygomycota* (Table 1.4). *Briefly, the descriptions for the life cycle of many taxonomic groups are limited to only sexual or clonal phase or only one species.*

TABLE 1.4

Estimation on the number of fungal species, for which life cycle is described.

Clade	Known cycle (no.)	Reference
Chytridiomycota *Albugo candida, Chytriomyces hyalinus, Lagenidium rabenhorstii, Monoblepharis polymorpha, Olpidium viciae, Saprolegnia, Nowakowskiella ramosa, Polyphagus euglenae, Phytophthora infestans, Pythium debaryanum, Rhizophydium couchii, Synchytrium endobioticum*	12 (0.65%)	Alexopoulos and Mims (1979)
Oomycota *Allomyces arbusculus, A. macrogynus*	2	Webster and Weber (2007)
Myxomycota *Physarum polycephalum, Ceratiomyxa fruticulosa, Rhizidiomyces apophysatus*	3	Webster and Weber (2007)
Zygomycota *Rhizopus stolonifer, Phycomyces blakesleeanus, Basidiobolus ranarum, Mucor* sp,	4 (0.38%)	Webster and Weber (2007)
Basidiomycota	25 (0.10%)	

Table 1.4 contd. ...

...Table 1.4 contd.

Clade	Known cycle (no.)	Reference
Ustilaginomycotina *Tilletia caries, T. tritici, Ustilago maydis*	3	Begerow et al. (2014)
Dacrymycetes *Dacrymyces deliquescens, D. stillatus*	2	Oberwinkler (2014)
Tremellomycetes *Tremella mesenterica, Filobasidiella neoformans*	2	Webster and Weber (2007)
Agaricales *Agaricus bisporus, Exidia glandulosa, Coprinus,* *Flammulina velutipes, Lentinula edodes,* *Schizophyllum commune*	6	Webster and Weber (2007)
Pucciniomycotina		
Cystobasidiomycetes *Xanthophyllomyces dendrohous*	1	Webster and Weber (2007)
Microbotryomycetes *Sporidiobolus salmonicolor*	1	Webster and Weber (2007)
Pucciniomycetes *Gymnosporangium juniperi virginiana,* *Cronartium ribicola, Gymnoconia peckiana,* *Melampsora larici-epitea, Puccinia coronata,* *P. graminis, P. helianthi, P. malvacearum,* *Uromyces striatus, Hemilieia vastatrix,* *Rhodosporidium sphaerocarpum*	10	Aime et al. (2004)
Ascomycota	31 (0.04%)	
Taphrinomycotina *Taphrina deformans*	1	Alexopoulos and Mims (1979)
Saccharomycotina *Saccharomyces cerevisiae, Schizo-saccharomyces* *octosporus, Dipoda-scopsis uninucleatus,* *Saccharomycodes ludwigii*	4	Alexopoulos and Mims (1979)
Pezizomycotina		
Pezizomycetes *Tuber melanosporum*	1	Pipenbring (2015a, b)
Dothideomycetes *Venturia inaequalis, Elsinoe veneta,* *Mycosphaerella tulipiferae*	3	Pem et al. (2021)

Table 1.4 contd. ...

...Table 1.4 contd.

Clade	Known cycle (no.)	Reference
Eurotiomycetes *Penicillium chrysogenum, Aspergillus fumigatus, Talaromyces vermiculatus, P. vermiculatum*	4	Ghany and El-Sheikh (2016)
Leotiomycetes *Botryotinia fuckeliana, Rhytisma acerinum, Sclerotinia sclerotiorum, Monilinia fructicola*	4	Amselem et al. (2011)
Sordariomycetes *Sordaria fumicola, Neurospora crassa, N. sitophilia, Nectria cinnabarina, Neonectria ditissima, Fusarium graminearum, F. solani, Magnaporthe oryzae, Claviceps purpurea*	9	Kuck et al. (2009)
Laboulbeniomycetes	1	Pipenbring (2015a, b)
Hyphomycetes *Plasmopara viticola, Phoma ligulicola*	3	Jackson (2008), Wurzbacher et al. (2011), Pethybridge et al. (2008)
Coelomycetes *Coelomomyces psorophorae, Erynia conica*	1	Whisler et al. (1975)
Total	77 (0.07%)	

2

Spatial Distribution

Introduction

Life has existed in oceans longer than on land. Fossils reveal the existence of prokaryotic bacteria over 3.7 and 3.1 billion years ago (BYA) in the oceans and on land, respectively. On land, the greater variety of environmental niches provides a better scope for specialization, which has led to larger occupation of space and in turn, to more biological niches. Species richness on land seems to have overtaken that in the oceans ~ 125 million years ago (MYA) (Costello and Chaudhary, 2017). This has coincided with diversification of flowering plants, which comprise 79% of photosynthetic eukaryotes (see Pandian, 2022). Flowering plants have increased terrestrial productivity (52% global productivity, Field et al., 1998) to almost equal to that in the aquatic systems. To complete recycling of matter in the trophic dynamics, the need is obvious for the emergence and diversification of decomposing organisms like eukaryotic fungi and prokaryotic bacteria. Being sessile and acquiring micronutrients alone via the body surface, the fungi are characterized by (i) structural simplification, (ii) limited number of tissues, (iii) heterogonic but direct cycles and (iv) mode of dispersal mostly through spores alone. Hence, they differ from those of animals and plants. As complements, they help us to expand our understanding of the phenomenon of life and different evolutionary strategies.

2.1 Habitat Distribution

Oceans cover 70% of the Earth's surface with 97% of its water. Freshwater systems, however, cover only 1% and hold as small as 0.01% of its water. The remaining 29% of the Earth is covered by land. Holding water mass amounting to 1.3 billion km^3 (*jbutler@uh.edu*), the oceans provide 900 times more livable volume of space than that on land. It must, however, be noted that ocean and land masses are not isolated entities. There are a lot of exchanges between them. On a large scale, for example, the Nile, prior to the construction of dams, exported 1,820 million m^3 of water along with alluvium (see Pandian, 1980), spores and mycelia of several fungi (cf LeBrun et al., 2018) into the Mediterranean Sea. On the other hand, some 64,600 species of insects emerge (carrying aquatic fungi attached to their body) from freshwater source after their larval sojourn (see Pandian, 2021a). The dragonfly may serve as an example for mini-scale exchange between land and the limnic habitats. A sum of 7.3 million eggs is imported into the Idumban Pond (Palani, Tamil Nadu, India) through oviposition by *Brachythemis contaminata* and 23,990 are exported into the land by emerging adults (Mathavan and Pandian, 1977). Over 80 million metric tons (mmts) of fish (carrying aquatic fungi) are imported into almost all parts of the land through capture fisheries (see Pandian, 2015).

At this juncture, two points should be noted. 1. In water, penetrability of solar radiation diminishes to zero between ~ 200 and ~ 400 m depth. The absence of light and photosynthetic activity at > 400 m depth eliminates the existence of plants but allows that of filter- and sediment-feeders and carnivorous animals up to 10,000 m depth, albeit at drastically reduced number of species and individual (indi)/species. For example, the reductions for sponges on the continental shelf of the Bahamas are from 10 species and 35 indi/species at the surface to 5 species and 2 indi/species at the depth of 450 m (see Pandian, 2021b). They are also reflected in the fungal flora. More importantly, most fungi are propagated by spores. The dispersal of 97% clonal spores is achieved externally in the air. *Surprisingly, this fact - not recognized by mycologists - has limited the number of fungal species in aquatic habitats to < 4% (p 65).* Of the global biomass of 550 gigaton carbon (GtC), marine fungi constitute just 0.03 GtC or 0.05%, in comparison to 12 GtC or 2.2% terrestrial fungi (Vargas-Gastelum and Riquelme, 2020). This fact also supports the influence of mega-level difference in the density of the media and their effects on spore dispersal and fungal standing biomass. *A second important factor – also not recognized by mycologists – is the obligate need for external digestion in fungi, which has limited their aquatic diversity. All free-living decomposing fungi have to obligately secrete and apply digestive enzymes on their food prior to digestion and absorp micronutrients via the body surface. The scope for faster dilution of these enzymes in aquatic milieu is greater than that in terrestrial habitats.* This is elaborated in Chapter 4. The Third reason is their highest tolerance to the lowest water potential, often found in dry land (see p 74).

Among fungi, the motile zoospores are engaged by (i) the 1,840 speciose chytrids, 500 speciose Oomycotes and 260 speciose Neocallimastigomycota (see Table 1.1), as well as the vestigial relics by Hyphomycetes for clonal multiplication alone (e.g., Figs. 1.42, 1.43). In Myxomycota *Physarum polycephalum*, the swarmers are limited to sexual reproduction (Fig. 1.24). In all, not more than 2,600 species (or 2.4% fungi) may engage the zoospores for clonal and/or sexual propagation(s). Notably, the size of the chytrid zoospores is limited between 1 and 6–7 µm (Wurzbacher et al., 2011). Consequently, their dispersing ability may be limited to short durations and shorter distances of a few micrometers. In contrast, the urediniospores of wheat yellow rust *Puccinia striiformis* were lifted by air from South Africa to Australia at an altitude of 1,200 m, covering a distance of 9,700 km in 5 d (Nagarajan and Singh, 1990). Another air-borne urediniospores of the coffee rust *Hemileia vastatrix* were dispersed across the Atlantic from African Angola to South American Brazil at 1,500–2,000 m altitude in ~ 6 d (Bowden et al., 1971, see also Table 9.4). Incidentally, the reproductive brown alga *Ascophyllum* attached to wood was carried from the east coast of South America to the west coast of Africa at the speed of 13 km/d over a period of 430 d for its passage of 5,500 km (see Pandian, 2022). *Note the difference of dispersal speed is at ~ 1,000 km/d by air but 13 km/d through water. Briefly, dispersal via air is cheaper and faster than through water.*

Marine fungi: Like most organisms, fungi have also originated from the seas. For reasons discovered and described in the earlier pages, denser sea water has not facilitated fungal evolution and speciation. Since the discovery of the first aquatic ascomycote *Phaeosphaeria typharum*, efforts have been made to update the number of marine fungi to 530 species in 209 genera, 56 families and 26 orders (Jones et al., 2009) and the erection of the new superphyla Opisthosporidia (Pang and Jones, 2016). The number of species described increased from 1 in 1850s to 43 in 1980s and to 145 during 1990s prior to declining to 43 during 2010 (Jones et al., 2009), i.e., the number increased from 4.3 species/y during 1980s to 14.5 species/y during 1990s and returned to the original level. Hence, the number of fungal species from the marine habitat may not exceed 1,100 (Ramesh et al., 2021) more precisely 1,112 species (Sarma, 2019, Vargas-Gastelum and Riquelme, 2020). These 1,112 species are constituted by the following groups:

Phylum	(no.)	(%)	Phylum	(no.)	(%)
Chytridiomycota	26	2.3	Unidentified filamentous fungi	44	4.0
Zygomycota	3	0.3	Others including yeasts	213	19.2
Basidiomycota	21	1.8			
Ascomycota	805	72.4			

Spore size is an important factor that determines the ability of fungi to recolonize the marine habitats. For example, the smaller sporangiospores of 20–50 µm size (see Wurzbacher et al., 2011) are unable to recolonize the marine habitats. With little larger sized spores, a few Basidiomycota are mostly limited to the littoral area (see Pang and Jones, 2016) or mangroves (Ramesh et al., 2021). With spore sizes up to > 200 μm (see Wurzbacher et al., 2011), a large number of ascomycotes have successfully recolonized different marine habitats.

Ramesh et al. (2021) indicated the existence of 625 fungal species in worldwide mangroves. Of them too, 278 are ascomycotes and 30 are basidiomycotes. In a relatively smaller Mandovi mangroves of Goa, India, Sarma and Raghukumar (2013) cataloged 45 fungal species including 31 for Ascomycota and 2 for Basidiomycota. For more details on their species-specific distribution in an array of marine habitats, Pang and Jones (2016) may be consulted. In the islands like Andaman and Nicobar, Niranjan and Sarma (2008) provided a checklist for 446 fungal species in 216 genera, 96 families, 44 orders and 10 classes. Kohlmeyer and Kohlmeyer (1977) were the first to detect the existence of *Allescheriella bathygena* from 1,722 m and *Periconia abyssa* from 5,315 m depths. With the advent of molecular tools, a large number of fungi, which were collected from oceans, could be identified. *Nigrospora* sp, *Apergillus sydowii* were collected from 800 and 3,000 m depth and cultured at the National Institute of Ocean Technology, Chennai, India (Arumugam et al. 2014, Ganesh Kumar et al., 2021). Table 2.1 lists the most prevalent unicellular yeasts (seven genera) and multicellular filamentous fungi (10 genera) from the deepest Mariana Trench (10,900 m). Vargas-Gastelum and Raquelme (2020) also listed the most common fungal genera in million counts on the sediments of varying depths from 0.1 m to 30 m. Notably, the filamentous fungi *Acremonium*,

TABLE 2.1

The most common fungal genera in deep sea and sediments. Genera common to the trench and sediment are indicated by bold letters (compiled from Vargas-Gastelum and Raquelme, 2020).

Fungi prevalent in deep sea		Conserved fungi in sediments
Filamentous	**Yeast**	
Acremonium	*Candida*	Filamentous
Alternaria	**Cryptococcus**	*Acremonium*
Aspergillus	*Malassezia*	*Aspergillus*
Aureobasidium	*Pichia*	*Cladosporium*
Cladosporium	*Rhodosporidium*	*Fusarium*
Exophiala	**Rhodotorula**	*Penicillium*
Fusarium	*Trichosporon*	Yeasts
Hortaea		**Cryptococcus**
Penicillium		*Malassezia*
Trichoderma		**Rhodotorula**

Aspergillus, Cladosporium, Fusarium and *Penicillium* and unicellular yeasts *Cryptococcus* and *Rhodotorula* are found both in the Marina Trench and sediments of varied ages. From the Antarctic Ocean, *Acremonium fusidioides, Penicillium alliisativi, P. chrysogenum, P. palitanis, P. solitum* and *Pseudogymnoascus verrucosus* were identified and are found culturable (Ramesh et al., 2021). No information is available for fungi from hydrothermal vents and cold seeps. However, some details on microsporidia from methane seeps are available (see Sapir et al., 2014).

Freshwater fungi: Freshwater is relatively less dense than seawater but still a lot denser than air. Since the first description for 6 species during 1950s, the number is increased at the rate of 2.1 species/y (El-Elimat et al., 2021). Grossart et al. (2021) reported the updated species number for the following taxonomic groups. Hence, the chytrids and ascomycotes constitute

Phylum	Species (no.)	Species (%)
Chytridiomycota	1,340	45.5
Zygomycota	198	6.7
Neocallimastigomycota	260	8.8
Blastocladiomycota	220	7.5
Basidiomycota	120	4.0
Ascomycota	560	19.0
Others	247	8.4

the majority off freshwater fungi. Within Ascomycota, Dong et al. (2020) listed 77 species in 46 genera for Dothideomycetes. Regarding habitat and mode of their life, 317 species are found in peat swamps, 270 are lichenized (e.g., *Dermatocarpon fluviatile*), 226 are endomycorrhizals and 40 are parasites (Grossart et al., 2021). Incidentally, it is difficult to comprehend how Grossart et al. included Neocallimastigomycota, which are symbiotic in terrestrial herbivorous mammals, as freshwater fungi. The species number reported by Grossart et al. for ascomycotes is 560, against 738 by El-Elimat et al. (2021). Hence, the number of species for freshwater fungi is to be taken as 2,863 after deleting the 260 species for Neocallimastigomycota, and after the addition of 178 species to Ascomycota. On the whole, aquatic fungi may constitute 3,975 species including 2,863 species from freshwater and 1,112 species from marine habitats. In any case, not more than 4,000 species or 3.7% of fungi may be aquatics.

In freshwater systems, three major types of fungi are recognized: (i) the transient terrestrial fungi that are passively introduced by runoff or wind, which may not survive in water (e.g., *Aspergillus* spp, *Penicillium* spp, see El-Elimat et al., 2021), (ii) amphibious aero-aquatic fungi with one stage of their life cycle in water and another in air-water boundaries. These aero-aquatics do not sporulate underwater but require air exposure to complete their life cycle and (iii) indwellers or ingoldian fungi that are characterized by their ability to sporulate on submerged plants. From the site point of view, freshwater fungi may be considered under (i) standing water in lentic systems and (ii) running water in lotic systems. In running rivers (e.g., LeBrun et al., 2018) and streams (Gorniak et al., 2013), the transport of fungi or their spores is unidirectional, i.e., from the origin head to the sea or local ponds. To retain their site position, the lotic fungi are structurally adapted; they produce large tetraradiate (e.g., *Articulospora tetracladia*) or long sigmoid (e.g., *Anguillospora longissimi*) or filiform branched spores alone. A few others achieve it not as spores but as mycelia or dormant structures associated with plants detritus. Within a distance of 200–1,000 m, all leaves carrying spores or mycelia are generally entrained (see Magyar et al., 2016). In lentic waters, the fungi display an additional feature for dispersal. For example, the totally submerged ingoldian hyphomycetes also produce relatively large multi-radiate, sigmoid or spherical conidia with tips covered with sticky mucilage to facilitate attachment and colonization of a specific substrate. The ascomycotes form microscopic ascomata (< 0.5 mm) contain ascospores characterized by several sheaths or wall ornamentation covered by a gel-like sticky material that facilitate spore dispersal and attachment to new substrate. Briefly, *the spores are structurally adapted in running waters but chemically also (with sticky mucilage) in lentic waters.*

The undermentioned list provides a comparative picture on the proportion (%) of eukaryotes distributed in marine, freshwater and terrestrial habitats.

Kingdom	Marine	Freshwater	Terrestrial
Fungi	1.0	2.7	96.3
Plantae	4.8	7.2	88.0
Metazoa	15.1	7.8	77.1
Protozoa	74.0	20.0	6.0

For reasons described earlier (p 61), fungi are more terrestrial. In contrast, Protozoa are more aquatic, as they are able to exist and diverse only in moisture-dependent soil. In the absence of light, not more than 5% plants occur and thrive in marine and 7% in freshwater habitats. On the other hand, filter- and sediment-feeding and carnivorous animals are distributed up to 10,000 m depth.

2.2 Terrestrial Spore Dispersal

Terrestrial fungi may produce **dry xerospores** or **wettable slimy gloiospores**. According to Magyar et al. (2016), they have developed different strategies for spore dispersal, which involves three stages: (i) liberation, (ii) transport and (iii) deposition. Ballistospore discharge is a trademark feature of basidiomycotes, whereas those of pezizomycotinous fungi are shooters; this explosive discharge is the fastest among fungi. To be air-borne following the discharge, the spore must cross the laminar layer, in which drag is created by spore surface. The layer is thickened at night times. The discharging speed required to detach the spore over the laminar layer ranges from 0.4–2.0 m/s in *Alternaria alternata* to 1.8–2.8 m/s in *Puccinia striiformis* (see Magyar et al., 2016). From their analysis, Oneto et al. (2020) suggested that timing of spore liberation may be finely tuned to maximize fitness during airborne transport. Spores released during the day fly for several days, whereas those released at night return to the ground within a few hours. These differences are caused by intense turbulation during the day and the weak ones at night. Tropical fungi appear to release spores at night.

Transport: Following launching into air, wind plays a key role in the transport of fungal spores. Airborne spores are primarily present in the troposphere. Dispersal of airborne spores is closely related to their (i) size, (ii) shape, (iii) roughness, (iv) density and (v) electrostatic charging of individual spore as well as air (a) viscosity, (b) convection, (c) layering, (d) wind, (e) turbulence, (f) wind gradients close to the ground and (g) pattern of atmosphere circulation. Fungal spores may be dispersed by wind over distances of thousands of kilometers. Some examples are cited in Table 9.4.

Deposition: Airborne spores are removed from air by (i) sedimentation, (ii) washout or (iii) impaction. Spores heavier than air tend to sediment rapidly. The heavier spores settle faster than the lighter ones. Therefore, gravity determines the range of spore dispersal. Rain removes spores from air by impaction. During prolonged rain, the xerospores are washed off. The wettable gloiospores become incorporated in rain drops and spread as a film across a wettable surface or drip from non-wettable one. Dense vegetation facilitates spore removal from the air. An individual tree can filter ~ 73% of the airborne spores passing through its canopy.

Magyar et al. (2016) also elaborated on spore dispersal by animals. The vectors engaged by fungi for this mode of dispersal can briefly be described under the following headings: 1. **Inadvertent dispersal**: (i) many lichens are dispersed by animals, as soredial spores readily adhere electrostatically to the cuticle of insects or to the feather, fur or extremities of vertebrates. (ii) The sticky drops of conidiospores of *Acremonium*, *Fusarium* and *Verticillium*, for example, developed in different orientations, are particularly efficient to touch by small insects. (iii) The slime molds, especially the phallales, which have lost the ballistosporic mechanism (see p 24) are dispersed by beetles, which inadvertently feed on their fruiting structures. (iv) The digestion-resistant sporocarps of *Enteridium* are dispersed through feces defecated by beetles. 2. **Dispersal by pollinators**: (v) In plant pathogenic fungi, insects serve as prominent vectors. *Microbotryum violaceum* is a causative fungus of the anther smut disease prevalent among Caryophyllaceae. The pollinating visitors also transfer the infectious spores of *M. violaceum* to benign flowers. The infected male flower develops an anther-like structure filled with spores instead of pollen grains. (vi) The rust fungus *Puccinia monoica* causes the production of pseudoflowers with a pungent smell but with exuding sugar-rich fluid to attract more vectors to the host plant *Arabis*. 3. **Hypogenous fungi** lack spore discharging mechanism. They engage small mammals as vectors, in addition to insects. Their spores are a regular part of the diet of these vectors. Their indigestible spores are dispersed through defecation. 4. **Mutuals**: leaf-cutter ants (e.g., *Atta, Acromyrmex*) cultivate fungal gardens. The ants carry and maintain a selected fungal species (e.g., *Leucoagaricus*) to inoculate the piles of harvested leaf pieces to provide the cellulose-digested leaf food for their larvae. For more information, Fisher et al. (1994) may be consulted. Other fungi like *Termitomyces* are associated with termites (e.g., *Termes*). The ambrosia beetle *Xyleborus* cultivate *Ambrosiella* spp to feed their larvae. For more examples, Magyar et al. (2016) could be referred to. 5. **Man-made spore dispersal**: The immature gall of the corn smut is a delicacy; as it fetches higher prices, the US Department of Agriculture allows the artificial infection of corn with the smut.

2.3 Harsh Habitat – The Deserts

Fungi grow and thrive, wherever water and carbon are available. Fungi and plants draw water in its liquid form from the surrounding *milieu* or acquire it from soil moisture enriched by precipitation and/or fog. Chamber's Dictionary of Science and Technology defines fog as minute water droplets suspended in atmosphere. Webster's New Collegiate Dictionary states moisture as liquid diffused or condensed in relatively small quantity in air. A cloud is a visible mass of water particles existing as liquid or solid ice, which occurs in the form of mist or haze floating at altitudes in air. Lichenized fungi and rootless epiphytic plants acquire liquid water directly from precipitation and/or cloud. All other fungi and plants may have to acquire water in its liquid form from the soil (as in irrigated crops) or moisture content enriched by precipitation and/or fog. *Irrespective of fog or moisture, fungi and plants can accept water only in its liquid form.*

Water is available in solid frozen form in the Arctic (14.0 million km^2) and Antarctic (14.5 million km^2), in which biological activity is at the least. The arid deserts are characterized by evaporation exceeding precipitation. Consequently, water availability in its liquid form in the deserts is scarce. Ward (2009) named as many as 23 deserts, which span over an area of 28.5 million km^2 and cover 9% of land surface, leaving only 20% land area with relatively more production. In some of these deserts, day temperature may shoot up to 57°C (e.g., Libya) but may sink to freezing point at night (e.g., Gobi Desert). In them, precipitation occurs not in pulses but as flash floods and its level ranges from < 25 cm/y in the Sonoran Desert of Arizona to < 4 cm/y in the Atacama-Sechura Desert of South America. From flash floods, water is retained as moisture by desert soil and/or fog in the elevated zone; from soil moisture and/or fog, water is absorbed and stored by plants, especially the succulents, and part of it, is eventually transferred to animals. Not surprisingly, 2,308 plant and 476 vertebrate species are reported from the rainier Sonoran Desert. With the least precipitation in the Atacama-Sechura Desert, the number of plant and vertebrate species is only 1,930 and 146, respectively (see Pandian, 2021b, 2022). Irrespective of their relatively greater dependence for liquid water, the decomposing fungi may also exist to complete recycling of matter in trophic dynamics. This account presents an almost contrasting picture of the existence and diversity of fungi in the Arabian and Atacama-Sechura Deserts. The latter is the oldest and most arid desert on Earth (Goncalves et al., 2016). It extends from 10°S to 30°S and covers an area of 10,400 km^2 and receives 1.5 to 4.5 cm/y precipitation at an average temperature of 17°C (*enclima-data.org*). The Saudi Arabian Desert, Asia occupies an area of 2,300,000 km^2, experiences 55°C and receives precipitation of 10 cm/y (see Pandian, 2022).

Irrespective of extreme temperatures, prolonged desiccation, lowest availability of water and nutrients and high level of UV radiations, extensive researches by Goncalves et al. (2016) have brought to light the existence of 29 species of Ascomycota, of which *Cladosporium halotolerans*, *Penicillium chrysogenum*, *P. citrinium* occur more frequently at different altitudes. In a more elaborate account, Santiago et al. (2018) reported the incidence of ~ 200 lichenized fungi from the Atacama Desert. Notably, *Neucatenulostroma* is a gypsum-inhabiting fungus. Thanks to the support for science by the Saudi King, very useful data are reported not only for the taxonomic distribution of desert plants but also fungi and for the altitudinal effect on area coverage by flora but not by fungi.

Information on fungi associated with the Arabian Desert soil is available from publications of Ameen et al. (2021a, b), who reported the existence of 302 species including *Aspergillus* (32 species), *Alternaria* (10), *Chaetomium*

(12), *Cladosporium* (18), *Fusarium* (27), *Glomus* (23), *Penicillium* (93) and others (87). Of them, some 44 species are arbuscular mycorrhizal fungi and another 8 species are lichenized fungi. About 25 fungal species are culturable. Murgia et al. (2018) reported the existence of 11 colonial filamentous fungi belonging to Sordariomycetes (*Albifirmia, Chaetomium* and *Fusarium*) in the Arabian Desert of Jordan. Unfortunately, no information is yet available on the fungal distribution in the Arabian mounts.

Considering some aspects of fungi from widely distributed habitats, two points should be considered. The first one is associated with fungi that are grown on the montane zone of the arid desert. On plotting the number of lichenized fungal species as a function of altitude of the Atacama-Sechura Desert, an interesting trend became apparent (Fig. 2.1A). Both the number for species and genus peaked at 700 m altitude in correspondence with fog distribution and density, and subsequently began to progressively decline up to 4,500 m altitude. For want of corresponding information on fungal distribution as a function of altitude in the South Arabian Desert, plant coverage – an indirect approximate indicator of diversity – is considered. Whereas the fog distribution and consequent plant coverage is limited to 900 m altitude in the Atacama-Sechura Desert (Fig. 2.1B), they are reported up to > 2,000 m altitude in the Saudi Arabian Desert (Fig. 2.1C). In contrast to the aridity in the former, the latter with more rain holds soil moisture and fog over > 2000 m with the consequent plant coverage progressively increasing from 25 to 75% (Fig. 2.1C). Receiving ~ 150 cm precipitation/y (*wikipedia*), the number of plant species in the Mount Congo is nearly 10 times greater (Fig. 2.1D) than the relatively drier (68 cm precipitation/y, *climateknowledge-portable.worldbank.org*) Mount Kenya (Fig. 2.1E). Presumably, the flora on Mount Congo may entirely depend on soil moisture supporting more herbaceous species, while those on Mount Kenya may depend on both soil moisture and fog.

The second one is related to a few fungal species, that exist in deep trenches and sediments, on one hand, and those growing at extremely hot temperatures in arid deserts on the other. Optimum temperature for release of fungal α-amylases ranges between 50° and 70°C for *Aspergillus niger* (see de Souza and de Oliveira-Magalhaes, 2010). Likewise, the optima for the release of proteases are reported as 55°C for *A. oryzae* (Visvanathan et al., 2009), 60°C for *Trichoderma rosei* (Savitha et al., 2011) and 40°C for *Penicillium citrinium* (Chrzanowsa et al., 1995). Interestingly, *A. niger* and *A. oryzae* (Ameen et al., 2021a, b) are reported to grow at 55°C in the Saudi Arabian Desert. Representative species of *Aspergillus* are also known from the cold Mariana Trench and sediments. Briefly, the enzyme release required to digest cellulose and proteins is continued by some of the fungi, irrespective of their existence in cold or hotter temperature.

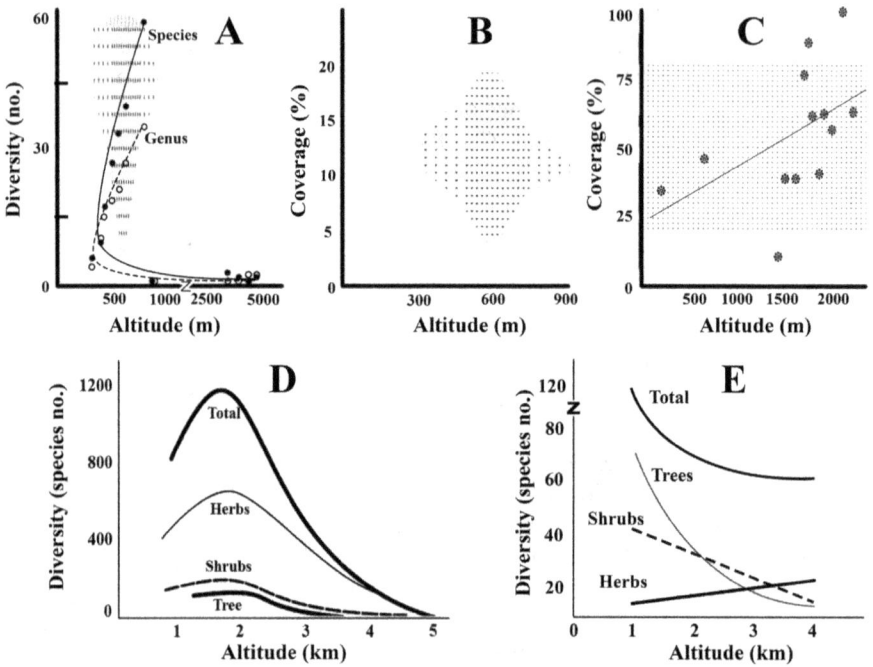

FIGURE 2.1

(A) Number of lichenized ascomycote species and genus that are found in the Atacama-Sechura Desert (compiled from Santiago et al., 2018). Note the superimposed distribution of fog and consequent fungal presence as a function of altitude. (B) Fog distribution and consequent plant coverage as a function of altitude in the Atacama-Sechura Desert. (C) Fog distribution and consequent plant coverage on the montane Al Baha region of Saudi Arabian Desert. Species diversity and plant forms as a function of altitude on Mount Kenya (D) and (E) Mount Congo (modified and redrawn from Pandian, 2022).

3

Fungi: Modes of Life

Introduction

Fungi exist as (i) free-living decomposers, (ii) commensals, (iii) symbionts, (iv) parasites or (v) antagonists. Regarding commensalic fungi, there are only sporadic incidences (e.g., Limon et al., 2017). They do not make up any reasonable number. So are antagonists. Kiss (2003) quantified their number to 40 species. For the key antagonistic role played by *Trichoderma* spp on biocontrol, Verma et al. (2007) may be consulted. Quantitative estimates have revealed that 14.4% (Table 3.5) and 40.0% (Table 3.9) fungi are symbionts and parasites, respectively. The remaining < 46% fungi play a key role as decomposers in terrestrial ecosystem. Sheerly by number of species (< 4% of fungi), fungi may not play as important a role as bacteria for decomposition of organic matter in aquatic system.

3.1 Free-living Decomposers

In selected terrestrial ecosystems, the estimated average quanta are (16 + 14 + 8 =) ~ 38 ton (t) Carbon (C)/hectare (ha)/y for plant productivity and (341 + 147 + 15 =) ~ 503 t C/ha for standing plant biomass to be decomposed by fungi (Table 3.1). For more details on their distribution as above-ground, below-ground, litter layer and soil organic matter, Sanderman and Amundson (2003) may be consulted. But for the inconvertible services rendered by the fungal decomposers, the land area may be covered by stinking hillocks of litter.

TABLE 3.1

Productivity (t C/ha/y) and biomass (t C/ha) that are to be degraded in forests and grassland (condensed from Sanderman and Amundson, 2003).

Tropical forest	Temperate forest	Temperate grassland
11–21 = 16 t C/ha/y	4–24 = 14 t C/ha/y	2–14 = 8 t C/ha/y
209–473 = 341 t C/ha	13–281 = 147 t C/ha	2–27 = 15 t C/ha

In decomposition of the non-consumed terrestrial biomass, fungi play a major role in the biogeochemical cycle of matter. In general, leaf tissue accounts for > 70% of the above-ground litter fall in forests and the rest is composed of stem, twigs and propagative structures. Climatic factors like temperature, precipitation and/or humidity exert a great deal of influence on the rate of decomposition (Krishna and Mohan, 2017). Litter decomposition proceeds through many routes, especially heterotrophic consumption and degradation of organic compounds. It consists of mainly three concurrent processes: (i) communition or fragmentation, (ii) leaching of water-soluble compounds and (iii) microbial degradation. The fragmentation of the coarse layer by soil invertebrates significantly enhances the accessibility to microbial invasion and colonization (Wolters, 2000), and leaching of organic compounds (Reddy and Venkataiah, 1989). Briefly, the soil fauna fragment, partially solubilize plant residues and thereby facilitate the establishment of decomposing fungi and bacteria (Sanderman and Amundson, 2003). The fragmentation is a rapid process. Following the shift to recalcitrant component of litter, the decomposition rate slows down, for example, the pine needle litter and paper birch foliage litter (Fig. 3.1). The non-glucan polysaccharides and glucans are degraded more rapidly than lignin. Within a period of 30 mo, the components remaining to be degraded increase from 42% for non-glucan polysaccharides to 66% for glucans and 79% for lignin (Fig. 3.1 window).

In terrestrial habitats, by number and biomass, fungi along with actinomycetes form three – four times larger components than bacteria (Table 3.2). With increasing decomposer size, their number and biomass decrease from fungi ($10^5 - 10^6$ no./g and 800 g/m^2) to Collembola (1 – 10 no./g, 0.5 – 1.5 g/m^2). Yet, the role played by fauna cannot be underestimated. In woodland forests, the leaf litter fall decreases from 5.5 – 15.0 t/ha to 2.5 – 3.5 t/ha and ~ 0.5 t/ha in the tropic, temperate and alpine-arctic zones, respectively. The pastures containing no earthworm in New South Wales (UK) accumulate surface mats at the rate of 4 cm/y. Consuming 27 mg litter/worm/d, the Indian earthworms decompose 3 t litter/ha/y.

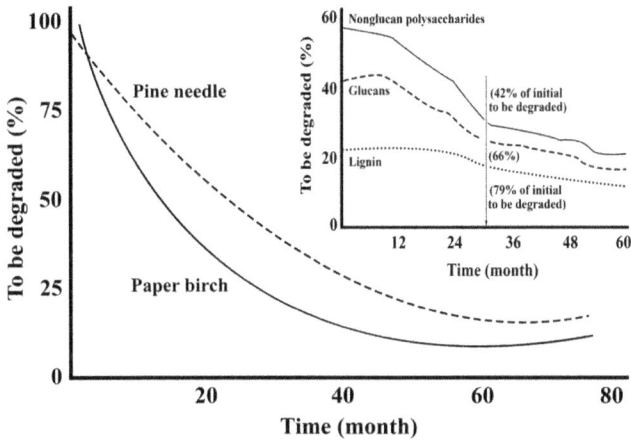

FIGURE 3.1

Proportion of initial biomass remaining to be degraded in the red pine needle litter and paper birch foliage litter as a function of time (compiled from Melillo et al., 1989, Aber et al., 1990). In the window: proportion of plant components to be degraded in the Scots pine needle litter as function of time (modified from Berg et al., 1982).

Being ecosystem engineers, their services need no emphasis (see Pandian, 2021b).

TABLE 3.2

Relative abundance and biomass (live weight) as well as size of soil fauna and flora within 15 cm soil surface (condensed and compiled from Swift et al., 1979, Brady and Weil, 2002).

Organisms (body width)	Number/g	Biomass (g/m²)	
		Range	Mean
Microflora			
Bacteria (1–2.5 µm)	10^8–10^9	40–500	270
Acitomycetes	10^7–10^8	40–500	270
Fungi (3–6 µm)	10^5–10^6	100–1,500	800
Algae	10^4–10^5	1–50	25
Fauna			
Protozoa (10–128 µm)	10^4–10^5	2–20	11
Nematodes (3.5–9.0 µm)	10–10^2	1–15	8
Earthworms (1.8 µm–40 mm)		10–150	80
Mites (100 µm–1.5 mm)	1–10	0.5–1.5	1
Collembola (130 µm–1.5 mm)	1–10	0.5–1.5	1
Diplopoda/Isopoda/Coleoptera (2–60 mm)		1–10	5

Water availability, as measured in units of water potential and water film thickness, determines the limits of the activity of decomposers. Accordingly, seed germination ceases already at –1.5 MPa (Pandian, 2022). Activity of soil inhabiting Protozoa also ceases at –0.03 MPa water potential and 4.0 µm water film thickness (Table 3.3). These values are –4.0 Mpa and < 3.0 nm for bacteria (e.g., *Bacillus*) but go far lower to range between –10.0 MPa and < 1.5 nm for *Fusarium* and –40.0 MPa and < 0.9 nm for *Penicillium*. The fungal ability to sustain their decomposing activity even on such low water containing substrates is something phenomenal. It is much greater for fungi than that for bacteria. *This is another reason why 96% fungi are distributed in land* (cf p 61).

Table 3.4 lists the decomposing activity by fungi and bacteria on different substrates. Of a dozen substrates, fungi degrade all of them but bacteria cannot degrade tannin, humic- and fulvic-acids. The number of fungal genera degrading cellulose, hemicellulose and lignin are greater than that of bacteria. Only the number of bacterial genera is greater than fungi in degrading starch, pectin and inulin.

TABLE 3.3

Microbial tolerance to water stress (modified from Hartel, 1999).

Water potential (MPa)	Water film thickness	Limit of activity
–1.5		Seed germination ceased
–0.03	4.0 µm	Protozoa
–0.01	1.5 µm	Zoospores, bacteria
–4.0	< 3.0 nm	Bacterial growth (*Bacillus*)
–10.0	< 1.5 nm	Fungal growth (*Fusarium*)
–40.0	< 0.9 nm	Fungal growth (*Penicillium*)

TABLE 3.4

Fungal and bacterial genera that act on different substrates (compiled from Crawford, 1981, Eriksson et al., 1990, Jin et al., 1990).

Substrate	Fungi	Bacteria
Cellulose	*Alternaria, Aspergillus, Chaetomium, Coprinus, Fomes, Fusarium, Polyporus, Myrothecium, Penicillium, Rhizoctonia, Rhizopus, Trametes, Trichoderma, Trichothecium, Verticillium, Zygorynchus* (16)	*Achromobacter, Angiococcus, Vibrio, Bacillus, Celfalcicula, Cellulomonas, Cellvibrio, Clostridium, Cytophaga, Polyangium, Pseudomonas, Nocardia, Sporocytophaga, Streptomyces, Sorangium, Micromonopora, Streptosporangium* (17)
Hemi-cellulose	*Alternaria, Aspergillus, Fusarium, Fomes, Rhizopus, Trichothecium, Coriolus, Penicillium, Zygorynchus, Polyporus, Chaetomium, Helminthosporium,* (12)	*Bacillus, Achromobacter, Cytophaga, Pseudomonas, Sporocytophaga, Lactobacillus, Vibrio, Streptomyces* (8)

Table 3.4 contd. ...

...Table 3.4 contd.

Substrate	Fungi	Bacteria
Lignin	*Clavaria, Clitocyle, Collybia, Flammula, Hypholoma, Lepiota, Mycena, Pholiota, Arthrobotrys, Cephalosporium, Humicola* (11)	*Pseudomonas, Flavobacterium* (2)
Chitin	*Fusarium, Mucor, Mortierella, Cytophaga Trichoderma, Aspergillus, Gliocladium, Penicillium, Thamnidium, Absidia,* (10)	*Achromobacter, Bacillus, Beneckea, Micrococcus, Pseudomonas, Nocardia, Chromobacterium, Flavobacterium, Streptomyces, Micromonopora* (10)
Cutin	*Penicillium, Rhodotorula, Mortierella* (3)	*Bacillus, Streptomyces* (2)
Tannin	*Aspergillus, Penicillium* (2)	0
Humic acid	*Penicillium, Polystictus* (2)	0
Fulvic acid	*Poria* (1)	0
Starch	*Aspergillus, Fomes, Fusarium, Polyporus, Rhizopus* (5)	*Achromobacter, Bacillus, Cytophaga, Chromobacterium, Clostridium, Nocardia, Micromonopora, Streptomyces* (8)
Pectin	*Fusarium, Verticillium* (2)	*Bacillus, Clostridium, Pseudomonas* (3)
Inulin	*Penicillium, Aspergillus, Fusarium* (3)	*Pseudomonas, Flavobacterium, Beneckea, Micrococcus, Cytophaga, Clostridium* (6)

On counts of number and biomass, *the ability of fungi to continue lytic activity is greater with decreasing water availability. Spreading lytic activity over a dozen substrates, the fungi play a much bigger role in degradation of plant litter in terrestrial habitat than bacteria.*

3.2 Symbionts

Fungi have a mutually beneficial symbiotic association with other fungi, plants and animals. As plants and fungi have no immune system, fungi may freely associate with them at external or internal level. However, only a few fungi have developed a mechanism for immune reception and resistance systems to symbiotically associate with animals. For the first time, an attempt has been made to estimate the number of symbiotic fungal species. For lower phyla and the Basidiomycota, relatively precise data became available or could be estimated (e.g., Lyophyllaceae, Atractiellomycetes, Table 3.5). However, it was a task to do it for the Ascomycota. Repeated computer searches with the keywords, 'symbiotic Dothideomycetes' yielded no information. This led one to consult *Systematics and Evolution: The Mycota* by McLaughlin and Spatafora (2014). This series is devoted to explain phylogenesis and systematics of Ascomycota. However, a few, who have authored chapters, have made passing remarks on species number and their modes of life. With explosive number of descriptions for new species, genera and so on for the Ascomycota, the required information is in a more fluid state. For example, the Class Sordariomycetes is divided into 14 orders by Barr (1990), 11 major + 1 minor orders by Samuels and Blackwell (2001) and 16 orders by Eriksson (2005). Consequently, many authors (during the period from 1990 to 2005) found it difficult to place the newly erected species, genera and families into one or another order. For example, the plant parasite *Fusarium* is placed as Sordariomycetes (by Zhang and Wang, 2015) but as Coelomycetes (by *eagri.org*). In fact, the reported number of *Fusarium* species ranges from 20 (Early, 2009) to 200 (*onezoom.org*) and to > 1,000 (*wikipedia*). Many authors consider a few of their species as subspecies, while others as strains. Therefore, the values listed in Table 3.5 may not be precise, except for those such as Arthoniomycetes (Schoch and Grube, 2015) and Lecanoromycetes (Gueidan et al., 2015). The value 113 for mycorrhizas of Eurotiomycetes was reached from hints of their incidences in a few genera (Geiser et al., 2015). Similarly, the value 1,140 for symbionts with ants was obtained from the hint that all the members of Myrmecophytes are ant symbionts and their species number is drawn from Chomicki and Renner (2015). According to Tedersoo et al. (2006), only 45 pezizomycete species are mycorrhizals.

TABLE 3.5

Survey on the number of symbiotic fungi. * = compromised, † = estimated. Myc = Mycorrhizas, Herbi vert = Herbivorous vertebrates, Term = Termites. ‡6,000 is suggested by a reviewer.

Phylum	Species (no.)		Remarks	Reference
	Total	Symbionts		
Chytridiomycota	1,840	0		
Neocallimastigomycota	260	260	Herbi vert	Gruninger et al. (2014)
Glomeromycota	230	230	Myc	Kolmer et al. (2009)
Zygomycota	1,065			
Zoopagomycotina		65	Amoebae	Margulis et al. (2009)
Myxomycota	500	?		

Table 3.5 contd. ...

...Table 3.5 contd.

Phylum		Species (no.)		Remarks	Reference
		Total	Symbionts		
Basidiomycota		**23,975**	‡		
Ustilaginomycotina	1,000				
Agaricomycotina	14,559				
Dacrymycetes	*329*				
Tremellomycetes	1,117				
Agaricomycetes	*8,500*				
Lyophyllaceae	35		35	Term	*wikipedia*
Phallales			172		
Pucciniomycotina	8,416				
Atractiellomycetes	50		6	Myc	Aime et al. (2014)
Ascomycota		**78,532**			
Taphrinomycotina	*140*		0		Kurtzman and Sugiyama (2015)
Saccharomycotina	1,309		0		
Pezizomycotina	*77,083*				
Pezizomycetes	1,684		45	Myc	Kirk et al. (2008) Tedersoo et al. (2006)
Orbiliomycetes	288				
Dothideomycetes*	19,000		500	Myc	Schoch and Grube (2015)
Arthoniomycetes	1,500				
Eurotiomycetes[†]	3,000		113	Myc	Geiser et al. (2015)
Myrmecophytes	1,140		1,140		Chomicki & Renner (2015)
Sordariomycetes	10,000		0		Zhang and Wang (2015)
Leotiomycetes	14,714			*Rhizoscypus*	Fehrer et al. (2018)
Lecanoromycetes	14,199		12,779	Myc	Gueidan et al. (2015)
Laboulbeniomycetes	2,486		0		
Coelomycetes	10,000		0		
Hyphomycetes	160		0		
Mycelia sterilia	52		0		
	Total	**106,761**	**15,345 (14.4%)**	Mycorrhizas = 13,673 (90.1%) Others = 1,500 (9.9%)	

*266 lichenocolous fungi, [†]6,755 lichenocolous fungi.

Irrespective of these inconsistencies, reasonable values could be estimated for Ascomycota. *In all, of 106,761 fungi species, 15,345 or 14.4% of fungi are symbionts. Of them, 13,673 or 90.1% of symbiotic fungi are mycorrhizas.* Hence, the majority of fungal symbionts have opted for the mycorrhizal mode unlike the Neocallimastigomycota, which are symbiotic in herbivorous vertebrates and Myrmecophytes symbiotic with ants (Table 3.5).

Fungal symbiosis involving ants and termites is already reported on p 66–67. As mentioned earlier, fungi may freely be associated externally or internally with plants or fungi, which have no immune system. *Not surprisingly, all endophytic symbiosis with mycorrhizas is spread over 89% flowering plants (Table 11.1), while all mutualistic association of fungi with insects is limited to ectosymbiosis alone.* Lichen-forming mycorrhizas are named as mycobionts and their plant counterparts as photobionts. The former protects the latter from UV radiation, temperature extremes and to a certain extent from desiccation. In return, they receive carbon, mostly in the form of glucose. In aquatic symbiosis, > 90% photobionts are chlorophytes such as *Asterochloris, Chlorella, Trebouxia or Trentepohlia.* In addition to providing water and nutrients, the cyanobionts also directly fix atmospheric nitrogen (see Gueidan et al., 2015). Some of these aspects are elaborated in Chapter 11.

Two types of mycorrhizas are recognized: (i) ectomycorrhizas (EC) and (ii) endo- or Arbuscular-Mycorrhizas (AM). Of them, arbuscular mycorrhizas are more ancestral and associated with a vast majority of plants inclusive of bryophytes. According to a survey by Wang and Qui (2006), 89% of the flowering plant species symbiotically engage in arbuscular mycorrhizas. Yet, the mycorrhizas themselves cannot directly fix atmospheric nitrogen to benefit the photobionts. But they stimulate the bacteria in the rhizosphere to fix nitrogen. "Incidentally, rhizosphere is the soil region that surrounds the surface of roots. Sugars, oligosaccharides, organic acids, vitamins, nucleotides, flavonoids, enzymes, hormones and volatiles are diffused from the roots that stimulate microbial activity in the rhizosphere. For example, the range of microbial densities (CFU/g soil) in the rhizosphere are 2.4×10^3 for Protozoa, 5.0×10^3 for algae, 1.2×10^8 for ammonifiers and 1.2×10^9 for bacteria (Vega, 2007). The reason for the circuitous route (mycorrhiza → bacteria), through which nitrogen is obtained, is traced to the fact that mycorrhizas also enhance the uptake of nutrients and water along with nitrogen (Dighton, 2009, see Pandian, 2022)".

3.3 Parasites

Structural simplicity, i.e., possession of six- seven tissue types and the need for economization of externally released digestive enzymes have driven many fungi toward parasitism. In fact, the cornerstone of fungal life style is parasitism. As much as beneficial fungi (see Table 4.5), there are also harmful parasitic fungi. Table 3.6 lists some of them. They cause (i) death and diseases

TABLE 3.6

Death, disease and cost of harmful fungi. bil = billion, mil = million, $ = USD.

Particulars	Death/disease/cost	Reference
Man	1.6 mil death – 1 bil people suffer – costs $ 7.2 bil on disease management and research	Anonymous (2017a, b), Benedict et al. (2018)
Amphibians, bats	Driven to extinction	Casadevall (2017)
Cereals	15–20% loss on yield of cereals Fungal diseases on rice, wheat and maize cost $ 60 bil/y	Figueroa et al. (2018), Rozewicz et al. (2021) *sciencedaily.com*
Fruits & vegetables	20% loss enough to feed 600 mil people annually	Anonymous (2017a, b)
Mycotoxin	25% global loss amounting to 1 bil metric ton of cereal products	*apsnet.org*
Fungicide	$ 14 bil	*alliedmarketresearch.com*

in man, (ii) near extinction of a few amphibians and chiropterans, (iii) yield loss of cereal and horticulture crops, (iv) mycotoxins-induced spoilage of food and dairy products, (v) fungicidal management cost and so on. As many as 625 fungal species are reported to infect vertebrates (Fisher et al., 2020). Approximately, 400 fungal species are known to cause diseases in man (Boddy, 2016). Under diseases called mycoses, Almeida et al. (2019) list 10 diseases caused by fungi. Of 250 species of *Aspergillus*, fewer than 40 caused disease in humans. For details on human diseases caused by *Aspergillus* and *Candida* and the consequent hospital costs, Benedict et al. (2018) may be consulted. The amphibian chytrid fungi *Batrychochytrium dendrobatidis* and *B. salamandrivorans* have driven frogs to the point of near extinction (see Casadevall, 2017). Dean et al. (2012) named the following fungal parasites that cause 15–20% losses on yield of crops: *Magnaporthe oryzae, Botrytis cinera, Puccinia* spp, *Fusarium graminearum, F. oxysporum, Blumeria graminis, Mycosphaerella graminicola, Colletotrichum* spp, *Ustilago maydis* and *Melamspora lini*. Analyzing a 30 years survey of fungal diseases in French forests, Vacher et al. (2008) traced the direct effect of exogenous factors like climate as responsible for fungal infection. Fungal diseases may cause (i) hypertrophy or abnormal growth by hyperplasia (excessive cell division) or etiolation (elongation) or gall formation, (ii) abscission, i.e., premature defoliation and fruit drop, (iii) host tissue replacement, especially reproductive elements like pollens leading to host sterility, (iv) necrosis cell death and/or, (v) wilting due to destruction or blockage of vascular tissues (Gould, 2009).

Though internalized, the parasites continue to extracellularly release a number of digestive enzymes (Table 3.7). Some amylases and lipases produce their respective monomers, which serve as food for the parasites. Others like cellulases, cutinases and pectinases serve to (i) soften, (ii) facilitate penetration and (iii) expansion of the parasitic area on the host's substrate. The functions of hemicellulases and proteases are not yet known.

TABLE 3.7

Functions of the digestive enzymes, produced by plant parasitic fungi (rearranged from Gould, 2009).

Parasitic enzymes	Substrate/site in host	Function in parasites
Amylase	Stored polysaccharides	Hydrolyzed glucose used as food
Lipase	Major component of plasma	Hydrolyzed fatty acids used as food
Cellulase	Skeletal component of cell wall	Softens cell wall – Increases area of parasitization
Cutinase	Wax + cellulose, the main component over chitin	Facilitates direct penetration of host's cuticle – Necessary to enhance parasitization
Pectinase	Component of cell wall + middle lamella	Facilitates penetration and colonization of the host
Hemicellulase	Skeletal component of cell wall	Their role is not yet known
Ligninase	Major component of cell wall and middle lamella	
Protease	Component of enzymes	

For the first time, a census was also made to estimate the number of parasitic fungal species. In it, random reports on suspected facultative parasitic species were not considered. As in symbiosis, the systematics and number of species included in different taxonomic hierarchy are known a little better for lower fungi and basidiomycotes than for the Ascomycota, in which the taxonomy is in a more fluid state and the species number for different hierarchical groups keeps changing. For lower fungi and the 23,975 speciose Basidiomycota, computer searches indicated no parasitic incidence in (i) Neocallimastigomycota, (ii) Glomeromycota, (iii) the 329 speciose Dacrymycetes and (iv) 14,559 speciose Agaricomycotina, with the exception of the 300 speciose Boletales and 402 speciose Auriculariales (Table 3.8). In contrast, all members of (i) the 322 speciose Entomophthoromycotina within

TABLE 3.8

Survey on the number of lower fungi and basidiomycotic species parasitic on fungi, plants or animals. * = compromised, † = estimated.

Phylum	Species (no.)		Hosts	Reference
	Total	Parasites		
Chytridiomycota*	1,840	1,104	Algae, animals	Table 1.3
Neocallimastigomycota	260	0		
Glomeromycota	230	0		
Zygomycota	1,065			
Entomophthoromycotina		322	Vertebrates	*wikipedia*
Myxomycota	500	45	Fungi, plants	*Biologydiscusssion.com*
Basidiomycota	23,975			
Ustilaginomycotina, 1,000		1,000	Plants	
Agaricomycotina, 1,4559				
Dacrymycetes, 329		0		
Tremellomycetes, 1,117		23	Animals	Akapo et al. (2019)
Agaricomycetes				
Boletales, 300 Gomphidiaceae		50	Fungi, host specific	*wikipedia*
Auriculariales, 402*		201	Plants	
Pucciniomycotina, 8,416				
Pucciniomycetes, 7,800		7,800	Plants	
Non-pucciniomycetes†, 616		434	~ 350 species (1.3%) are parasitic on animals, 10,629 are parasitic on plants + fungi	
Subtotal	27,870	10,979 (39.4%)		

Zygomycota are parasitic on insects. (ii) But, the 1,000 speciose Ustilaginomycotina and (iii) 7,800 speciose Pucciniomycetes are all plant parasites. Relevant information was also available for the number of parasitic species of Tremellomycetes (Akapo et al., 2019) and Boletales (Gomphidaceae). For the non-pucciniomycetes, an estimate was made for seven classes (see Table 1.3). Only the 332 speciose class Agaricostilbomycetes, consisting of saprotrophic or parasitic species, posed a problem. From available indications by Aime et al. (2014), the number of saprotrophic species belonging to 8 genera was estimated (*onezoom.org*). Briefly, of 616 non-pucciniomycetes, 434 were found as plant parasites. From available information, 6 of 10 chytrids were parasites. Hence, 60% or 1,104 chytrid species are taken as parasites. Likewise, a compromise of 50% was also taken for Auriculariales. In all, of 27,870 species of lower fungi and Basidiomycota,

10,979 species (or 39.4%) are parasites; of them, only 350 species (3.2% of parasitic species or 1.3% of fungal species) are parasites on insects and vertebrates. Notably, immune reception and resistance are developed randomly by a fewer number of lower and basidiomycote fungal species, which parasitize insect or vertebrate hosts.

TABLE 3.9

Survey on the number of ascomycotic species parasitic on fungi, plants or animals. * = compromised, † = estimated.

Phylum	Species (no.)		Hosts	Reference
	Total	Parasites		
Taphrinomycotina	140	140	Plants	*wikipedia*
Saccharomycotina	1,309	10	Plants, man	Kurtzman and Sugiyama (2015)
Pezizomycotina	77,083			
Pezizomycetes	1,684	251	Plants	Pfister (2015)
Orbiliomycetes	288			
Dothideomycetes	19,000	3,789	Plants	Schoch and Grube (2015), *indexfungoram. org,onezoom.org*
Arthoniomycetes	1,500			
Eurotiomycetes	3,000	~ 1,000		Geiser et al. (2015), Li and Zhang (2014)
Sordariomycetes	10,000	6,183	Fungi, plants, insects, man	Zhang and Wang (2015), Rodriquez et al. (2021), Early (2009)
Leotiomycetes	14,714	714	Plants, obligate	
Lecanoromycetes	14,199	0		Gueidan et al. (2015)
Laboulbeniomycetes	2,486	2,486	Insects	Zhang and Wang (2015)
Coelomycetes	10,000	10,000	Fungi, Plants, Vertebrates	*eurofinsus.com*
Hyphomycetes	160	160	Fungi, plants	*biodiversitylibrary.org*
Mycelia sterilia	52**	unknown		
Subtotal	78,532	24,733 (31.5%)	20,163 plant (81.5%) and 4570 animal parasites (18.5%)	
Lichenocolous parasites		7,021	For all fungi: 20,163 plant (18.9%) and 4570 animal parasites (4.3%) in ascomycota	
From Table 3.8		10,979		
Total	106,761	42,733 (40.0%)	For basidiomycota + ascomycota: 350 + 4570 = 4920 (11.5%) animal parasites. 10629 + 20163 + 7021 = 37813 (88.5%) plant parasites.	

Unlike in lower fungi and basidiomycotes, *parasitism is spread over all the three subphyla of Ascomycota (Table 3.9)*. *Of 78,532 ascomycotes, 24,733 species or 31.5% are parasites. Hence, parasitism is more prevalent among basidiomycotes + lower fungi complex than ascomycotes. Of 24,733 ascomycotes, 4,570 species or 18.5% (as compared to 1.3% for lower and basidiomycote fungi) are parasites on animals, indicating that more numbers of ascomycotes have developed mechanisms against the immune system and can parasitize animals, including man. In all, 42,733 or 40.0% fungal species are parasitic on fungi, lichens (7,021 species), plants and animals. Of them, 4,920 (i.e., 350 + 4,570) or 11.5% fungal species are animal parasites. Among animal hosts, most fungal species are ectoparasites.* Not more than 100 fungal species, which have developed mechanisms for immuno-reception and resistance, are endoparasitic in animals. Being parasites or symbionts, *54.4% fungi (40.0% parasites + 14.4% symbionts) cannot serve as decomposers.*

About 7% animals parasitize other animals (Pandian, 2021b); similarly, 1.2% plants parasitize other plants (Pandian, 2022). Unusually, fungi parasitize fungi, plants, lichens and animals. Whereas the heterotrophic slow motile acellular protozoans constitute 33.5% parasites, 40% of the tissue grade heterotrophic sessile fungi are parasites. Hence, *structural simplicity seems to drive organisms towards parasitism.*

3.4 Hyperparasites

Interestingly, Gleason et al. (2014) drew attention to the incidence of hyperparasitic chytrids, i.e., parasites on other parasites. From their long list (i) facultative hyperparasites (e.g., *Phytophthora cinnamomi*) and (ii) facultative parasites (e.g., *Allomyces arbusculus*) as well as (iii) parasites (e.g., *A. macrogynus*), for which host names are not mentioned, are not listed in Table 3.10. Though they are considered under Chytridiomycota, Blastocladio-mycota and Oomycota, they are all complex within Chytridiomycota (Table 1.1). In all, there are 23 chytrid obligate hyperparasitic species on 26 chytrid parasites, to which many plants and fewer animals serve as hosts. The number of these hosts decreases in the following order: 22 plants > 9 algae > 3 nematodes > 3 fish > 1 chytrid. The chytrid *Chytridium parasiticum*, is remarkable - a hyperparasite on another chytrid *Septosperma rhizophydii*, to which a third chytrid serves as a host. Notably, *Saprolegnia ferax* can only be an ectoparasite in fish, which have a fairly well-developed immune system. Of 23 hyperparasites, four of them, i.e., *Pythiella vernalis, Rozella cuculus, R. polyphagi, Rhizidiomyces japonicus* engage one of the two chytrid parasites as hosts; *Phythium oligandrum* and *Woronina pythii* engage one species of the five chytrid parasites (Table 3.10). In contrast, the parasites *Pythium aphanidermatum, Py. monospermum, Py. irregulare* may serve as a host for three hyperparasitic species; others like *Py. ultimum, Phytophthora megasperma* and *Saprolegnia ferax* hosts one of two hyperparasites. Hence, there is a wide choice to serve as a host by eight parasites or select one of the few parasitic species as a host for six hyperparasites.

TABLE 3.10

Obligative chytrid (including oomycetes) hyperparasites, parasites and their respective host species (modified, rearranged from Gleason et al., 2014).

Hyperparasite	Parasite	Host
Rozella marina	*Chytridium polysiphoniae*	Red algae
R. rhizophlyctii	*Rhizophydium globosum*	Diatoms, algae
R. achlyae	*Dictyuchus anomalus*	Fish
R. laevis	*Pythium gracile*	Green algae
R. barrettii	*Phytophthora cactorum*	Plant
R. pseudomorpha	*Lagenidium rabenhorstii*	Green algae
Chytridium parasiticum	*Septosperma rhizophydii*	Chytrid
Phlyctochytrium synchytrii	*Synchytrium endobioticum*	Plant
Septosperma anomala	*Phlyctidium bumelleriae*	Xanthophyceae
Canteriomyces stigeoclonii	*Phytophthora megasperma*	Plant
Olpidiopsis incrassata	*Saprolegnia ferax*	Fish
Rhizophydium pythii	*Py. monospermum*	Nematode
Pythiella pythii	*Py. dictyosporum*	Green algae
Pythium monospermum	*Phytophthora megasperma*	Plant
Woronina polycystis	*Saprolegnia ferax*	Fish
Ro. polyphagi	*Polyphagus laevis*	Euglena
	Po. euglenae	Euglena
Ro. cuculus	*Py. intermedium*	Plant
	Py. monospermum	Nematode
Rhizidiomyces japonicus	*Ph. megasperma*	Plant
	Ph. erythroseptica	Plant
Pythiella vernalis	*Py. aphanidermatum*	Plant
	Py. gracile	Green algae
Sorodiscus cokeri	*Py. irregulare*	Plant
	Py. undulatum	Plant

Table 3.10 contd. ...

...Table 3.10 contd.

Hyperparasite	Parasite	Host
Rhizophydium carpophilum	*Synchytrium fulgens*	Plant
	S. macrosporum	Plant
	S. linariae	Plant
Woronina pythii	*Py. aphanidermatum*	Plant
	Py. debaryanum	Plant
	Py. irregulare	Plant
	Py. monospermum	Nematode
	Py. ultimum	Plant
Pythium oligandrum	*Py. aphanidermatum*	Plant
	Py. irregulare	Plant
	Py. mamillatum	Plant
	Py. sylvaticum	Plant
	Py. ultimum	Plant

4

Acquisition of Micronutrients

Introduction

Both animals and fungi are heterotrophs. Whereas the former are motiles, the latter are sessiles. The combination of motility and intake of food followed by internal digestion (except the anus-less planarian, see Pandian, 1975) required the development of structural complexity up to the organ and system levels in animals. Conversely, that of sessility and external digestion has led to the retention of structural simplicity up to the tissue level alone in fungi. *Although many mycologists have hinted at external digestion by fungi, none has ever associated it with their structural simplicity.* Like fungi, most bacteria are also heterotrophs and sessiles. Earlier, botanists considered fungi as plants that lack chlorophyll (e.g., Evert and Eichhorn, 2013). At present, many microbiologists include fungi along with bacteria. However, there are succinct differences between fungi and bacteria. 1. Bacteria are prokaryotes, that have opted for unicellularity, while the fungi are eukaryotic multicellular hyphae. Depending entirely on extracellular release of digestive enzymes and external digestion, the fungi have evolved toward terrestrial habitat. In contrast, evolution of bacteria, which partially depend on direct uptake of Dissolved Organic Matter (DOM) from aquatic *milieu*, has led toward aquatic habitats. Briefly, microbiology may be concerned with prokaryotic bacteria, viruses and the like. But mycology shall be an independent subject dealing with fungi alone. However, a few fungi are indicated as predators (e.g., *Zoophagus tentaculum* growing as epiphytes on *Nitella* capturing rotifer, Karling, 1936, microsporidians predating nematodes, Sapir et al., 2014). The acquisition of micronutrients through the body surface has required the fungi to have (i) expansive hyphal surface area, (ii) adaptation in the cell wall at structural and functional levels, (iii) extracellular release of large molecule sized proteinaceous enzymes to facilitate external digestion, (iv) an array of enzymes and their economization in terrestrial habitats and (v) release of copious quantum of one or more exceptional enzymes to meet the specifications of industries. This chapter elaborates some of these aspects.

4.1 Hypha – Surface Area

Containing a full gamut of organelles typical of eukaryotes, the hyphae exhibit extraordinary developmental versatility, phenotypic plasticity and diverse functionality (see Read, 2017). Growing at the rate of 20 μm/min (Steinberg, 2007), an individual hypha grows to 29 m/d but can also achieve a kilometer length in a day (*csus.edu*). Rare estimates on the hyphal distribution of *Trichoderma viridae* indicate that the total hyphal length is 220 km/g dry mycelium and 3 km length/g dry soil in birch forest (Hanssen, 1974). A climax is the occupation of 965 ha by the largest known fungus *Armillaria ostoyae* (*scientificamerican.com*).

At this juncture, three points should be noted: 1. In the radial expansion of a mature colony of filamentous fungi like *Coprinus sterquilinus* (opposite figure), a few hyphae in the inner core may grow toward each other and anastomose to form an interconnected network. However, all the hyphae on the outer zone avoid each other (negative tropism) and remain isolated to eliminate interhyphal competition for nutrients, especially where they are actively growing. 2. Irrespective of the intense demand for expansion of surface area, the shape of most hyphae remain cylindrical/tubular rather than square or rectangular in shape, which may have provided a relatively larger surface area. However, *penetrability seems to be more important rather than surface area expansion*. In a large number of fungal parasites and endophytic mycorrhizal symbionts, the penetration into plant tissues is required. It is also needed among decomposers. Playing a prominent role in degrading a decalcifying molluscan shell, the hyphae penetrate to ramify and degrade the proteinaceous shell matrix and thereby increase losableness of its mineral content (Lauckner, 1983). 3. The saccharomycotines (1,309 species) and Neocallimastigomycota (260 species) have reverted to non-motile and motile unicellularity, respectively and thereby have increased the surface area per unit volume. The critical diameter for induction of budding process in *Saccharomyces* is ~ 8 μm (Zakhartsev and Reuss, 2018).

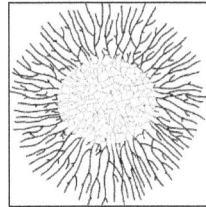

4.2 Cell Wall: Structure and Function

Unlike those of animals and plants, the fungus cell wall is a more dynamic organelle. Its composition (Table 4.1) and functions are greatly influenced by modes [(i) free-living, (ii) symbiotic, (iii) parasitic] of life and external *milieu-*

TABLE 4.1

Chemical composition of cell wall of fungi (modified from Webster and Weber, 2007).

Group	Chitin	Glucan	Protein	Cellulose	Lipid
Chytridiomycota	58	16	10	0	?
Ascomycota: *Fusarium*	39	29	7	0	6
Basidiomycota: *Coprinus*	33	50	10	0	?
Zygomycota	9	44	6	0	8
Basidiomycota: *Schizophyllum*	5	81	2	0	?
Ascomycota: *Saccharomyces*	1	60	13	0	8
Oomycota	0	65	4	25	2

imposed stress. It determines the fungal shape, morphogenesis and differentiation. Essentially, it consists of fibrous and gel-like carbohydrate polymers that form a tensile and robust core of scaffolds, to which a variety of proteins and other superficial compounds are added. In most fungi, it is a layered structure and consists of (i) the innermost, that surrounds the plasma membrane, a rather conserved skeletonous layer and (ii) the outer hexagenous layer tailored to suit the functions of a particular fungus. The former consists of a core of covalently attached branched β 1,3 glucans with 3–4% interchain and chitin (Fig. 4.1). The glucan and chitin form intrachain hydrogen bonds. They can be assembled into fibrous microfibrils that form a basket-like scaffold around the cell (Gow et al., 2017). According to Webster and Weber (2007), chitin microfibrils of the Ascomycota and Basidiomycota, are a bundle of linear β-(1,4) N-acetylglucosamine chains, which are synthesized and extruded into the growing cell wall around the apical dom. In Zygomycota, the chitin fibers are modified (after their synthesis) by partial or complete deacetylation to produce the poly β-(1,3) and β-(1,6) bonds. Proteins represent the third component of fungal cell walls. Many of them can be modified by glycosylation, i.e., the attachment of oligo-saccharide chain to polypeptide. As an aldohexose, mannose is the main component of such proteins that are called mannans. The branched β-(1,3): β-(1,6) glucan is bound to protein (e.g., *Candida* yeast, *Histoplasma*, Fig. 4.1A, B), or other polysaccharide like mannan (e.g., *Pneumocystis*, Fig. 4.1E) or galactomannan (e.g., *Aspergillus fumigatus*, Fig. 4.1C) or amorphous polymerized phenolic compound, the melanin (e.g., *A. conidium*, Fig. 4.1D). Briefly, the exoskeleton is the load-bearing structure of the wall that resists the internal hydrostatic pressure exerted on cytoplasmic membrane. For example, the turgor pressure exerted at the penetrating hyphal tip can be as high as 0.01–0.1 µN/µm (N = Newton, the unit of force), which is adequate to penetrate the stiff 8% (wt/vol) agar (Gow et al., 2017). Therefore, the wall defines cellular structure, provides rigidity and protects the cell contents (Garcia-Rubio et al., 2020).

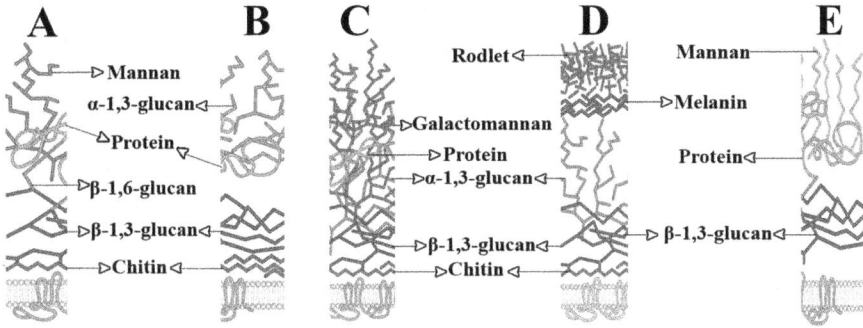

FIGURE 4.1

Structural organization of the cell wall of selected fungi: (A) *Candida* yeast, (B) *Histoplasma capsulatum*, (C) *Aspergillus fumigatus*, (D) *A. conidium*, (E) *Pneumocystis* spp (modified and redrawn from Gow et al., 2017).

The ability of fungi to produce copious quantities of extracellular large molecular enzymes has been recognized for many decades. Their highly efficient secretion of hydrolytic enzymes, for example, account for recycling of the most abundant polysaccharides, especially cellulose, lignin, keratin and so on. They transport large molecules of biological significance including (i) proteins, (ii) lipids, (iii) polysaccharides and (iv) pigments (e.g., astaxanthin, Rodriguez-Saiz et al., 2010) across the plasma membrane (pm) and cell wall (cw). Active research during the last 40 years have brought to light the role played by the vesicular mode of transport through pm and cw. According to the most well-studied pathway, protein excretion involves vesicular migration from the Endoplasmic Reticulum (ER) to the *trans*-Golgi face of the Golgi, (ii) subsequent loading into a complex network of vesicles, and (iiia) the *trans* Golgi network of transportation of vesicles that immediately move to fuse with pm and (iiib) release their contents by exocytosis (Glick and Malhotra, 1998). The rate, at which vesicles are generated and fused with the pm is one of the fastest processes. For example, Collinge and Trinci (1974) estimated the fusion of 38,000 secretory vesicle/min with pm of a single growing hypha of *Neurospora crassa*. The presence of the cell wall implies the existence of trans-cell mechanisms for release of these large molecules into an extracellular space. Fungal endosomes seem to play an important role in trans cell wall transport. Many studies have shown the vesicular mode of transport across the cell wall in the basidiomycote *Cryptococcus neoformans*. This model of bilayered vesicular transport occurs *in vitro* and *in vivo* of some ascomycotes like *Histoplasma capsulatum, Candida albicans, C. parapsilosis, Sporothrix schrenckii* and *Saccharomyces cerevisiae* (Rodrigues et al., 2008, 2011).

Another characteristic of the fungi is their ability to uptake organic or inorganic solutes from extremely dilute solutions from the *milieu* exterior. Thereby, they can accumulate them > 1,000 folds against the concentration gradient. For this uptake process, the main barrier is the plasma membrane. Webster and Weber (2007) named two systems namely (i) System I involves the selective process and is mediated by the proteinaceous pores called **channels**, which facilitate the diffusion of a solute according to its concentration. (ii) System II, called **porters** require ATP to accumulate the solute against the concentration gradient across the plasma membrane. For more details, consult Webster and Weber (2007).

4.3 Enzymes and Functions

The main characteristic of the fungi is to synthesize and apply hydrolytic and oxidative enzymes on their substrate and uptake the 'digested' micronutrients. Some fungi are capable of synthesizing and releasing > 20 g of a single enzyme or enzyme group/l of culture medium (Webster and Weber, 2007). This amazing characteristic holds potential for biotechnological and pharmaceutical application. The classification and functions of these enzymes in degradation of live or dead organisms, which consist of carbohydrates, proteins and lipids are described here. Accordingly, the description deals with 1. Carbohydrases, 2. Proteases and 3. Lipases.

1. Carbohydrases: Organisms store their carbohydrate resources usually in the form of starch. The pathways of glycolysis and metabolic routes are well known. In defense of predation, plants, however, also synthesize a large amount of secondary plant substances like cellulose, lignin and so on. In general, the cell wall of plants is constituted by (i) cellulose (glucose, 44%), hemicellulose (xylose, 30%) and lignin (phenol, 26%) (see Raghuwanshi et al., 2016). Cellulose is a complex carbohydrate or polysaccharide consisting of 3,000 or more glucose units. It is an organic compound with the formula $(C_6H_{10}O_5)_n$, a polysaccharide consisting of a linear chain of several hundred to over thousand β (1,4) linked D glucose units (Crawford, 1981). Hemicellulose is arranged in forms of strips that are tightly linked with cellulose and lignin in the plant cell wall (Kumar and Chandra, 2020). Hence, the role of fungal degradation of plants is considered under (i) cellulases, (ii) hemicellulases and (iii) ligninases.

In fungi, **(i) *cellulases*,** belonging to the Glycoside Hydrolase (GH) family of enzymes, hydrolyze the glycosidic bond between two or more carbohydrates or between a carbohydrate and a non-carbohydrate group (Carbohydrate Active Enzyme Database, CAZy). The International Union of Biochemistry and Molecular Biology (IUBMB) classifies cellulases into three types: (i) endocellulases (EC 3.2.1.4), (ii) exocellulases and (iiia) cellobiohydrolases (EC 3.2.1.91) and (iiib) cellobasises or β-glucosidases (BGL 3.2.1.21). The majority of cellulases have a two-domain structure, of which the first represents the Cellulose Binding Domain (CBD) and the second one Catalytic Domain (CD). The cellulose is degraded at three levels by (a) endoglucanases, (b) cellobiohydrolases and (c) β-glucosidases (Panchapakesan and Shankar, 2016). Endoglucanases hydrolyze the β-1,4 glycosidic linkages randomly within the cellulose chain. This hydrolyzation produces a soluble long chain cellodextrin or insoluble cellulose fragment. The fungal cellobiohydrolases act in a progressive manner to break down the reducing and non-reducing ends of cellulose polysaccharide chain to liberate cellobiose, which is subsequently acted by β-glucosidases. The third type of cellulases degrade cellobiose to glucose and form a vital component of the cellulase system. β-glucosidases are thus crucial for breaking down of the cellulose, as it completes the saccharification by hydrolysis of cellobiose and small cello-oligosaccharides into glucose molecules. However, industrial biochemists look at these reactions in the more commonly used terms. Endoglucanases are more commonly known as α-amylases, the enzymes (e.g., De Souza and de Oliveira-Magalhaes, 2010, Saranraj and Stella, 2013) that catalyze the hydrolysis of internal α-1,4 glycosidic linkage within the cellulose chain of starch to low molecular products like glucose, maltose and others. They are present in fungi, plants and animals. Another group of these enzymes are β-amylases, which are synthesized by fungi and plants but not by animals.

(ii) *Hemicellulases*: Hemicelluloses are a highly branched and mostly non-crystalline heterosaccharides (e.g., D-xylose, L-arabinose, D-mannose, D- and L-galactose, L-fucose, uronic acids). They comprise 15–35% of plant biomass. Of them, xylan and glucomannan are the most important hemicelluloses. The former constitutes a major structural hetero-polysaccharide in the plant cell and is also found at the interface between the lignin and cellulose. The complex structures of hemicelluloses require synergistic action of different types of hemicellulases for degradation. Like cellulases, xylanases are divided into endo-acting xylanases (EC 3.2.1.8) and exo-acting xylanases (EC 3.2.1.156). The other hemicellulases are classified under glucoside hydrolases and carbohydrate esterases. Fungal xylanases are generally active at 40–60°C at slightly acidic pH. The following are some examples:

Endo-acting xylanases	Exo-acting xylanases
Endo-1,4-β xylanase (EC 3.2.1.8) hydrolyzes the glycosidic bonds in the xylan backbone and reduces them to xylo-oligomers	Exo-1,4-β-D-xylosidase (e.g., xylohydrolase, EC 3.2.1.37) hydrolyzes only xylobiose
Mannase (EC 3.2.1.78) catalyzes mannan-containing hemicellulases	Mannosidases (EC 3.2.1.25), galactosidases (EC 3.2.1.23) and arabinofuranosidases (EC 3.2.1.55) degrade hemicelluloses into monomeric sugars

(iii) Ligninolytic enzymes: Lignin is a complex aromatic polymer and characteristic of the cell wall of vascular plants. Constituting 26% of lignocellulose, it is also an abundant polymer but is the most highly recalcitrant to either chemical or biological degradation. It remains so recalcitrant that > 79% lignin remain to be decomposed in the Scots pine even after 80 months (see Fig. 3.1 window). In fact, it resists decomposition and remains as mummified conifer wood to be mineralized since ~ 25 MY (see p 172).

According to Sette et al. (2013), ligninolytic enzymes comprise oxidases and peroxidases. As an enzyme, oxidase catalyzes the oxidation-reduction reaction, involving molecular oxygen (O_2) as an electron acceptor. In it, oxygen is reduced to water (H_2O) or hydrogen peroxide (H_2O_2). As an enzyme, peroxidase catalyzes the transfer of oxygen from H_2O_2 to a suitable substrate and further the oxidation of lignin. The fungi-secreted ligninolytic peroxidases include (i) lignin peroxidase (LiP), manganese peroxidases (MnP) and the Versatile Peroxidase (VP). The activities of these enzymes are enhanced by various mediators and cofactors (Adarsh and Ram, 2020). Table 4.2 lists these enzymes, their cofactors and reactions in ligninolytic degradation by fungi.

TABLE 4.2

Ligninolytic enzymes and their reactions (condensed from Hatakka, 2001).

Enzyme	Cofactor	Substrate, mediator	Reaction
Lignin peroxidase (LiP)	H_2O_2	Veratryl alcohol	Aromatic ring oxidized to cation radical
Manganese (MnP)	H_2O_2	Mn, organic acids as chelators, thiols, unsaturated fatty acids	3 Mns oxidizes phenolic compound to phenoxyl radicals
Versatile peroxidase (VP)	H_2O_2	Mn, veratryl alcohol, compounds similar to LiP and MnP	Mn(ii) oxidized to Mn(iii); phenolic and non-phenolic compounds and dyes oxidized
Laccase (Lac)		Phenols, mediators, e.g., ABTS	Phenols oxidized to phenoxyl radicals; other reactions in the presence of mediators

As lignocellulose degraders, the basidiomycotes form an interesting group. Some enzymes, arising from different basidiomycote species, exhibit specialized catalytic reactions (Table 4.3) (Maciel et al., 2010).

TABLE 4.3

Special enzymes arising from basidiomycotes and their catalytic reaction.

Fungi	Enzyme	Catalyzed reactions
Trametes versicolor	Laccase	Phenol oxidation
Phanerochaete chrysosporium	Manganese peroxidase	Phenol oxidation from Mn^{2+} to Mn^{3+}
	Cellobiose-quinone oxireductase	Quinone reduction; cellobiose degradation
	Glyoxal oxidase	H_2O_2 production
	Manganese independent peroxidase	Activity on aromatic substrates
	Cellobiose dehydrogenase	Dispose manganeses (MnII) from manganese oxide
	Lignin peroxidase	Phenol polymerization
Pleurotus saborcaju	Aryl alcohol oxidase	H_2O_2 production
Pleurotus sp	Versatile peroxidase	Oxidizes Mn^{2+}; high redox-potential aromatic compounds

2. Lipases: Due to their high catalytic activity, fungal lipases stand out as the major sources of enzymes (Table 4.4). They are ubiquitous, especially in soils contaminated with oils and dairy product wastes (Singh and Mukhopadhyay, 2012, Mehta et al., 2017). Lipases of triacylglycerol hydrolases (EC 3.1.1.3) catalyze both the hydrolysis and synthesis of the long chain acylglycerol. In fact, they can catalyze a variety of chemical reactions, which include esterification, transesterification, acidolysis and aminolysis. A few lipases-mediated reactions are explained hereunder:

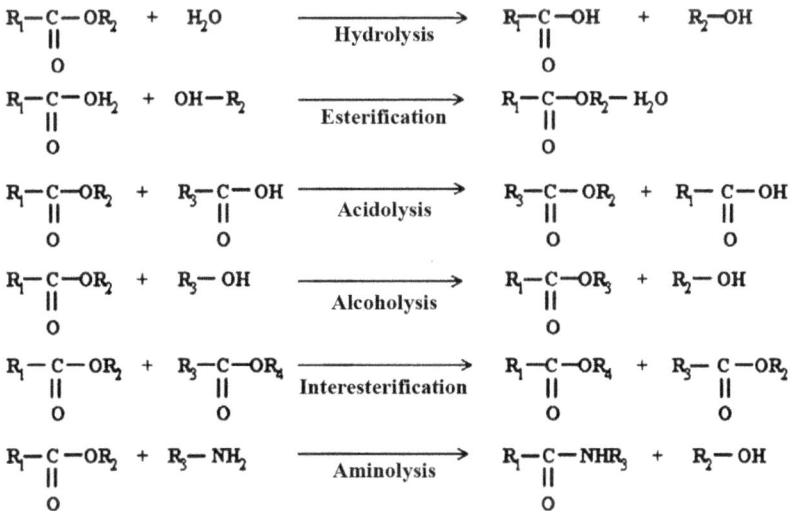

3. Proteases constitute a large group of enzymes that catalyze the hydrolysis of proteins (de Souza et al., 2015). Their occurrence in some fungi is listed in Table 4.4. Depending on the site of action, they are divided into two major groups: (i) endopeptidases (EC 3.4.21-99) and (ii) exopeptidases (EC 3.4.11-19). The former preferentially act at the peptide bonds of the inner regions of the polypeptide chain, whereas the latter cleaves the peptide bond at the N or C terminus. Both these groups are further divided into each with four subgroups. Their characteristics are described hereunder:

		(i) Endoproteases
A.	Berine proteases (EC 2.4.21)	Active at neutral or alkaline pH (7–11) - Consist of low molecular mass (18–35 kDa), e.g., trypsin
B.	Aspartic proteases (EC 2.4.23)	Acidic proteases – Depend on aspartic acid residue for catalysis, e.g., pepsin
C.	Cysteine proteases (EC 2.4.22)	Active only in the presence of reducing agents like cysteine, e.g., papain
D.	Metallo proteases (EC 2.4.24)	Require a divalent metal ion for activity
		(ii) Exoproteases
i.	Aminopeptidases (EC 3.4.14)	Act at a free N terminus of the peptide chain - Liberate a single amino acid residue, a dipeptide
ii.	Carboxypeptidases	Act at a free C terminus of the polypeptide chain - Divided into three major groups based on the nature of amino acid: (ii) serine peptidases (EC 2.4.16), (iii) metallopeptidases (EC 2.4.17) and (iv) cysteine peptidases (EC 2.4.18)

As emphasized earlier, almost all fungi release enzymes, digest their food externally and absorb micronutrients through their body surface. From a limited survey, this account could assemble relevant information on copious (commercial scale) release of the digestive enzymes from 121 fungal species (Table 4.4). In any case, their species number may not exceed ~ 1,400 (cf Mandels and Sternberg, 1976). *1. Hence, > 98% of fungi (especially endosymbionts and endoparasites) economize and avoid wasteful secretion and release of these enzymes. 2. For 121 species, relevant information is available for 51 (43%), 49 (40%) and 21 (17%) species that produce amylases, proteases and lipases, respectively. It would imply (i) that the greater number of fungal species produce relatively larger quantities of amylases and proteases than that of lipases* and/or (ii) Collection and purification of amylases and proteases from cultured fungi are more profitable than that of lipases. 3. Among these fungi, species belonging to *Aspergillus, Fusarium, Mucor, Penicillium* and *Rhizopus* serve as a source for all the three enzyme groups. For example, *Penicillium* includes 250 accepted species with many of them finding application in food and pharmaceutical industries (Panchapakesan and Shankar, 2016). 4. Yet, *Aspergillus niger* and *Thermomyces lanuginosus* are the only fungal species that copiously produce all the three enzymes, *A. oryzae* produce both amylases and proteases, but

P. camemberti, *P. roqueforti* and *R. oryzae* produce both proteases and lipases. 5. In a few species, strains are also identified; for example, three in *A. oryzae*, two each in *Pynoporus sanguineus* and *Th. lanuginosus* for amylases; similarly, six for *A. oryzae* and three for *T. lanuginosus* for proteases. Hence, *A. oryzae* and *Th. lanuginosus* are the most important fungi that copiously produce amylases and proteases. 6. However, Mandels and Sternberg (1976), who compared the cellulolytic ability of 1,400 fungal species, found that not one of them came even remotely close to *Trichoderma rosei*. Table 4.4 shows the proportions of taxonomic fungal groups that contribute to the industrial level of enzyme production, decrease in the following descending order: Ascomycota (84, 69.4%) > Basidiomycota (12, 9.9%) > Zygomycota (18, 14.9%) > Neocallimastigomycota (7, 5.8%).

TABLE 4.4

Survey on the number of fungal genera and species (in numbers) that are known to secrete abundant (commercial scale) amylases, proteases and lipases (compiled from de Souza and de Oliveira-Magalhaes, 2010, de Souza et al., 2015, Mehta et al., 2017, Panchapakesan and Shankar, 2016, Raghuwanshi et al., 2016, Saranraj and Stella, 2013, Singh and Mukopadhyay, 2012). *includes some species with strains. The numerals indicate the number of species in the genus.

Amylases				Proteases			
Neocallimastigomycota				**Zygomycota**			
Anaeromyces	1	*Chaetomium*	1	*Rhizomucor*	1	*Botryis*	1
Caecomyces	1	*Aureobasidium*	1	*Humicola*	1	*Clanostachys*	1
Orpinomyces	2	*Saccharomyces*	1	*Mucor*	1	*Cordyceps*	2
Neocallimastix	2	*Sporotrichum*	1	*Rhizopus*	2	*Cronostachys*	1
Piromyces	1	*Talaromyces*	2	*Thermomucor*	1	*Engyodontium*	1
Zygomycota		*Thermomyces**	2	*Conidiobolus*	1	*Metarhizium*	1
*Mucor**	2	*Trichoderma*	6	**Basidiomycota**		*Myceliophthora*	1
Rhizopus	1	Total	51	*Phanerochaete*	3	*Ophistoma*	1
Basidiomycota		**Lipases**		**Ascomycota**		*Penicillium*	7
Bjerkandera	1	**Zygomycota**		*Fusarium*	1	*Thermoascus*	2
Coriolus	1	*Mucor*	2	*Graphium*	1	*Thermomyces**	1
Cryptococcus	1	*Rhizopus*	5	*Hirsutella*	1	*Trichoderma*	2
Phanerochaete	2	*Humicola*	1	*Aspergillus**	12	Total	49
*Pycnoporus**	1	**Ascomycota**		*Beauveria*	3	**Grand Total: 121**	
Schizophyllum	2	*Aspergillus*	4	Amylases : 43%, Proteases : 40%,			
Volvariella	1	*Candida*	2	Lipases : 17%			
Ascomycota		*Fusarium*	1	Neocallimastigomycota:	7	(5.8%)	
Fusarium	1	*Geotrichum*	1	Zygomycota:	18	(14.9%)	
Penicillium	2	*Penicillium*	5	Basidiomycota:	12	(9.9%)	
*Aspergillus**	15	Total	21	Ascomycota:	84	(69.4%)	

4.4 Enzymes: Germination/Development

In eukaryotes, reproductive multiplication is accomplished by (i) spores in fungi, algae and tracheophytes, (ii) seeds in flowering plants and (iii) eggs in animals. In them, structurally complex amylose, proteins, lipids and others are tightly packed to sustain germination/development. This requires the digestive enzymes to release smaller molecules or monomers for their own structural construction and sustain metabolism. Being a gray area, not many publications are available. Although some of them fall outside the scope of this book, they are briefly summarized for the benefit of teaching. Accordingly, cellulase, amylase, proteases and others are produced by fungal spores during their germination (Fernandes et al. 2012). Spores of ferns also produce isocitrate lyase and malate synthase (DeMaggio and Greene, 1980). Seeds synthesize amylase, protease, β 1,3-glucanase during germination (Joshi, 2018). Muntz et al. (2001) reported on the presence and function of endopeptidases and carboxypeptidases during germination of dry quiescent seeds. In chicks, a dozen proteases are produced from the allantoic fluid to digest the albumin and egg yolk (Da Silva et al., 2017). Combinations of proteases and phosphatases as well as cathepsin B and phosphatase are reported from developing eggs of moths (Oliveira et al., 2013) and parasitic tick (Zhang et al., 2019), respectively.

4.5 Beneficial Fungi

Approximately, 22,500 biologically active compounds have been obtained from fungi (see Ganesh Kumar et al., 2019). Some like astaxanthin fetches US$ 2,500/kg (Rodriguez-Saiz et al., 2010). Of ~ 1% fungi that copiously secrete and externally release one or more enzymes, many are beneficial. They may be considered in the following four categories: 1. Food and food products, 2. Wine and beverages, 3. Pharmaceutical and 4. Industrial products (Table 4.5). Despite repeated computer searches, not many values are available for industrial products, although Singh and Mukhopadhyay (2012) listed the use of fungi in paper, textile, leather, cosmetic and other industries. Many authors like Sivakumar et al. (2014) elaborated on the use of *Aspergillus flavus* in color removal from textile industrial wastewater, but provided no information on their market value. Others like de Souza et al. (2015) and Arnau et al. (2020) dumped relevant information on industrial enzymes that are produced by both fungi and bacteria. Yet others like Arumugam et al. (2014) described bioactive metabolites by the marine *Nigrospora* sp from 800 m depth. Similarly, Ganesh Kumar et al. (2021) reported on the capability of degrading spent engine oil by *Aspergillus sydowii* collected from sediments at 3,000 m depth. As they are all from ongoing researches, no information on their market value is yet estimated. Table 4.5 lists the global market values for some beneficial fungi. Impressive is the benefit arising from some of these fungi amounting to > 1,000 billion US$. Not surprisingly, there are numerous publications, reviews (e.g., Raghuwanshi et al., 2016) and some books (e.g., Kemboi et al., 2020) devoted entirely to (i) culture of the fungi, (ii) collection and purification of a specific enzyme. Therefore, this account shall be limited to highlight one or two aspect(s) of them. Remarkably, thermostable proteases of some fungi act at temperatures of 55–60°C. For example, that of *Rhizopus oryzae* acts at 60°C (Kumar et al., 2005), *Trichoderma harzianum* at 60°C (Savitha et al., 2011), *Aspergillus oryzae* at 55°C (Vishwanathan et al., 2009) and *Engyodontium album* at 45–60°C (Chellappan et al., 2006). Similarly, the optimum pH can be as low as 3.0 for *A. niger* (O'Donnell et al., 2001) or as high as 11.0 for *E. album* (Chellappan et al., 2006). Both fungi and bacteria release amylase extracellularly to perform external digestion. However, fungi are better adapted for solid state fermentation. Their hyphal mode of growth, tolerance to low availability of water and high osmotic condition render them for bioconversion of solid substrate (Saranraj and Stella, 2013).

TABLE 4.5

Some examples for global economic cost of fungi. bil = billion, mil = million.

Product	Estimated cost (in US$)	Reference
Food and food products		
Mushroom	17 bil	*emergenresearch.com*
Baking	311 bil, 1.8 mil jobs	*blog.oup.com*
Yeast products	1 bil	*blog.oup.com*
Mycoprotein**	552 mil	*researchlandmarkets.com*
Soy sauce†	40 bil	*fortunebusinessinsights.com*
Wine and beverages		
Wine	220 bil, 1.7 mil jobs	*blog.oup.com*
Brewing (e.g., beer)	311 bil, 2.2 mil jobs	*blog.oup.com*
Pharmaceutical products		
Antibiotics (fungus)	10 bil	Meyer et al. (2020)
Cephalosporins	14 bil	*alliedmarketresearch.com*
Cyclosporines	2 bil	*verifiedmarketresearch.com*
Yeast insulin	15 bil	*blog.oup.com*
Industrial products		
Plant degrading enzymes	5 bil	
Citric acid‡	3.6 bil	Cairns et al. (2018)
Phytase	150 mil	

Acremonium chrysogenum, **Fusarium venenatum*, †*Aspergillus soyae, Candida versatilis*, ‡*A. niger*.

A few fungal species have attracted much research activities during the last 40 years. For example, the number of publications on them increased from < 100s in 1978 to ~ 520, 400, 150, 200 and 75 in 2017 for *A. fumigatus*, *A. niger*, *A. nidulans*, *Fusarium graminaerum* and *P. chrysogenum*, respectively. The cumulative number of publications for these fungi are 9,825, 7,879, 4,416, 1,755 and 1,456 (Fig. 4.2). In an interesting publication, Cairns et al. (2018) narrated how fungi, especially *A. niger* has shaped biotechnology. The following landmark discoveries in fungi opened new avenues for industrial production. Through his seminal publication, Currie (1917), a food chemist reported that when grown in a sugar medium, any strain of *A. niger* produced large quantities of citric acid. Within 2–3 y of this discovery, *A. niger* was exploited by 10 biochemical fermentation industries. As these fungi are active at broad abiotic ranges (pH, temperature, see above) and emergence of new developments in techniques for fermentation and protease purification, the subsequent 20 y witnessed further exploitation of *A. niger* for diverse products by many food, detergent and pharmaceutical companies (Table 4.6). Another level of complexity in *A. niger* was uncovered by pioneering researches, which have led to the discovery of heterogeneity of fungal secretion at the cell (de Bekker et al., 2011), hyphal (Vinck et al., 2005) and mycelium (Levin et al., 2007) levels. Since 2003, the hallmark studies on fungal genomes were published for *A. nidulans*, *A. oryzae* and *A. fumigatus*. With genome sequencing and editing of *A. niger* and others, a new era has commenced for more beneficial and profitable production of fungal enzymes.

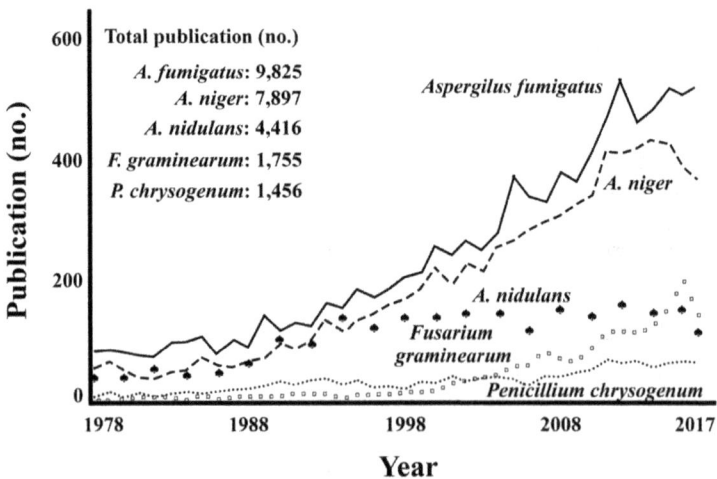

FIGURE 4.2

Number of publications for selected fungi (modified from Cairns et al., 2018).

To provide an idea of the magnitude of global fungal market, the country wise number of companies and their products are listed in Table 4.6. Of 10 countries, the Netherlands and Denmark stand out in production of 14 and 10 enzymes. However, the volume of production for each enzyme may vary, especially American companies may produce several times larger volumes of a few enzymes than that manufactured by other countries. Still, from the point of number of enzyme production, the rank descends for the countries in the following order: The Netherlands > Denmark > USA > Japan > Germany > China > India > UK. Citric acid, Glucoamylase and Hemicellulase are the more profitable enzymes that are produced from six countries. The descending order for these enzymes is: Citric acid = Glucoamylase = Hemicellulase > Catalase = Cellulase = β-Galactosidase > Pectinase = Protease > Glucose oxidase = Phytase > Lactoferrin = Lipase. Developing countries like India, may have to encourage research institutions and manufacturing companies to produce a greater number of enzymes from its vast natural resources.

TABLE 4.6

Number of companies from selected 10 countries, which exploit *Aspergillus niger* for industrial products (rearranged from Fiedler et al., 2013).

Country	Company (no.)	Product (no.)	Name of products
The Netherlands	2	14	Arabinase. Asparaginase, Catalase, Cellulase, β-Galactosidase, Glucoamylase, Glucose oxidase, Hemicellulase, Lactoferrin, Lipase, Pectinase, Phytase, Proteases, Xylanase
Denmark	2	10	Asparaginase, Catalase, Chymosin, β-Galactosidase, Glucoamylase, Hemicellulase, Lipase, Pectinase, Phytase, Proteases
The USA	6	9	Citric acid, Cellulase, Hemicellulase, β-Galactosidase, Glucoamylase, Glucose oxidase, Catalase, Inulinase, Proteases
Japan	3	8	β-Galactosidase, Glucoamylase, Glucose oxidase, Hemicellulase, Proteases, Arabinase, Catalase, Cellulase
Germany	4	5	Citric acid, Glucoamylase, Lactoferrin, Hemicellulase, Phytase
China	5	4	Citric acid, Glucoamylase, Cellulase, Pectinase
India	1	3	Cellulase, Hemicellulase, Pectinase
Israel	1	1	Citric acid
Switzerland	1	1	Citric acid
UK	1	1	Citric acid

Citric acid	6	Catalase	4	Pectinase	4	Phytase	3
Glucoamylase	6	Cellulase	4	Protease	4	Lactoferrin	2
Hemicellulase	6	β-Galactosidase	4	Glucose oxidase	3	Lipase	2

5

Mycology: Teaching – Degrees

Introduction

Like animals, fungi are heterotrophs but sessile like plants. The peculiar combination of heterotrophy and sessility has driven them to acquire micronutrients after external digestion by the externally released digestive enzymes. As of today, the number of described species for fungi is around 107,000 only, in comparison to 374,000 plant species and 1,543,196 animal species. Their structural simplicity and small size have slowed the discovery and description of existing fungal species. Hence, many mycologists estimate that fungi could reach up to 1.5–8.0 million species. Some of them are beneficial to man (Table 4.5) but others are harmful to him, wildlife and crops (Table 3.6). The economic importance of them has greatly attracted the attention of biologists for a long time.

Not surprisingly, there are ~ 100 research journals, exclusively dedicated to fungi and 366,282 articles are published. In terms of research, the USA, India, Japan, Brazil and Canada are some leading countries, where maximum researches related to fungi are carried out (*omicsonine.org*).

Table 4.5 shows that 26 multimillion US$ companies from 10 countries profitably produce 56 enzyme products. There are many other companies involved in production and marketing of antibiotics, cyclosporin and other pharmaceuticals. Hence, the potential for jobs for mycologists is great and is expected to grow at the rate of 13% (*onlinedegree.com*).

Apart from employment opportunities, the international level of recognition for fungal research is also high. For example, the following fungal specialists were awarded the Nobel Prize (Hohmann, 2016): In comparison, only a couple of zoologists were awarded the Nobel Prize.

Mycologists		
1. Alexander Fleming, Ernst, B. Chain and Howard W. Florey	1945	Discovery of antibiotics from *Penicillium*
2. Edward Tatum and George W. Beadle	1958	Regulation of chemical events by genes of *Neurospora*
3. Yoshinori Ohsumi	2016	Autophagy in yeast

Irrespective of their economic importance, employment opportunities and the highest recognition for fungal research, it is sadly disappointing to note that mycology is not taught as a subject leading to recognized degrees in any academic institutions. Rarely, the University of Exeter offers Medical Mycology as an optional subject in the medical syllabus. In India, a couple of agricultural universities have included mycology and plant pathology as one of the many subjects leading to a M.Sc. in Agriculture Science. The AIMST University, Malaysia is the only one offering a course in Mycology. Within the discipline of biology, zoology, botany and microbiology have been taught for a long time as independent subjects leading to undergraduate and postgraduate degrees in several colleges and universities. Understandably, these are subjects concerned with animals, plants and microbes, respectively. There are succinct differences between fungi and bacteria. For example, the former are eukaryotes but the latter prokaryotes. It is difficult to comprehend that they are considered as microbes; unlike bacteria, many fungi are macroscopic and are visible to the naked eye. In recent years, molecular biology and biotechnology, which are tools for the study of organisms rather than a proper subject, are also taught at the undergraduate and postgraduate institutions. The reasons for not teaching Mycology as an independent subject are not known. In these days, when innovative courses are being searched for, an independent subject Mycology leading to degrees would be welcome.

6

Clonal Multiplication

Introduction

A clone is defined as an offspring derived from a single parent (Bisognin, 2011) and is genetically identical at all loci with the parental genome (de Meeus et al., 2007). Clones are produced agametically and are genetically identical descendants with potential for independent existence and reproduction. Clonality, the ability to clone, provides an assurance for propagation and avoids time and resources associated with sexual reproduction. It provides a means by which adaptive genotypes can be sustained and/or rapidly replicated to successfully colonize new substrates/habitats, where sexual reproduction cannot establish it. Given the clear advantages of sexual reproduction, and the means by which clonality minimizes the costs and resources, most fungi harvest the benefits by accommodating both phases in their life cycle.

6.1 Clonal Bridge

In fungi, sexual reproduction is resorted, only when there is a need to (i) purge deleterious genes, (ii) repair DNA and/or (iii) escape from depletion of specific nutrient(s). Hence, it is difficult to estimate the required minimum number of clonal cycles to trigger sexual reproduction. In many of them, the clonal cycle may be continued and sustained for many generations. For example, a budding yeast is expected to sexually reproduce only once in 10,000 generations (Nieuwenhuis and James, 2016). However, the clonal and sexual reproductive cycles are independent from each other. Only in Leotiomycetes (Fig. 1.39) and some Sordariomycetes (Fig. 1.40B), each sexual cycle depends on the participation of the mating type, the microconidiospore, arising from the clonal cycle. It is the fusion of the microconidiospore (MAT-1) from the clonal cycle with MAT-2 from sexual cycle in the Leotiomycete *Botryotina fuckeliana*, and microconidiospore α and mating type A from sexual cycle in the Sordariomycete *Neurospora crassa* that initiates the subsequent events in the sexual cycle. Hence, their sexual cycle is bridged by a clonal cycle in about 18,688 species (17.5% fungi), i.e., 14,714 Leotiomycete species + approximately 50% of 3,974 Sordariomycete species. The same holds true for the Ustilaginomycotina (Fig. 1.28), Pucciniomycetes (Fig. 1.34) and a single species *Coelomomyces psorophorae* (Fig. 1.42), which are all parasites. In fact, their clonal cycle is independent and can continue for any number of generations but each sexual cycle needs to depend on the clonal cycle.

6.2 Budders and Fragmenters

Clonality is a hallmark feature of fungi. However, the agaricale complex consisting (8,500 + 300 + 1,767 + 172 + others =) 10,911 species and 6 species belonging to the yeast genus *Saccharomycodes* (Fig. 1.35C) are not clonals. The remaining 95,845 species or 90% do multiply clonally (see Table 7.3). In fungi, the clonal process may involve (i) sporulation, (ii) budding and/or (iii) fission. Based on whether clonal progenies are produced by simple conversion of parental body, or growth and development of new progeny by budding from parental resources, clonals that sporulate or bud may be considered together as budders, whereas fragmenters comprise those which multiply by fission or fragmentation. A survey has been made to identify and quantify fragmenting and budding clonal fungi (Table 6.1). Unfortunately, Alexopoulos and Mims (1979), Lee et al. (2010), Webster and Weber (2007) and Margulis et al. (2009) have not identified whether the Chytridiomycota, Oomycota, Myxomycota and Zygomycota multiply by fragmentation or budding. However, the fact that motile zoospores of the solitary chytrids and oomycotes (Fig. 1.23) arise from sporangium indicates that in them, clonal multiplication is accomplished by fragmentation, as in Metazoa, in which all fragmenters are motile solitaries (Pandian, 2021b). Also in Myxomycota , clonal myxoamoebae are solitary fragmenters, as most amoebae are known to multiply by either binary or multiple fission/fragmentation (see Pandian, 2023). In solitary motile zoospores of the 260 speciose Neocallimastigomycotes too, fragmentation may take place in the sporangium (Fig. 1.27). However, zygomycotes generate immotile sporangiospores from sporangium (Fig. 1.25). Interestingly, the 322 speciose entomophthoromycotines generate clonals by budding in cycles 1 and 3 (see Fig. 1.26). Hence, zygomycotes are considered as budders. In fact, the switching to budding commences from zygomycotes onwards. Yet, Oberwinkler (2014) proposed clonal multiplication by either microconidial budding or hyphal fragmentation in Dacrymycetes (Fig. 1.29). In view of the fact, that a vast majority of basidiomycotes multiply by clonal budding, the 329 speciose Dacrymycetes are considered as budders. Within Saccharomycotina, the 6 speciose (Helston et al., 2010) *Schizosaccharomyces* are reported as fragmenters (Fig. 1.35B). The fact that isidia of Lecanoromycetes are a detachable outgrowth (e.g., *Parmelina tiliacea*), they are likely to be budders. However, soredia of *Pertusaria flavicans* are granules (see p 166). Hence, it is not known whether these spores are generated by fragmentation or budding. As vast majority of ascomycotes are clonal budders, Lecanoromycetes are considered as budders (Table 6.1). In all, *of 94,420 species, for which the data have been assembled, 3.0% clonal fungi are fragmenters but 97.0% (including zygomycotes, basidiomycotes and ascomycotes) are budders. In fungi, evolution has proceeded from fragmenters in lower fungi to budders in basidiomycotes and ascomycotes. Being the cheapest clonal mode of multiplication, 97% of fungi have opted for budding mode of clonality.*

TABLE 6.1

Estimation of fragmenting and budding cloners of fungi.

Group	Reference	Species (no.)	
		Fragmenters	Budders
Chytridiomycota	Fig. 1.23A	1340	–
Oomycota	Fig. 1.23B	500	–
Myxomycota	Fig. 1.24	500	–
Neocallimastigomycota	Fig. 1.27	260	–
Glameromycota	-	230?	–
Schizosaccharomyces	Fig. 1.35B	4	–
Subtotal		2834 (3.0%)	–
Zygomycota	Figs. 1.25, 1.26	–	1065
Ustilaginomycotina	Fig. 1.28	–	1000
Dacrymycetes	Fig. 1.29	–	329
Tremellomycetes	Fig. 1.30	–	1117
Non-agaricale complex	Fig. 1.32	–	2202
Pucciniomycetes	Figs. 1.33, 1.34	–	8416
Subtotal			13064 (13.8%)

Table 6.1 contd. ...

...Table 6.1 contd.

Group	Reference	Species (no.)	
		Fragmenters	Budders
Taphrinomycotina	Fig. 1.36	–	140
Saccharomycotina	Fig. 1.35A	–	1299
Pezizomycetes/Orbiliomycetes	-	–	1972
Dothideomycetes/Arthoniomycetes	Fig. 1.37	–	20500
Eurotiomycetes	Fig. 1.38	–	3000
Leotiomycetes	Fig. 1.39	–	14714
Sordariomycetes	Fig. 1.40A, B	–	10000
Lecanoromycetes	see p 50	–	14199
Laboulbeniomycetes	Fig. 1.41	–	2486
Coelomycetes	Fig. 1.42	–	10,000
Hyphomycetes	Figs. 1.43, 1.44	–	160
Mycelia sterilia	–	–	52?
Subtotal			78522 (83.2%)
Total			94420
Fragmenters: 2,834 (3.0%); Budder: 91,586 (97.0%)			

6.3 Clonality and Tissue Types

From an innovative approach, Pandian (2021b, 2022) found a correlation between clonality and tissue types. In eukaryotes, clonality decreases with increasing number of tissue types. For example, it gradually decreases from 100% in algae among plants, which are constituted by < seven tissue types, to ~ 24% in angiosperms consisting of < 60 tissue types (Fig. 6.1). But it drastically decreases from 100% in sponges and cnidarians with six – seven tissue types to < 1% in flatworms (e.g., *Planaria*) and polychaete worms (e.g., *Nereis*) and 0% in mammals with > 200 tissue types. A reason for this difference between plants and animals is traced to the fact that whereas the animals undergo irreversible differentiation at fertilization or early embryonic stage, the reversible differentiation takes place in plants; in them, the differentiation is limited to reproductive organs alone and is expressed at a later stage of life. Hence, plants have retained more stemness in their stem cells, responsible for greater percentage of clonality than animals. Apart from it, clonality is also 100% in the acellular grade Protozoa, and 90% in the tissue grade Fungi with not more than seven tissue types.

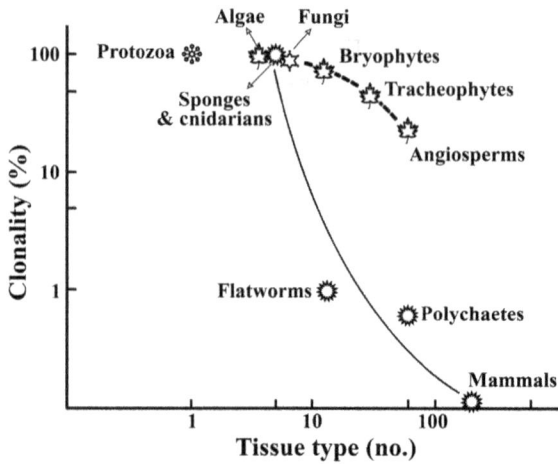

FIGURE 6.1

Clonality as a function of tissue type in eukaryotes (drawn from data reported in Pandian, 2022, 2023, this volume).

6.4 Anamorphs – Retention of Sex

The structurally simpler unicellular Protozoa and tissue grade Fungi incur 10 and 16% (see Table 7.1) secondary loss of sex, respectively. They are the anamorphic clonals with no recognizable morphological reproductive structure(s). They multiply clonally alone. However, a word of caution is made by Nieuwenhuis and James (2016). According to them, molecular tools have provided much insight to recognize that some of these anamorphs are indeed sexuals. For example, Dyer and Gorman (2011) showed the retention of sex in *Aspergillus* and *Penicillium*, which were previously considered as anamorphs (for *Fusarium oxysporum*, see p 111). Similarly, the yeasts *Candida* spp are also now shown as sexuals. In *Coccidioides immitis*, *Fusarium oxysporum*, *Coenococcum geophilum*, with the life cycle involving no meiosis, molecular evidences show the incidence of recombination at population level (see Sherwood and Bennett, 2009, Nieuwenhuis and James, 2016). Hence, the forthcoming molecular evidences may bring a paradigm shift from anamorphism to teleomorphism in more and more anamorphic fungi. With slowing of the rate of Loss Of Heterozygosity (LOH), sexual division in *Saccharomyces paradoxus*, for example, occurs once every 1,000 clonal divisions (Tsai et al., 2008). With the prolonged slowdown of LOH, some of these anamorphs may have retained sex, but express it only after a few thousand clonal generations.

7

Sex – Mating Type – Sexuality

Introduction

Sex is a luxury, and costs time and energy (Carvalho, 2003). For example, clonal protozoans may divide and multiply 5–100 times faster than their sexual counterparts (see Pandian, 2023). True sexual reproduction involves meiotic gametogenesis to reduce ploidy and recombination at fertilization to restore diploidy. Whereas mitosis is accomplished in a simple cycle, the completion of meiosis involves a complex cycle. In meiosis, the first chromosomal division is reductional with homologous chromosomes segregating from one and other (meiosis I) and the second meiotic division is equational with separation of sister chromatids (meiosis II). In sexually reproducing organisms, new gene combinations are gained through (i) random mutation and (ii) meiosis and (iii) recombination. Plants provide excellent examples for the speciation rate in sexless cyanobionts as well as sexualized algae-bryophytes and flowering plants. The number of years required to evolve a species is 0.76 million years (MYs) for clonal cyanobionts, which solely depend on random mutation to generate new combinations. The value is 0.24 MYs for chlorophytes, in which the scope for generation of a new gene combination is limited, as their sexual reproduction involves mitotic gametes. In contrast, the flowering plants require just 390 years to generate a new species, as their sexual reproduction involves meiotic gametes and fertilization (see Pandian, 2022). As benefits arising from meiosis and recombination outweigh the costs of time and energy, sex is successful and has been manifested as early as 1.6–2.0 BYA (Butlin, 2002). Gametes are expressed as mating types in prokaryotes, eukaryotic fungi and a few lower plants (e.g., chlamydomonads) but sexuality as female and male phenotypes in higher plants and metazoans, which generate gametes in the form of immotile eggs and motile sperms.

7.1 Fungal Features

Most random mutations are deleterious rather than beneficial. Fungal lineages continue to accumulate them until they are purged through sexual reproduction (Ni et al., 2011). Hence, many authors (e.g., Wallen and Perlin, 2018) considered the emergence of sexual reproduction as a protective measure from (random mutation) environmental stress and to repair DNA damage (e.g., Michod et al., 2008). It is possible that meiosis itself emerged as DNA repair mechanism. Regarding stress, for example, the depletion of the nitrogen source in a culture medium leads to differentiation of vegetative cells into somatogamic gametes in the unicellular green alga *Chlamydomonas reinhardtii* (Sager and Granick, 1954). During the checkered history of evolution, sex and sexuality have emerged as mating types in the ancient single-celled plants (e.g., *C. reinhardtii*), protozoans (e.g., ciliates, see Pandian, 2023) and single celled/a few tissue-typed fungi. Subsequently, higher eukaryotic plants and animals have discovered female and male phenotypes. Chamber's Dictionary of Science and Technology defines "mating types as groups of individuals within a species, which cannot breed among themselves but which are able to breed with individuals of other such group(s)". Mating types are genetically defined incompatible groups that as haploids precisely regulate successful mating only between complementary mating types (Nieuwenhuis and James, 2016). To establish and sustain dual or multiple mating types, a member of a mating type in a fungal species needs a mechanism to recognize complementary member of its species. The discovered mechanism by fungi is as simple as a pheromone produced by one mating type, which is recognized by complementary member from another mating type (Wallen and Perlin, 2018).

It is an established fact that inbreeding limits genetic diversity, whereas outcrossing promotes it leading to acceleration in evolution and speciation. In most eukaryotes, outcrossing is preferred mode of breeding. For example, despite being monoecious, plants have developed structural and/or chemical strategies to minimize inbreeding/selfing. At this juncture, it is necessary to distinguish outcrossing as different from non-outcrossing as well as outbreeding from inbreeding. Figure 7.1 explains the fine differences in usage of these terms, especially regarding breeding strategies of fungi. In a limited sense, outcrossing may mean heterothallism or dioecy and inbreeding homothallism or monoecy.

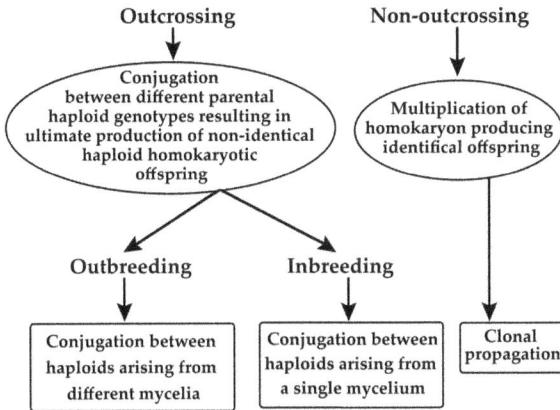

FIGURE 7.1

Terminology used for describing breeding strategies of fungi
(modified and redrawn from Rayner and Boddy, 1988).

Despite being unifactorial, a fungal species may have large numbers of
alleles within each mating type at the population level (Watkinson et al.,
2015). Mating occurs successfully between any two haploid mycelia/yeast
cells that are complementary as mating types (e.g., A_1 and A_2). Following
meiosis, two of the haploid spores will be of one mating type, while the
other two are of another mating type. This system is called **bipolar**, which is
common to ascomycotes and some basidiomycotes (see Fig. 7.2A). In them,
inbreeding potential is limited to 50%, as (i) any diploid mycelium yields
two mating types and (ii) many alleles exist within each mating type at
the population level; hence almost all encounters with non-sibling mycelia
alone can be compatible. But many basidiomycotes have a mating system
with two unlinked mating type traits named as AB. Alleles both the A and
B loci must differ to ensure successful mating. For example, A genes control
the development of clamp connections, which maintain the two parental
nuclei in each cell. The B genes regulate exchange of the nuclei between both
partners, nuclear migration from one mycelium to the other and *vice versa*,
and encode for pheromonal and receptor systems. Such systems with four
factors one with *a, b* and the other with *A, B* are known as **tetrapolar** (see
Fig. 7.2B). When haploid mycelia, derived from the same basidium are
crossed in all possible pair combinations, only 25% of the crosses will mate
successfully; they will have different alleles at both mating loci, i.e., the
inbreeding potential is reduced to 25% (see Watkinson et al., 2015).

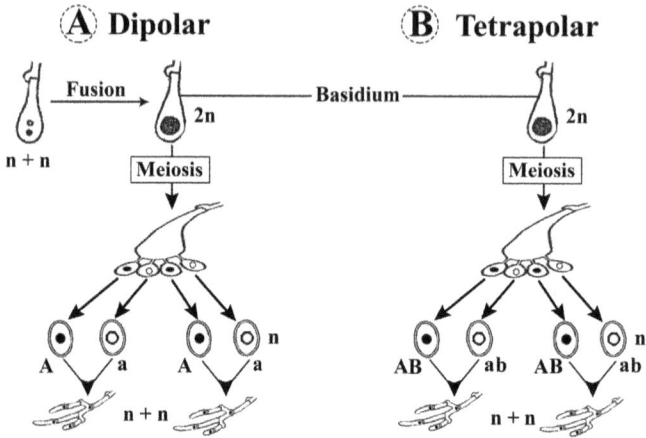

FIGURE 7.2

The sequence of meiosis and fusion in (A) dipolar and (B) tetrapolar basidiomycotes.

The sexualized chytridiomycotes may have emerged after the discovery and manifestation of sex as early as ~ 2.0 BYA. Having been sexualized, their motile zoospores, similar in structure to the vegetative stage in clonal multiplication, copulate (e.g., *Allomyces* spp, Fig. 1.23A) or conjugate (e.g., *Olpidium viciae*) or oogamically fertilize (e.g., *Albugo candida*). Of 1,840 chytrid-oomycote species, for which 14 life cycles are described, some (including Myxomycota) have explored the following modes of sexual reproduction:

Chytridiomycotoa	1840	Myxomycota	500
Sexualized thalloidal oogamy	7/14 = 920		
Zoosporic copulation	5/14 = 657		
Subtotal	1,577		500
Zoosporic conjugation	1/14 = 131		
Subtotal	131		
Total 1,708 + 500 = 2208			
Secondary loss of sex	1/14 = 131		1157
Conjugation = 131 species, 6%		**Fusion = 2,077 species, 94%**	

Although these zoospores arise from clonal zoospores, as 'vegetative gametes' (e.g., *Chytriomyces hyalinus*), the fact that conjugation (*O. viciae*) or oogamic fertilization occurs in 920 species clearly indicates that except in *Nowakowskiella ramosa* (which have secondarily lost sex), the gametic zoospores are characterized by an element of mating type. So are the 1,309 speciose Saccharomycotina. Though their somatogametes arise vegetatively from parental yeasts (Fig. 1.35), they are characterized by one or another mating system (Sherwood and Bennett, 2009, Ni et al., 2011, Wallen and Perlin, 2018).

Once manifested, the structurally complex organ/system grade organization in plants and metazoans do not incur loss of sex, except when parasitically castrated or sterilized (see Table 22.2, Pandian, 2021b). But the structurally simpler unicellular Protozoa and tissue grade Fungi are unable to maintain the manifested sex and secondarily lose it. For example, 10% of the unicellular Protozoa have secondarily lost sex (see Pandian, 2023). Even with elevation to tissue grade organization, Fungi are also unable to sustain the manifested sex. Clearly, *the minimum of organ level organization is required to sustain the manifested sex.*

A few authors hint that ~ 20% fungi may have secondarily lost sex and exist only as clonal anamorphics (e.g., Dyer and Gorman, 2011). Jones et al. (2009) noted the existence of marine anamorphs in 94 species in 61 genera. However, no survey has been thus far made to quantify the incidence of the sex loss across the taxonomic groups. For the first time, this account has made an attempt to quantify them (Table 7.1). Sex is lost in the symbiotic Neocallimastigomycota and Glomeromycota (however, see Taylor et al., 2015) at the phylum level as well as Coelomycetes, Hyphomycetes and Mycelia sterilia at class level. Repeated computer searches have revealed that no sex loss is incurred by Myxomycota, Zygomycota, Lecanoromycetes, Leotiomycetes and Laboulbeniomycetes. However, surprisingly there is the no loss of sex also in the unicellular non-motile 1,309 speciose Saccharomycotina.

This estimate is based on the names of anamorphic genera in Pucciniomycotina by Aime et al. (2014) and in four classes in Ascomycota by the authors listed in Table 7.1. The estimate does not include *Aspergillus* and *Penicillium* (Sordariomycetes), which were considered as anamorphs. Based on the prevalence and expression of functional sex-related genes, distribution of mating type genes and detection of recombination from population genetic analyses, Dyer and Gorman (2011) showed the retention of sex in these fungi, albeit expressed not frequently. Among the 20 speciose genus *Fusarium*, Rana et al. (2017) noted the retention of sex in 20% species, i.e., 4 of 20 species. Genes, known to be associated with sexual reproduction in other fungi, are also retained intact in the arbuscular mycorrhizal fungus *Rhizophagus intraradices* (see J.W. Taylor et al., 2015, for more information, see p 162), albeit may not be expressed. For the other taxonomic groups also, adequate care has been taken to quantify the number of species as precisely as possible to each anamorphic genus. Similarly, there may be more numbers of genera and species that are anamorphic but are not mentioned by the authors listed in Table 7.1. With these constraints, the following may be inferred: (1) *Approximately, 16,899 species or 15.8% of fungi have secondarily lost sex.* (2a)

Among the 1,605 speciose Zygomycota, which engage hyphae as mating types, none has lost sex. (2b) So are the 13,113 speciose Agaricomycetes, which also engage hyphae as mating types (Fig. 1.31). (2c) This is true as well for the 1,117 speciose Tremellomycetes, which employ yeast as a mating type. (2d) The sex loss occurs only in a few Pucciniomycotines, which engage basidial spores for mating. *In all, none among Zygomycota, Agaricales and Tremellomycetes, but only 204 species or 0.85% of other Basidiomycota incur secondary loss of sex.* (3) In contrast, *of 77,083 Ascomycota species, which employ sexualized spores and mostly spermatization mode of fusion to accomplish sexual reproduction, 16,064 species or 20.8% of them have lost the sex. Briefly, employing*

TABLE 7.1

Estimate on anamorphic fungi. *Nowakowskiella ramosa.*

Group		Species (no.)		Reference
		Total	Anamorph	
Chytridiomycota		1,840	131*	Table 1.3
Zygomycota		1,065	0	
Neocallimastigomycota		260	260	Gruninger et al. (2014)
Glameromycota		230	230	*Wikipedia*
Myxomycota		500	0	see Webster and Weber (2007)
Basidiomycota		23,975	204	Aime et al. (2014)
Ascomycota		78,532		
Taphrinomycotina	140		7[†]	Kurtzman and Sugiyama (2015), Schisler et al. (2010)[†]
Saccharomycotina	1,309		0	Dujon and Louis (2017)
Pezizomycetes/Orbiliomycetes	1,972		565	Pfister (2015)
Dothideomycetes/Arthoniomycetes	20,500		3,098	Schoch and Grube (2016), Videira et al. (2016)
Eurotiomycetes	3,000		129	Geiser et al. (2015)
Sordariomycetes	10,000		2,053	Zhang and Wang (2015), Vega et al. (2012)
Leotiomycetes	14,714		0	
Lecanoromycetes	14,199		0	Miadlikowski et al. (2008)
Laboulbeniomycetes	2,486		0	
Coelomycetes	10,000		10,000	
Hyphomycetes	160		160	Margulis et al. (2009)
Mycelia sterilia	52		52	
	Total	106,402	16,889 (15.8%)	

[†]of which 2 are *Geotrichum candidum, G. citri-aurantii.*

hyphae or yeasts as the mating type, Zygomycota and Basidiomycota discourage the secondary loss of sex, whereas that of sexualized spores by Ascomycota tend to allow the loss of it.

7.2 Homothallism vs Heterothallism

Fungi have discovered two paradigmatic sexualities: (i) heterothallism, which avoids selfing but requires two compatible individuals to accomplish reproduction and (ii) homothallism, which involves a single individual and in many of them selfing is possible. Both share common key features namely meiosis and fertilization (Ni et al., 2011). Gametically controlled heterothallism was first discovered by Albert Blakeslee in the Zygomycota *Phycomyces blakesleeanus* (see Watkinson et al., 2015). In fungi, an attempt to quantify them encounters the following hurdles (see also Chapter 10): 1. Both modes of thallism can be found concomitantly in different species within a genus (e.g., *Fusarium*) and sometimes even within a species. For example, the wood-decaying basidiomycote *Stereum sanguinolentum* have 30 distinct inbreeding (homothallic) populations in Northwestern Europe but the outbreeding (heterothallic) population in Austria and North America (Watkinson et al., 2015). 2. Sporadic occurrence of homothallics among the predominantly heterothallic taxonomic groups, as in *Cochliobolus neoformans* among basidiomycotes, *Fusarium* among Eurotiomycetes, *Sordaria* among Sordariomycetes and 3. Switching between mating types is reported in *Saccharomyces cerevisiae*, *Schizosaccharomyces pombe* and *Kluyveromyces lactis*. 4. In the absence of a compatible mating type, heterothallics can switch to homothallism. For example, the heterothallic basidiomycote *C. neoformans* can switch to homothallism and become self-fertile. 5. The existence of three paradigmatic sexualities namely homothallism, pseudohomothallism and heterothallism in *Sordaria macrospora*, *Podospora anserina* and *Neurospora crassa*. 6. For many taxonomic groups, the life cycle is known only for a single species (*Auricularia auricula-judae*). It may be debatable to generalize the same life cycle for 402 species of Auriculariales. Irrespective of these exceptional hurdles and other constraints, this account marks the beginning of a survey to quantify homothallism and heterothallism in fungi. Accordingly, the relevant information has been assembled in Table 7.2. As far as possible, the information was drawn from the life cycles illustrated in Figs. 1.23 to 1.42 and other sources. The number of anamorphic species (Table 7.1) is subtracted from the total number for the relevant taxonomic group. For Pezizomycetes/Orbiliomycetes and Lecanoromycetes, no distinct life cycle is yet available. However, piecemeal information was collected from different sources to arrive at the values. The survey reveals *the existence of 26,926 homothallic species or 30% fungi and 62,987 heterothallic species or 70% fungi*. Being the very first attempt, this account does not claim precision of the values assembled in Table 7.2. However, *the proportions between 30% homothallism and 70% heterothallism may remain valid*. Being the ancestral lineage, Chytridiomycota along with Oomycota exist as homothallic in 1,709 species and heterothallic in 394 species. From this basic predominantly aquatic lineages, fungal evolution has proceeded to homothallism in 26,926

TABLE 7.2

Estimate on homothallism vs heterothallism and conjugation vs fusion in fungi (compiled from Figs. 1.23 to 1.42, Table 1.5).

	Fertilization	Group	Species (no.)	Figure
		I. Homothallism		
1.	Zoospore fusion	Chytridiomycota/Oomycota	1,709	1.23
2.	Basidiospore fusion	Dacrymycetes	329	1.29
3.		Tremellomycetes	1,117	1.30
4.	Aeciospore fusion	Pucciniomycetes	7,800	1.34
		Subtotal	**10,955**	
5.		Ustilaginomycotina	1,000	1.28
6.		Cystobasidiomycetes	25	1.33 A
7.		Microbotryomycetes	200	1.33 B
8.	Yeast conjugation	Non-pucciniomycetes	191	
		Subtotal	**1,416**	
9.		Taphrinomycotina	133	1.36
10.		Saccharomycotina	1,309	1.35 A,B
		Subtotal	**1,442**	
11.	Hyphal conjugation	Agaricomycetes	**13,113**	1.31, 1.32
		Total	**26,926**	
		II. Heterothallism		
1.	Zoosporic fusion	Chytridiomycota/Oomycota	394	
2.	Amoebic fusion	Myxomycota	500	1.24
3.	Microconidial spore + ascospore fusion	Leotiomycetes	14,714	1.39
4.		Sordariomycetes	3,974	1.40 B
5.	Ascospore fusion	Lecanoromycetes	14,199	
		Subtotal	**33,781**	

Table 7.2 contd. ...

...Table 7.2 contd.

	Fertilization	Group	Species (no.)	Figure
6.	Hyphal conjugation	Zygomycota	**1,065**	1.25, 1.26
7.		Pezizomycetes/Orbiliomycetes	1,407	
8.		Dothideomycetes/ Arthoniomycetes	17,402	1.37
9.		Eurotiomycetes	2,871	1.38
10.	Spermatization	Sordariomycetes	3,974	1.40 A
11.		Laboulbeniomycetes	2,486	1.41
12.		Coelomycetes	1	1.42
		Subtotal	**28,141**	
	Total	62,987	*Grand total*	89,913 (84.2%)
	Homothallism = 26,926 (30.0%)		**Heterothallism = 62,987 (70.0%)**	

species in Basidiomycota and to heterothallism in 1,605 Zygomycota and 61,028 species in Pezizomycotina (not indicated, but added from Table 7.2), which are mostly terrestrial habitants, i.e., colonization of land by fungi proceeded via homothallism in basidiomycotes but through the successful heterothallism in more speciose ascomycotes, i.e., *the colonization of land by homothallism decelerated species diversity in basidiomycotes but heterothallism accelerated it in Pezizomycotina. To minimize selfing among homothallics, basidiomycotes have devised tetrapolar mating type. Interestingly, inbreeding potential or selfing is reduced to 50% in bipolar basidiomycotes, but to 25% in tetrapolar basidiomycotes* (see p 109).

7.3 Fertilization Site and Dispersal

Fungi reproduce mostly by producing clonal (e.g., Coelomycetes) and/or sexual (e.g., Agaricales) spores. *As parent : progeny ratios in sexual reproduction is limited to 1 : 2 in lower fungi and 1 : 4 in ascomycotes, fungi may invest more on clonal multiplication than on sexual reproduction.* Whereas the dispersal mode is external in the former, the fertilization site is internal in the latter. A survey was also made to quantify them. Surprisingly, 95,485 species (leaving some like Agaricales, which propagate solely by sexual reproduction) are clonals (Table 7.3); their life span is longer (except perhaps in ustilaginomycotines) and externally release clonal spores for dispersal by air or water (e.g., chytrids and oomycotes). *Air being 800-times less dense than water, 95,485 species or 88.5% fungi (Table 7.3) disperse their clonal spores by air.*

Clonal spores of 95,485 fungal species are released externally. Notably, fertilization site is also external in 14,714 in lower fungal species but it is internal in 54,112 species or 60.5% fungi, especially pezizomycotines (Table 7.3). In fungi, evolution has proceeded from the riskier and costlier external fertilizations in lower fungi to safer and cheaper internal fertilization in pezizomycotines. External fertilization may immediately link spore dispersal by the relatively less efficient ballistosporic mechanism. Though fertilization is internal, the pezizomycotines have a more efficient explosive discharge and the fastest launching gun-shooting mechanisms. These are elaborated in Chapter 9. With the involvement of sclerotium, the fertilization site is external only in the 14,714 speciose Leotiomycetes (Fig. 1.39). The values reported for fungal spore density in the atmosphere are in percentages (see Table 9.1). Yet, it is notable that fungal spores found in the atmosphere for 102,501 species (i.e., excluding aquatic fungal species), the basidiomycotes with 23,975 species or 23% share and ascomycotes with 78,526 species or 77% share, contribute 32 and 63% atmospheric spores, i.e., with safer internal fertilization site, the pezizomycotines require to produce ~ 20% less in number of spores than the externally fertilizing basidiomycotes.

TABLE 7.3

Life span (LS), dispersal of clonal spores and fertilization site (FS) of sexual gametes in fungi (compiled from Figs. 1.23 to 1.44). L = Long, S = Short, In = Internal, Ex = External, Dis = Dispersal, Aq = Aquatic, * = within the host/substratum.

Clade	Clonal			Sexual		
	LS	Dis	Species (no.)	LS	FS	Species (no.)
Chytridiomycota/Oomycota	L	Ex	1,840	S	Ex Aq	1,709
Coelomycetes	L	Ex	10,000	S	Ex Aq	1
Myxomycota	L	Ex	500	S	Ex	500
Neocallimastigomycota	L	Ex	260	–	–	0
Glomeromycota	L	Ex	230	–	–	0
Zygomycota	L	Ex	1,065	S	Ex	1,065
Ustilaginomycetes	S	Ex	1,000	L	Ex	1,000*
Dacrymycetes	L	Ex	329	S	Ex	329
Tremellomyces	L	Ex	1,117	S	Ex	1,117
Agaricale complex	L	–	0	S	–	10,911
Auricularial complex	L	Ex	2,202	S	Ex	2,202
Other pucciniomycetes	L	Ex	616	S	Ex	412*
Taphrinomycotina	?	Ex	140	S	Ex	133

Table 7.3 contd. ...

...*Table 7.3 contd.*

Clade	Clonal			Sexual		
	LS	Dis	Species (no.)	LS	FS	Species (no.)
Saccharomycotina	L	Ex*	1,303	S	Ex*	1,309
Leotiomycetes	L	Ex	14,714	S	Ex	14,714
	Clonal External = 35,316			**Sexual External = 35,396**		
Pucciniomycetes	L	Ex	7,800	L	In	7,800
Pezizomycete/Orbiliomycetes	L	Ex	1,972	S	In	1,407
Dothideomycetes/Arthoniomycetes	L	Ex	20,500	S	In	17,402
Eurotiomycetes	L	Ex	3,000	L	In	2,871
Sordariomycetes	L	Ex	10,000	S	In	7,947
Lecanoromycetes	L	Ex	14,199		In?	14,199
Laboulbeniomycetes	L	Ex	2,486	S	In	2,486
Hyphomycetes	L	Ex	160	–	–	0
Mycelia sterilia	L	Ex	52	–	–	0
	Clonal External = 60,169			**Sexual Internal = 54,112**		
Total			**95,485**			**89,508**

7.4 Modes of Fertilization

Prior to the description of different fusion modes, a preamble on a few aspects exclusively relevant to fungi and a few technical terms may have to be explained (Ni et al., 2011, Watkinson et al., 2015).

Gametangial fusion: The fusion between the gametangia and formation of the zygote, following pheromonally attracted complementary hyphae, e.g., Zygomycota (Fig. 7.3D, E).

Somatogamy: The fusion between 'vegetative' gametes, which are not morphologically different from the vegetative body. These gametes may be single-celled motile zoospores (e.g., *Polyphagus macrogynus*, Fig. 7.3A, B) or immotile yeasts (Fig. 7.3F) or even multicellular hyphae (Fig. 7.3D, E).

Gamete fusion: The fusion between iso- or aniso-gamic gametes arising from the oogonium and antheridium to form an oospore. They may be motile (e.g., chytrids) or immotile (e.g., Agaricales). In most cases, anisogamy leads to oogamy (Fig. 7.3C).

Spermatization: Penetration and fusion of antheridial nucleus with larger immotile oogonium to form oospores, after the development of the fertilization tube and its chemotropic growth, e.g., Ascomycota (Fig. 7.3G). In contrast, the trichogyne curves to receive 'male' gamete in *Neurospora* (Fig. 7.3H).

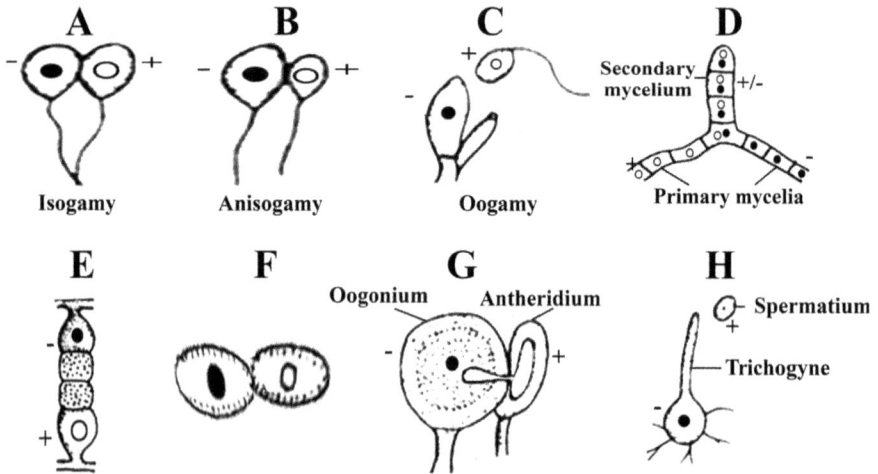

FIGURE 7.3

Modes of fertilization in fungi. Note: the gametangia serve as 'gametes' in (D, E) zygomytcotes and some (G) ascomycotes (modified and redrawn from *onlinebiologynotes.com*).

Sexual reproduction in fungi occurs in temporally separated three stages (see p 34). Firstly, plasmogamy involves the cytoplasmic fusion between compatible mating types. It is followed by karyogamy, in which the two haploid nuclei fuse. Finally, meiosis takes place to restore haploidy from the newly produced diploid status. The interval between plasmogamy and karyogamy may last for hours or days. Rarely, a few basidiomycotes may not encounter compatible mycelium for many weeks (wk) or years; for example, *Trametes versicolor* persist as monokaryon for several years in nature (see Watkinson et al., 2015).

The modes of fusion can be looked at from different points of view: (i) cellular (e.g., yeast conjugation in Pucciniales) vs. hyphal conjugation (e.g., Zygomycota). (ii) motile zoospores (e.g., chytrids, oomycotes) vs. immotile gametes/hyphae. (iii) fusion between iso- (e.g., *Physarum polycephalum*) and aniso-gamy (e.g., *Polyphagus euglenae*) and (iv) oogamic fusion. Another mode of grouping is (i) motile complementary mating types that move toward each other (e.g., *P. euglenae*), (ii) multicellular hyphae that are pheromonally attracted to grow toward each other (e.g., Zygomycota) and (iii) only + mating type is either dispersed (e.g., chytrids) or grown (e.g., ascomycetes) toward the – mating type. Notably, the yeast cells are engaged as 'gametes' in parasitic non-pucciniales; to them, resource may not be limiting. This may hold good for the free-living yeasts growing on assured substratum like bread. More importantly, based on the potential cost of bringing the 'gametes' together by different modes of fusion, the fungi may be divided into the following five groups: 1. The costliest is the aquatic gametes swum by flagella (e.g., chytrids and oomycotes) or amoeboid movement (e.g., Myxomycota). 2. The conjugation between adjoining unicellular 'gametes' in yeasts is costlier. 3. The relatively less costly is the multicellular hyphal conjugation (e.g., Zygomycota) by gametangia growing towards each other. 4. The cheaper is spermatization, in which the growing male gamete penetrates the immotile oogonium or trichogyne bending to receive 'male' spore (e.g., pezizomycotines) and 5. External fusion in the air is the least, albeit riskier; however, it also facilitates dispersal. The following shows the quantification of these five groups, for which values are drawn from Table 7.2.

Zoosporic fusion	1,709 + 394 + 500	2,603	(2.9%)
Yeast conjugation	1,000 + 25 + 200 + 191 + 133 + 1,305	2,854	(3.2%)
Hyphal conjugation	13,113 + 1,065	14,178	(15.8%)
Spermatization	1,407 + 17,402 + 2,871 + 3,974 + 2,486 + 1	28,141	(31.3%)
Spore fusion via dispersal by air	329 + 1,117 + 7,800 + 14,199 + 3,974 + 14,714	42,133	(46.9%)

Based on the potential cost of gametic fusion in fungi, the modes of fusion increases in the following ascending order: *zoosporic fusion (2.9%) > yeast conjugation (3.2%) > hyphal conjugation (15.8%), spermatization (31.3%) > spore fusion by aerial dispersal (46.9%)*. Briefly, the aquatic motile zoosporic mode of fusion decelerates species diversity. So is the conjugation by unicellular yeast or hyphal conjugation. Being oogamic fusion, spermatization ensures safety and reduces the cost of fusion by restricting the growth of 'male' gametes alone. However, *the spore fusion by aerial dispersal is the preferred mode of 'gamete' fusion by fungi; it seems cheaper to produce a large number of sexual spores to neutralize the risk and loss of spores than to meet the higher cost of moving or growing the 'gametes' toward each other.*

7.5 Life Cycle and Ploidy

Parasitic fungi are more by number among ascomycotes than basidiomycotes (see Tables 3.8, 3.9). Yet, life cycles for more numbers of basidiomycotes are known than that for ascomycotes. One reason may be that pucciniale and ustilaginomycotine parasites are more virulent and cause greater loss to commercial crops. Interestingly, the ploidy status of either clonal or sexual phases during the life cycle is altered in pucciniales (Fig. 1.34) but not in ustilaginomycotines (Fig. 1.28).

For the first time, an attempt has been made to assemble relevant information for the ploidy status during clonal multiplication and sexual reproduction of different taxonomic groups of fungi. Unfortunately, not many authors have indicated the ploidy status during different stages of clonal and sexual phases in the life cycle of fungi, especially for ascomycotes. For many life cycles, a few authors have hinted at the ploidy status but thankfully Lee et al. (2010) and Evert and Eichhorn (2013) clearly stated the ploidy status during the entire life cycle of *Allomyces arbusculus* (Fig. 1.23B) and *Melamspora larici-epitea* (Fig. 1.34). Yet, many authors have clearly located the stages, in which meiosis and fusion occur in the life cycle of different taxonomic group of Ascomycota. From these sources, relevant information on the possible ploidy status during the life cycle of fungi is presented in Table 7.4. Surprisingly, *clonal spores are all haploids except in aquatic chytridiomycotes/oomycotes, arboreal dacrymycetes and parasitic pucciniomycetes*; for the latter, an explanation is provided in Fig. 7.4A. As meiosis precedes fusion in Ascomycota, their sexual cycles pass through n → 2n → n, except in a single aquatic Coelomycete species, in which the ploidy pass via 2n → n → 2n (Fig. 1.42). Similarly, the aquatic chytridiomycotes/oomycotes also pass through 2n → n → 2n with an exception of the terrestrial parasitic *Synchytrium endobioticum*. Understandably, the terrestrial zygomycotes also go through n → 2n → n. It seems that *aquatic fungi pass through 2n → n → 2n ploidy status during the sexual phase*. It is for experts in phylogeny to trace the lineage of basidiomycotes passing through 2n → n → 2n status, as in aquatic chytrids.

It may be relevant to compare the life cycles of free-living and parasitic fungi involving one or two host(s). Figure 7.4 shows the cycle of free-living mushroom and parasitic fungi involving one and two host(s). The sustained availability of debris has assured nutrient supply to mushrooms, which

TABLE 7.4

Ploidy status during clonal and sexual phases of fungi. Ma = Myxoamoeba, Y = Yeast, Z = Zoospore, D = Dormant (compiled from Figs. 1.23 to 1.44).

Clade	Clonal		Sexual	
	Ploidy	Species (no.)	Ploidy	Species (no.)
Chytridiomycota/Oomycota	2n, Z	1,840	2n → n → 2n	1,709
Myxomycota	n, Ma	500	n → 2n → n	500
Neocallimastigomycota	2n?, Z	260	–	0
Glomeromycota	2n?	230	–	0
Zygomycota	n	1,065	n → 2n → n	1,065
Basidiomycota				
Ustilaginomycetes	n, Y	1,000	n+n → 2n → n+n	1,000
Dacrymycetes	2n	329	2n → n → 2n	329
Tremellomyces	αn	1,117	n → 2n → n	1,117
Agaricale complex	–	0	n+n → 2n → n+n	10,911
Auricularial complex	n	2,202	n+n → 2n → n+n	2,202
Pucciniomycetes	2n	7,800	n → 2n → n	7,800
The others	2n	616	n → 2n → n	4,112
Ascomycota				
Taphrinomycotina	n	140	n → 2n → n	133
Saccharomycotina	n	1,303	n → 2n → n	1,303
Pezizomycete/Orbiliomycetes	n	1,972	n → 2n → n	1,407
Dothideomycetes/Arthoniomycetes	n	20,500	n → 2n → n	17,402
Eurotiomycetes	n	3,000	n → 2n → n	2,871
Sordariomycetes	n	10,000	n → 2n → n	7,947
Leotiomycetes	n	14,714	n → 2n → n	14,714
Lecanoromycetes	n	14,199	n → 2n → n ?	14,199
Laboulbeniomycetes	n	2,486	2n → n → 2n	2,486
Coelomycetes	n	10,000	2n → n → 2n	1
Hyphomycetes	n	160	–	0
Mycelia sterilia	n	52	–	0

have limited the cycle to the sexual phase alone. On assumption of parasitism, the need for risky dispersal from one host to another has demanded the insertion of the clonal phase in the life cycle of pucciniales, e.g., *Uromyces striatus* and agaricomycotines, e.g., *Auricularia auricula-judae*. Instead of karyogamic filaments developing into sexual mushrooms, the clonally multiplying telial stage is inserted to compensate the potential loss involved in the transmission from one host to another. The involvement of two hosts in the cycles of many pucciniales is difficult to comprehend. The transfer of adjoining plasmogamic spore from the basidium to host 1 is riskier but is facilitated by air and/or by vectors like insects or birds. The karyogamic

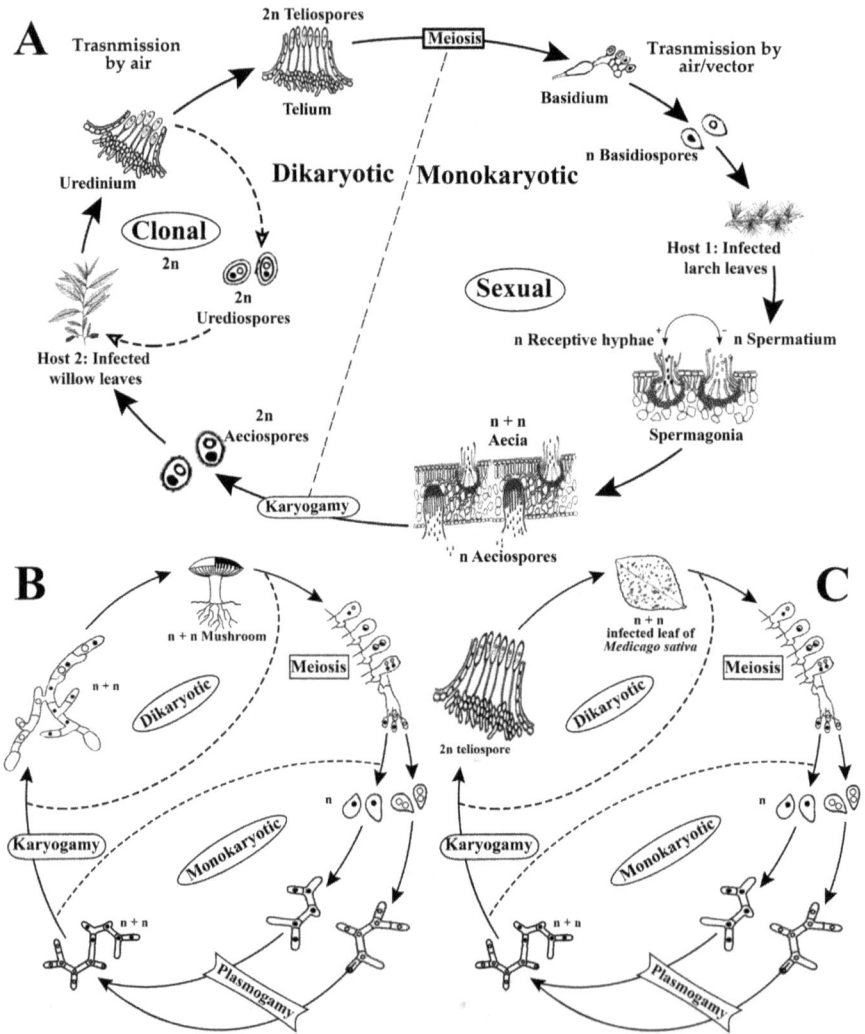

FIGURE 7.4

Life cycle of (A) pucciniale parasitic *Melamspora larici-epitea*, (B) free-living of an agaricale and (C) the parasitic *Uromyces striatus* on the leaf of *Medicago sativa* (A, B – copy from Figs. 1.31 and 1.34, C – drawn from different sources).

diploid aeciospore-urediniospore-telium-basidium to host 1 from host 2 seems to be riskier. Not surprisingly, a clonal phase is inserted at the uredinial level in host 2, the urediniospores are highly buoyant and when air-borne can travel long distances (p 62). As 11 pucciniale species are heteroecious and macrocyclic and cause huge crop loss, any control measure should target host 2, from which urediniospores multiply and travel far and wide. Incidentally, this leads one to quantify the internal or external dispersal of clonal and sexual spores.

8

Gametangia – Gametogenesis

Introduction

Gametogenesis is the process, through which gametes are generated. In Metazoa characterized by a minimum of 7–14 tissue types, distinct reproductive organs/systems are manifested and maintained. In them, the parent diploid oogonium and spermatogonium undergo meiosis to generate haploid oocyte and spermatozoa, respectively. In them, meiosis precedes fertilization and provides the base for recombination at fertilization (see Pandian, 2021b). In Protozoa, the imperfect gametogenesis ranges widely from one taxonomic group to another, indicating the trials and tribulations undergone by them. *Yet, the unicellularity in Protozoa has not let them establish true meiosis.* At tissue grade organization, most algae with three – seven tissue types are also unable to establish a distinct reproductive organ, except for the carpogonium in red algae with nine tissue types (see Pandian, 2022). However, sponges and cnidarians with six – seven tissue types do possess distinct reproductive tissues/organs. With ~ 14 tissue types, distinct reproductive organs begin to appear, for example, archegoniophore and antheridiophore in bryophytes or as reproductive systems in planarians. *Irrespective of animals or plants, the minimum number of tissue types required to manifest and maintain reproductive organs/systems seems to lie between 7 and 14 tissue types.* In this context, fungi are a fascinating group of organisms. In them, gametangial fusion ensures sexual reproduction in Zygomycota with < seven tissue types. Most fungi including Saccharomycotina, which have subsequently reverted to unicellularity, engage mating types as gametes to accomplish sexual reproduction. These mating types are chemically distinguishable but not morphologically. It is only in Ascomycota with a little more than seven tissue types, that gametes are generated from distinct organs namely ascogonium and antheridium to achieve sexual reproduction. Though distinct, these simple gonads can afford a maximum of parent : progeny ratio of 1 : 4 only. This chapter elaborates some of these aspects.

8.1 Chytrids – The Basal Lineage

For the 1,840 speciose Chytridiomycota, the life cycle is described for 12 species (Table 1.4). Being the basic lineage, they have representatives for different types of 'sexual' reproduction, like homothallism in 9 of 12 species, heterothallism in 2 of 12 species, generation of gametes from oogonium and antheridium in 4 + 1 species, conjugation in 2 species and so on. Unable to maintain sex, one species *Nowakowskiella ramosa* has secondarily lost it. It is conceivable that almost all other (phyla) lineages of fungi have emerged from the chytrids. In them, four haploid zoospores are generated following meiosis. Hence, the parent: progeny ratio is 2: 8 or 1: 4 (e.g., *Olpidium viciae*).

8.2 Gametangial Fusion: Zygomycota

Mating types are established in Zygomycota with < seven tissue types. On pheromonal attraction, the hyphae of complementary mating types grow to approach each other to commence the conjugation process (Fig. 8.1). Their haploid hyphal tips swell and gametangially conjugate to form the diploid zygote. Following meiosis, four haploid offspring out of the two conjugating hyphae. Notably, the conjugation between two complementary hyphae yields two offspring, i.e., the parent: progeny ratio is 1: 2.

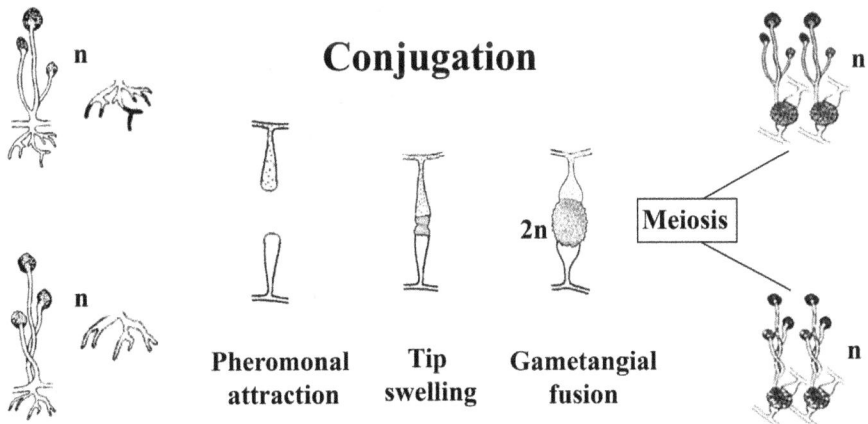

FIGURE 8.1

Conjugation and gametangial fusion in a representative Zygomycota.

8.3 Homothallic Basidium

In Basidiomycota, the basidium serves as a reproductive organ, which initially holds two haploid nuclei drawn from complementary mating types and subsequently fuse as diploid zygotic cell (Fig. 7.2). Thus, evolution seems to have explored this unique but not a very successful strategy (as Basidiomycota are not as diverged as Ascomycota) to introduce minimal heterozygosity in the basidiospores. In the basidium, the diploid spores undergo meiosis and the asymmetrically produced haploid basidiospores are released. Thereby, the integrity of the respective mating types is sustained, albeit introducing certain level of heterozygosity. To further increase heterozygosity, some basidiomycotes have explored shifting from bipolarity, i.e., unifactorial mating type (A, B) to tetrapolarity, i.e., bifactorial mating type (e.g., AB – ab). Notably, the fusion between the two haploid complementary mating basidiospores or yeasts yields four n + n dikaryotic progenies. Hence, their parent: progeny ratio remains 2: 8 or 1: 4 in Basidiomycota, except in some parasites.

8.4 Sexualized Ascomycota

Barring Saccharomycotina, the other ascomycotes are sexualized, i.e., their female and male mating types arise from the ascogonium and antheridium, respectively.

On the whole, *the number of offspring arising from a single reproductive organ in fungi is very limited*; it is 1: 2 in Zygomycota but 1: 4 in Chytridiomycota, Basidiomycota and Ascomycota. Hence, *to increase progeny number, the fungi must invest in a greater number of reproductive organs.*

In fact, the Pezizomycotina is another unusual group, in which a rare combination of mating type and sexuality co-exists. In contrast to the homothallic Basidiomycota, in which the mating types are meiotically generated, the heterothallic pezizomycotines mitotically generate 'gametes' from their haploid reproductive organs (Fig. 8.2). Briefly, *fusion between heterothallic mitotic gametes precedes meiosis in pezizomycotines, whereas meiosis*

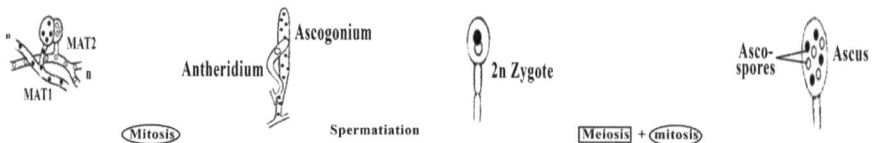

FIGURE 8.2

Gametogenesis in a representative Ascomycota.

precedes fusion between homothallic meiotic mating types in Basidiomycota. This provides a unique opportunity to know that within the system of the mating type, whether the homothallic meiotic mating types or heterothallic mitotic gametes accelerate species diversity. Certainly, *the combination of heterothallism and mitotic gametism accelerates species diversity, as the 77,083 speciose pezizomycotines are more diverged than the 23,975 speciose Basidiomycota.* Hence, *heterothallism is a more decisively important factor than meiosis in accelerating species diversity.* In fact, the introduction of sexualization into the system of the mating type may be a reason for the diversity of Ascomycota.

Incidentally, it may be interesting to compare the metazoans and angiosperms engaging the system of sexuality with that of fungi, which engage the system of the mating type. With 95% dioecy (heterothallism) in metazoans, the female and male gametes appear from distinct organs, the ovary and testis. Despite the complication by double fertilization, the 94% monoecious (homothallic) angiosperms do generate four pollens for every ovule. As structural and/or chemical mechanisms avoid self-fertilization in 76% of angiosperms (see also Table 15.15), they are functionally dioecious (Pandian, 2022). Metazoans too generate four spermatozoa/spermatogonium but one oocyte/oogonium. Most of them produce immotile oocyte and motile spermatozoa to ensure oogamous fertilization. In all of them, meiosis precedes fertilization. Hence, the combination of (heterothallism) dioecy and meiosis preceding fertilization has accelerated species diversity to 296,225 species in angiosperms and 1,543,196 species in Metazoa. The combination of dioecy (gonochorism) and meiosis preceding oogamous fertilization has accelerated the species diversity faster in metazoans than that of dioecy (heterothallism) and meiosis succeeding fertilization in pezizomycotines. Briefly, of three combinations that evolution has experimented namely (i) homothallism (monoecy/hermaphroditism) and meiosis preceding external fusion in Basidiomycota, (ii) heterothallism (monoecy/gonochorism) and meiosis succeeding oogamous fusion in Pezizomycotina and (iii) gonochorism (heterothallism) and oogamy succeeding meiosis in metazoans, species diversity is accelerated in the following ascending order: Basidiomycota > Pezizomycotina > Metazoa.

9

Sporogenesis – Quantity – Dispersal

Introduction

Fungal propagation involves sporogenesis, which includes the simpler and more ubiquitous clonal spores and/or more complex and less frequent sexual spores. These spores serve as vehicles for transmission of fungi through space and time. The capacity for production of spores by fungi and their efficiency for transmission can be appreciated from the following two examples: (i) Despite the initial impediments to become airborne, an estimated 50 million tons of fungal spores are annually dispersed into the atmosphere, corresponding to $> 10^{23}$ spores (see Money, 2016). (ii) It may be interesting to know that each human inhales $\sim 10^3$–10^6 conidial spore/d (see Shlezinger et al., 2017). Clonal spores include (i) sporangiospores and (ii) conidiospores and sexual spores comprise (iii) basidiospores and (iv) ascospores. This chapter describes sporogenesis, attempts to quantify them and elaborates on their dispersal.

9.1 Clonal Spores

In fungi, clonal spores carry identical gene copies of the parent. They can be diploids namely 2n (e.g., sporangiospores of chytrids, oomycotes, urediniospores and teliospores of pucciniomycetes) or haploid (e.g., pezizomycotines) (see Table 7.4). The simplest mechanism of clonal sporogenesis involves the differentiation of **preformed** mycelium. No information is yet available on the minimum number of somatic mycelial cells that are required to form a differentiated sporogenic cell. Yet, it is possible to infer from Fig. 3 of Magyar et al. (2011) that a minimum of ~ five somatic cells are acquired to support one conidium of the hypomycete *Pyrigemmula aurantiaca* (see Fig. 1.22B). Based on the mode of sporogenesis, three major types are recognized: **(i) Sporangium:** The diploid sporangiospores are formed inside a walled sporangium, from which motile zygospores are released into an aquatic *milieu*, as in chytrids, myxomycetes and neocallimastigomycetes or immotile sporangiospores are expelled into the air, as in Zygomycota. Figure 9.1 shows a few examples for the shapes of sporangia and sporangiospores. **(ii)** Glomeromycotes generate **sporocarps** (Fig 1.11B), but not much is known about them.

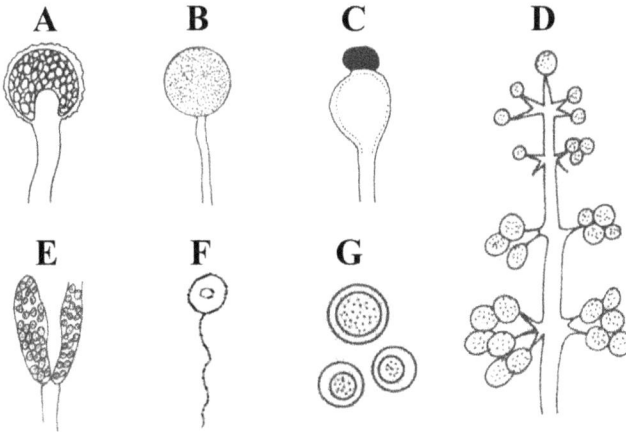

FIGURE 9.1

Diversity in morphology of sporangium: (A) *Rhizopus*, (B) *Pythium*, (C) *Pilobolus*, (D) *Mortierella*, (E) *Saprolegnia* and (F) chytrids zoospore, (G) sporangiospore of Zygomycota (free hand drawings based on Peberdy, 1980).

(iii) Conidium: According to Money (2016), more than 100 terms are used to discriminate between different types of conidia. These include annellospore, botryo-aleuriospore, closterospore, polarocularspore, stalagmospore and so on. However, the following four modes of development are important. Figure 9.2 illustrates a few common modes of conidiogeneses, which are explained hereunder:

Development mode	Explanation
Thalloida (Fig. 9.2A)	The conidial primordium does not enlarge prior to its separation from the conidiophore
Holoblastic (Fig. 9.2B)	The layers in the entire cell wall expand during conidiogenesis
Enteroblastic (Fig. 9.2C)	Only inner layer of the cell wall expands through an aperture in the outer wall of the conidiophore
Phialidic (Fig. 9.2D)	Conidium is formed by synthesis of a new cell wall within the neck of the phialide

Within thallospores, two subgroups are recognized: they are the **arthrospores** produced through fragmentation of hypha into compartments separated by the septa and **chlamydospore** produced, when the cell wall of a hyphal compartment is thickened. Phialides are vase-shaped cells that develop on conidiophores. They produce chains of conidia by extruding cell walls through the upper neck opening and injecting the uninucleate fraction of cytoplasm into the spore (Fig. 9.2E).

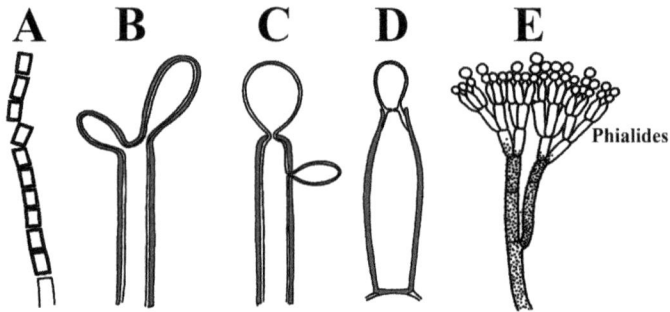

FIGURE 9.2

Modes of conidiogenesis: (A) thallic, (B) holoblastic, (C) enteroblastic, (D) phialidic and (E) conidiophore of *Penicillium* (free hand drawings based on Webster and Weber, 2007).

9.2 Sexual Spores

The sexual spores arise as haploid mating types/'gametes'. They are released as haploids and their karyogamic fusion may be accomplished externally in the air, as in most basidiomycotes or internally by spermatization, as in most Pezizomycotina; following meiosis, the haploid sexual ascospores are released.

Aquatic spores: Motile fungal spores called zoospores have a single posterior flagellum. They are produced by Chytridiomycota, Oomycota and Myxomycota. Except for the ploidy status, the sexual zoospores in the first two groups are not morphologically distinguishable from their clonal counterparts. The Ingoldian fungi consist of the aquatic submerged or emerging ones (Money, 2016). The broader span of these odd-shaped spores enhances the chances of their collision and attachment with submerged plants (Fig. 9.3).

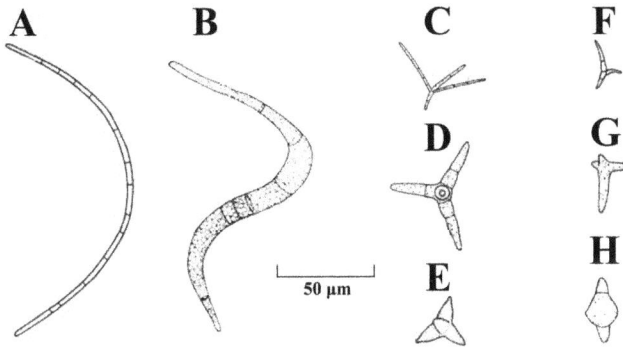

FIGURE 9.3

Selected representative examples of conidia of aquatic hypomycetes:
(A) *Anguillospora* sp., (B) *A. crassa*, (C) *Alatospora acuminata*, (D)
Lemonniera terrestrial, (E) *Clavatospora stellata*, (F) *Volucrispora* sp., (G)
Heliscus lugdunensis and (H) *Tumularia aquatica* (free hand drawings
based on Cole and Baron, 1996).

Terrestrial spores: The zygospores of Zygomycota are immotile. Being
small (< 50 µm, see Margulis et al., 2009), they may, however, be dispersed
by the wind or water. Regarding ploidy status and site of fusion of mating
types/'gametes', Basidiomycota and Ascomycota present a contrasting
picture. Following meiosis, the haploid basidiospores may or may not
plasmogamically fuse within the sporophyte but karyogamic fusion occurs
only in the air. But internal fertilization takes place between the haploid
'gametes' by spermatization in Pezizomycotina and their haploid ascospores
are expelled after meiosis. Though noted (see Table 6.1) by Watkinson
et al. (2015), this contrast is adequately recognized in this account. Of course,
there are exceptions. For example, external fertilization is accomplished
between haploid microconidiospore (MAT-1) arising from clonal cycle and
the haploid appearing from MAT-2 to produce sclerotium in Leotiomycetes
like *Botryotina fuckeliana* (Fig. 1.39) and in some Sordariomycetes like
Sordaria fimicola (Fig. 1.40B). Within Pucciniomycetes, plasmogamy (n + n)
occurs at the spermagonial stage, karyogamy at the aeciosporal stage, clonal
multiplication at the uredinial stage and formation of basidium following
the teliospore stage (Fig. 1.34). Briefly, internal plasmogamy is followed by
external karyogamy in the air.

9.3 Quantities of Spores

For spore producing capacity of fungi, available values are a few and widely
scattered. Some estimates are based on density or weight of cumulative

aerospores per unit volume of air, while others are species-specific. Yet, the limited species-specific values reveal the astonishing capacity of fungi for spore production. Fungi usually undergo clonal multiplication. Only when the nutrients are depleted, they switch to sexual reproduction. In them, sexual reproduction takes place once after a few hundred or thousand generations of clonal multiplications. For example, the estimated number of clonal generations per sexual reproduction ranges from 30,000 (see Magwene et al., 2011) to 50,000 (Ruderfer et al., 2006) and 60,000 (Farlow et al., 2015) for *Saccharomyces cerevisiae* and from 1,000 to 100,000 for *S. paradoxus* (Tsai et al., 2008).

Ingold's (1971) book on *Fungal Spores* is the major comprehensive source for structure and function of spores as well as their release and dispersal. Spore size ranges from 20–50 µm for sporangiospores of Zygomycota and 3–5 µm for basidiospores to 100 × 300 µm for the giant ascospores (Money, 2016). Figure 9.4 shows some examples of them. Airborne spores are subjected to rapid dehydration. Conidial- and basidio-spores lose viability within a few hours (Money, 2016). The thick-walled pigmented spores are more resistant to dehydration and UV radiation during aerial transport than thin-walled hyaline spores. They can be hydrophobic dry xerospores or wettable slimy gloiospores, which may be carried by rain droplets (Magyar et al., 2016). Aeciospores are produced in tightly packed chains and released on dissolution of intercalary cells and aerially disseminated, whereas teliospores remain attached to the host organ (Money, 2016).

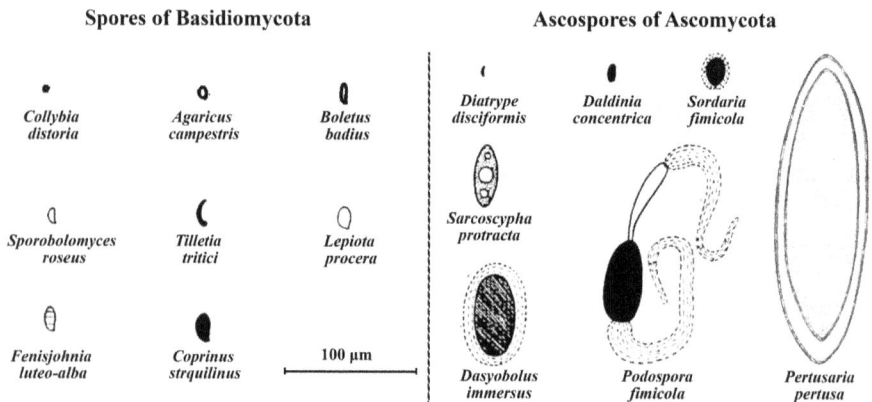

FIGURE 9.4

Examples for ballistospores and violently discharged ascospores (rearranged from Ingold, 1971).

Aerospores: Within aeroplankton, fungal spores account up to 45% of coarse particle mass (> 1 µm) in tropical rainforest air and up to 4–11% of fine particle mass (< 2.5 µm) in urban and rural air (Womiloju et al., 2003). At the global level, ~ 50 million tons of fungal spores are emitted annually, which in terms of number amounts to 10^{23} (Table 9.1). Of them, 10^{21} spores survive on their sojourn over air (see Oneto et al., 2020). On an average, the density and mass of fungal spores in the continental air are in the range of 10^3–10^4/m^3 and 1 µg/m^3, respectively (Elbert et al., 2007). In terms of species richness, the fungal spores hail from > 1,000 species (see Frohlich-Nowoisky et al., 2009). Further analysis has shown that 64, 34 and 2% identified fungal spores were from Ascomycota, Basidiomycota, Zygomycota and others (Table 9.1). Class level groupings revealed that within the 32% Basidiomycota, Agaricales dominated with 29.6% fungal spores or 87% within the Basidiomycota alone. Among Pezizomycotina, the dominance is, however, shared by Sordariomycetes (17.3%), Dothideomycetes (14.1%) and Eurotiomycetes (11.5%). In all, Agaricales (29.6%) and Sordariomycetes (17.3%) constitute nearly 50% aerospores of fungi.

TABLE 9.1

Density and weight of fungal spores.

Location	Density and weight	Reference
Global air	10^{23}; 50 million ton/y	see Money (2016)
Continental air	10^3–10^4/m^3; 1 µg/m^3	Elbert et al. (2007)
San Diego	200–80,000/m^3	see Money (2016)
Composition of spores (compiled from Frohlich-Nowoisky et al., 2009)		
Basidiomycota: 34%		Ascomycota: 64%			
Agaricomycetes	29.6%	Sordariomycetes	17.3%	Saccharomycetes	8.3%
Pucciniomycetes	1.7%	Dothideomycetes	14.1%	Lecanoromycetes	1.3%
Ustilaginomycetes	0.3%	Eurotiomycetes	11.5%	Orbiliomycetes	1.3%
Tremellomycetes	0.3%	Leotiomycetes	9.6%	Total	63.4%
Total	31.9%				

Discharge and dispersal: Spore discharge refers to the separation of fungal spores from their parent colonies/fruiting bodies and spore dispersal for their subsequent travel on air/water. The discharge can be passive, owing to air flow, rain and other disturbances, or active, where the mechanism is powered by (i) hydrostatic pressure, (ii) fast movement induced by cytoplasmic dehydration or (iii) utilization of surface tension force. Despite its importance, the passive spore release mechanism has not received much attention. Wind plays a major role in spore release. Wind speed of 0.4–2.0 m/s is adequate to ensure spore release (Gregory, 1961). However, the range of wind speed required to release the spores varies among species; for example, it is 0.4–1.0 m/s for *Alternaria alternata* (Rotem, 1994), 0.5 m/s for *Aspergillus fumigatus* (Pasanen et al., 1991) and 1.8–2.3 m/s for *Puccinia striiformis* (Geagea et al., 1997). Notably, gloiospores of *Stachybotrys chartarum*, which remain attached to the substrate, are not removable even at the wind speed of 0.3–1.6 m/s (Tucker et al., 2007).

The 'squirt gun' of *Pilobolus kleinii* is the best-known device based on hydrostatic pressure for spore discharge among Zygomycota. Its sporangiophore with 2–4 mm height and 0.5 m diameter holds 30,000–90,000 sporangiospores. The discharge occurs, when the sporangiophore tip fractures due to swelling and thereby propels the sporangium and jets sporangiophore fluid over a horizontal distance of up to 2–5 m at a launching speed of 9 m/s (see Money, 2016).

Most basidiomycotes are characterized by the ballistosporic mechanism to discharge spores from the gills, spines and other structures in mushrooms: The mechanism may hold good for yeasts of the phytopathogenic rusts and smuts. It is powered by the surface tension-based rapid expansion of fluid drop over the spore surface area. In them, a few seconds before the discharge, fluids begin to condense from the surrounding air on the hilar appendix called Buller's drop and adaxial drop on the adjacent spore surface (Fig. 9.5). The condensation induces the release mannitol and other osmotically active compounds. Due to ongoing condensation, the expansion of Buller's drop reaches a critical limit to discharge the spore along with adaxial drop. Yet, the ballistospory is capable of propelling spores only up to distances of < 2 m (see Money, 2016), which aids their escape from the initial drag of the boundary layer (see Magyar et al., 2016).

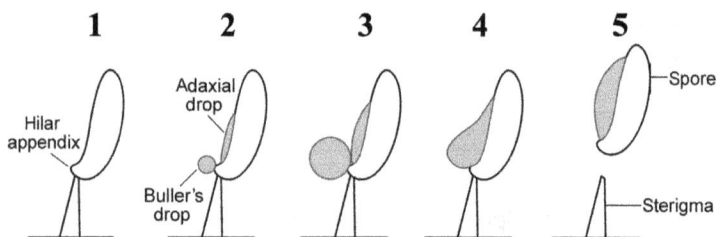

FIGURE 9.5

Mechanism of ballistospore discharge (free hand drawings thankfully based on Mark Fischer, Mount St. Joseph University, Cincinatti).

To increase pressure by solubilization of glycogen, another mechanism merits explanation. The basidiomycote *Sphaerobolus stellatus* is known for artillery mode propelling the large 1 mm diameter spherical spore-filled projectile called the gleba over distances up to 6 m (Fig. 9.6). Being unusually complex for a fungal organ, its spherical basidiome with < 2 mm diameter is covered by six-layers. At maturity, the basidium opens outward to expose the centrally sitting gleba. The palisade cells below the two layered peridium is osmotically pressurized by solubilization of glycogen to sugars. The consequent strain developed within these cells is relieved, when the peridium flips outward, propelling the gleba into the air at the velocity of 6 m/s (Burnett, 1976).

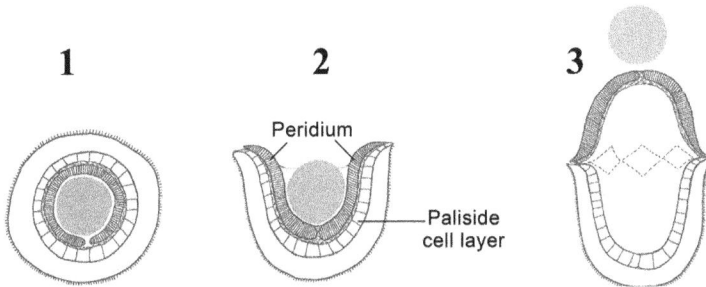

1 **2** **3**

Peridium

Paliside cell layer

FIGURE 9.6

The artillery fungus, *Sphaerobolus stellatus* slice through the center to show its multiple tissue layers (1) unopened fruit body, (2) open fruit body with black capsule bathed in fluid within the cup and (3) gleba jettisoned from the triggered cup (free hand drawings based on Burnett, 1976).

Explosive discharge of the Pezizomycotina is the fastest launching mechanism among the fungal spores. As small guns, they shoot the ascospore to 1 or 2 cm distance to let the spores cross the boundary layer, although launched at the velocity of 30 m/s. In *Ascobolus crenulatus*, over 1,800 spores in singles are deposited within 1 cm (see Ingold, 1971). A few turgor-driven mechanisms of conidial discharge have been described. The mechanism is quite widespread and important in determining the spread of crop diseases. Money (2016) reported many examples for the speed, at which the ascospores were launched and the maximum distance travelled by them prior to sedimentation. Sedimentation rates of spherical spores with a diameter of 5–10 μm range from 1 to 4 mm/s, i.e., they take 4–17 min to fall one meter through still air. Being larger (Fig. 9.4), the ascospores of many pezizomycotines settle faster and cover a maximum distance of < 2 cm. Some examples cited by Money (2016) are explained hereunder: Conidia of rice blast fungus *Magnaporthe grisea* are propelled up to 0.5 mm distance. *Podospora decipiens* fire the spores > 0.5 m distance. The coprophagous fungus *Ascobolus immersus* project the ascoma up to 1 mm to expel cluster of eight spores embedded in mucilage at the velocity of up to 18 m/s. Prior to discharge, the ascospores of *Neurospora tetrasperma* are bathed in fluid containing inorganic ions and alcohols that generate an internal turgor pressure of a few atmospheres. When the tip of the ascus bursts open, the spores are expelled at the speed of 16 m/s. Notably, the spores may travel in singles or in groups of 2 to 4. In them, 70% spores travel singly and cover 12 mm, in comparison to those travelling in groups of 2, 3 or 4 spores, connected by strings of ascus sap, travel further up to 15 mm.

Clonal and sexual spores: Values reported by different authors for species-specific spore production are in number of spores released per day (e.g., *Daldinia concentrica*) or per colony (e.g., *Penicillium*) (Table 9.2); some authors have not indicated whether the values are related to clonal or sexual spores (e.g., *Daldinia concentrica*). Hence, they are not comparable. Thanks to Goettel et al. (2005) and Wurzbacher et al. (2011), values associated with entomophthoromycotine (Zygomycota, see Jaronski, 2014) fungal conidiospore production is assembled in Table 9.2. However, they are

TABLE 9.2

Clonal and sexual spore production capacity of some fungi.

Species	Spore production	Reference
Clonal spores **Chytridiomycota: Motile 2n Zoospores**		
Rhizophlyctis petersenii	70,000	Sparrow (1960)
Lagenidium giganteum (Oomycota)	20,000 sporangium/mosquito larva	Couch and Romney (1973)
Zygomycota: Immotile n sporangiospore		
Choanephora	1/sporangium	*wikipedia.com*
Pilobolus kleinii	30,000–90,000/sporangium	see Money (2016)
Ascomycota: Conidiospores Laboulbeniomycetes		
Zoophthora phalloides	30,000/aphid	Glare et al. (1987)
Neozygites fresenii	30,000/aphid	Steinkraus et al. (1999)
	2.6×10^6/gypsy moth	Shimazu and Soper (1986)
Clavisporus clavisporus	10^4/mosquito larva	Cooper and Sweeny (1986)
Coelomycetes		
Coelomomyces sp	100–1,500 or 5,000/sporangium/ mosquito larva	Pillai and O'Loughlin (1972)
Pandora (Erynia) neoaphids	500,000/aphid	Glare and Milner (1991)
Eurotiomycetes		
Penicillium	400 million conidium/colony	Ingold (1971)
Sordariomycetes		
Daldinia concentrica	1 billion spore/d	Ingold (1971)

Table 9.2 contd. ...

...Table 9.2 contd.

Species	Spore production	Reference
Sexual spores **Basidiomycota: Immotile basidiospores** Agaricales		
Formitiporia ellipsoidea Sporophore of 10.8 m length x 85 cm width	450 million = 500 kg	*brittanica.com*
Agaricus campestris	2.7 billion basidiospore/d	see Watkinson et al. (2015)
Ganoderma applanatum	30 billion basidiospore/d or 5 trillion basidiospore/y	see Watkinson et al. (2015)
Calvatia gigantea	7 trillion/specimen	Ingold (1971)
Pucciniomycetes		
Tilletia caries	12 million teliospore/grain	Ingold (1971)
Ascomycota: Leotiomycetes		
Sclerotinia sclerotiorum	30 million ascospores	Ingold (1971)

estimated based on number of conidia per cadaver host of aphid, mosquito larva or moth. Conidial production is dependent on the host size and its nutritional quality as well as temperature and Relative Humidity (RH). In general, to infect a preferred host, 10^7–10^9 conidium/terrestrial host are generated. For the fungus *Neozygites rileyi* infecting *Helicoverpa armigera*, the number of conidial spores increases from 5 to 10-fold with increasing size of caterpillar from the 2nd instar to the 5–6th instar (Glare, 1987). *Zoophthora phalloides* sporulate on aphid only at 100% RH. However, not all entomophthoran fungi depend on RH. For example, *Entomophthora muscae* can also sporulate at 50% RH (Kramer, 1980). The values for conidial production ranges from 30,000 conidium/aphid in *Z. phalloides* to 500,000 conidium/ aphid in *Pandora neoaphids* and from 100–5,000 conidium/sporangium of mosquito larva in *Coelomomyces* sp. to 10^4 conidium/larva in *Clavisporus clavisporus*. These numbers for the haploid non-motile conidia (70,000 in *Rhizophlyctis petersenii*) are far greater than that for the diploid motile zoospores of Chytridiomycota. For free-living Ascomycota, Ingold (1971) assembled values ranging from 400 million conidium/colony in *Pencillium* to 1 billion spore/d in *Daldinia concentrica*.

For the sexual spore, only three values are available from free-living Agaricales, in which propagation is limited to sexual reproduction alone. They also generate million and billion of basidiospores. The values range from 2.7 billion spore/d in *Agaricus campestris* to 30 billion basidiospore/d in *Ganoderma applanatum*. As the latter continue to release spores for 6 mo in a year, *G. applanatum* can produce 5 trillion spore/y. The parasitic Pucciniomycetes and Leotiomycetes also produce a few million sexual spores.

Some fungi release their spores intermittently. But many others display finely-tuned timings for spore release to ensure maximum survival and fitness of their spores during air travel. Ingold (1971) described light- and temperature-dependent spore releases from *Sordaria fimicola*. According to Oneto et al. (2020), (i) spores released during the day, i.e., diurnal spores may fly for a few days, whereas nocturnal spores sediment within a few hours. (ii) Tropical fungi, however, are characterized by nocturnal spores to avoid the hotter day and desiccation. (iii) Species characterized by short longevity discharge spores, when there is strong turbulence to carry their spores for longer distances. (iv) The cosmopolitan fungi, for example, *Leptosphaeria maculans* follow different diurnal rhythms in different regions; for example, most of its spores are released at night in Canada or in the morning in England or early afternoon in Western Australia.

Longevity of spores: During dispersal, the flight duration of a fungal spore is determined by the balance between two opposing forces: gravity causes their downward sedimentation but turbulence keeps them afloat. Hence, the residence duration in the air may be shorter for the larger ascospores but longer for the smaller basidiospores. During the flight, the major threats limiting the longevity of spores are (i) exposure to UV radiation, (ii) harsh temperatures and (iii) humidity. Not surprisingly, longevity of the relatively larger ascospores is shorter and ranges between 1% during 12 hour (h) exposure to sunlight for *Botrytis cinerea* and 5% during a 4-d exposure for *Alternaria macrospora* (Table 9.3). The corresponding values for the smaller urediniospores of Basidiomycota are 10% during a 8 h exposure in *Puccinia striiformis* and 10% during a 8 h exposure or longer in darkness for *Mycosphaerella graminicola*. The spores of entomycorrhizal basidiomycotes and teliospores of pucciniomycetes survive over longer durations of > 4 y (Table 9.3). For example, teliospores of *Tilletia indica* can survive over 13–18 y.

TABLE 9.3

Longevity of some fungal spores (compiled from Money, 2016, Oneto et al., 2020 and others).

Species	Longevity
Chytridiomycota	
2n zoospore	5 h
Ascospore	
Sirococcus claviginenti-juglandacearum	< 10%, 3 h exposure
Conidia	
Botrytis cinerea	1%, 12 h exposure
Aspergillus niger	15%, 12 h exposure
Alternaria macrospora	5%, 4 d exposure

Table 9.3 contd. ...

...Table 9.3 contd.

Species	Longevity
Basidiomycota: Urediniospore	
Puccinia striiformis	10%, 8 h exposure
P. graminis	10%, 20 h exposure
Mycosphaerella graminicola	10%, 24 h exposure but 1–2 week in darkness
Entomycorrhizal basidiomycete spores	4 y
Teliospore	
	6 mon to 4 y
Tilletia indica	3–4% viable for 7 y or up to 13–18 y

Once airborne, the phytopathogenic basidiospores, especially the smaller urediniospores are at the mercy of turbulence and rarely by storms. Table 9.4 shows spectacular examples for fungal spores intercontinentally travelling at altitudes of 1,200–2,000 m over distances of 2,000–10,000 km within 5–6 d at a speed of 2,000 km/d. Others like *Phytophthora infestans* were transmitted through the export of the host potatoes from America to Europe. *Puccinia striiformis* was transmitted through clothes of men, who travelled from England to South Australia. Interestingly, ascospores of chestnut *Cryphonectria parasitica* (Sordariomycetes) causing the dreadful chestnut blight was transmitted by accidental man-made transmission from one place to other all the way from Japan to the USA. Notably, the spores of all these fungi are phytopathogens. For more details, Brown and Havmoller (2002) may be consulted.

TABLE 9.4

Distance travelled by some fungal spores by either airborne transmission or man-made transmission through export or cloths.

Oomycota: Zoospores			
Phytophthora infestans (1845, 1976)	Export transmission	USA to Belgium	Fry et al. (1993)
Pucciniomycetes: Urediniospores			
Hemileia vastatrix (1970)	6 d, airborne transmission	Angola to Brazil (~ 7,500 km)	Bowden et al. (1971)
Puccinia melanocephala (1978)	9 d, airborne transmission	Cameroon, Africa to Cuba + Florida (~ 10,000 km)	Purdy et al. (1985)
P. striiformis var. *tritici* (1979)	Transmission in ~ 15 d via clothes	England to South Australia (~ 9,700 km)	Wellings et al. (1987)
P. striiformis (1969)	5 d	South Africa to Australia (~ 10,400 km)	Nagarajan and Singh (1990)
Sordaiomycetes: Ascospores			
Cryphonecttria parasitica (1904)	Man-made transmission	Japan to USA	Rigling and Prospero (2018)

Aquatic spores: Chytridiomycota and some ascomycetes are aquatic fungi. The former release motile zoospores to accomplish clonal or sexual reproduction. The propulsion of zoospore is driven by high undulation frequency of the flagellum from the base to the tail at a velocity of 100 µm/s. A zoospore swimming at 25 µm/s for 5 h may travel 0.5 m distance. Except when chemically attracted by the host, the zoospores of free-living chytrids may not travel in a straight line. However, motile zoospores and cysts of chytrids may passively be dispersed over longer distances in water trickling through soils or carried by running water between aquatic habitats (see Money, 2016).

Many aquatic pezizomycotes decompose leaves of submerged and other Ingoldian emerging aquatic plants. Water being 800 times denser than air, the spore sedimentation rate becomes slower from µm/s to mm/min. The adaptive spores of aquatic pezizomycotes are of unusual shapes (Fig. 9.3). Of them, the tetraradiate spores are the most adaptive. The tetraradiate conidial spores produced by the Ingoldian *Brachiosphaeria troculis* are 0.4 mm in length and weigh ~ 40 µg each. The spore with a central hub and slender arm, in which the cytoplasmic investment is reduced, is 400 times lighter and sediment more slowly. Consequently, its chances of being trapped on its potential host leaf are the greatest. Webster (1959) measured the trapping efficiency of aquatic hyphomycete spores of different shapes in waters running at three speeds. From Table 9.5, the following may be inferred: (i) The optimum for the highest trapping efficiency is at the water speed of 25 cm/s for the radiate spores (except in *L. aquatica*). (ii) The efficiency is far higher for radiate spores than the others. Within radiate spores, the efficiency is higher for the tetraradiate spores at the running water speed of 15 cm/s.

TABLE 9.5

Trapping efficiency (%) of selected hyphomycetes spores of different shapes at three speeds of running water (values from Webster, 1959, figures and size from Cole and Baron, 1996).

Species	Figure	Size (µm)	Running water: speed (cm/s)		
			15	25	35
Tetraradiate spores					
Lemonniera aquatica		Triradiate: ~ 60/arm	0.35	0.13	0.04
Tetracladium marchalianum	–	–	0.28	0.37	0.27
Clavariopsis aqautica	–	–	0.30	0.40	0.04
Articulospora tetracladia		Tetraradiate: ~ 40/arm	0.35	0.51	0.33
		Mean	0.32	0.35	0.17

Table 9.5 contd. ...

...Table 9.5 contd.

Species	Figure	Size (μm)	Running water: speed (cm/s)		
			15	25	35
Long, sigmoid spores					
Anguillospora longissimi		> 200 length	0.11	0.10	0.10
Flagellospora curvula	–	–	0.06	0.14	0.14
Mean			0.08	0.12	0.12
Other spores					
Dactylella aquatica	–	–	0.004	0.009	0.008
Heliscus lugdunensis		~ 30 length	0.004	0.009	0.007
Mean			0.004	0.009	0.007

10

Sex Determination – Pheromones

Introduction

Sexual reproduction is central to eukaryotic evolution by providing scope for genetic diversity and elimination of deleterious mutations (Ni et al., 2011). It involves ploidy reduction through meiosis and gametic fusion at fertilization. During the sexual cycle, meiosis occurs prior to gamete fusion in Chytridiomycota and Basidiomycota but after fusion in Zygomycota and Ascomycota. In the majority of fungi, gametes are manifested as mating types, although in a few chytrids and most pezizomycotines (except in Leotiomycetes, see Fig. 1.39), the gametes arise from distinct sexual organs, the oogonium or ascogonium and antheridium. In fungi, the following sexual dichotomies can be recognized : 1. **Bipolar versus tetrapolar mating systems**: 2. **Gametangial mating types**: The zygomycotes are characterized by gametangial conjugation, which ensures sexual reproduction. This also holds true for many pezizomycotine groups (e.g., Eurotiomycetes, Fig. 1.38). 3. **Thallism**: Inbreeding/selfing (homothallism) versus outbreeding (heterothallism) modes of reproduction. In fungi, sexual reproduction is accomplished by mating types, which may not be equated with the gametes of higher plants and metazoans. Though both refer to sexual compatibility and asymmetry between gametes, mating type refers to genetic mechanisms that allows discrimination between gametes, independent of size dimorphism. Whereas the number of sexual phenotypes is limited to three, female, male and hermaphrodite in motile Metazoa and to six, monoecy, dioecy, gynomonoecy, andromonoecy and so on in sessile angiosperms (see Pandian, 2021b, 2022), there can be thousands of 'sexes' in mating types of fungi. Reduction in the number of sexual phenotypes to two or a few limits the average number of mating partners per individual to only 50% of the population. This limitation is avoided by the existence of thousands of 'sexes' within each mating type (see Nieuwenhuis, 2012, Wallen and Perlin, 2018). *Thus, evolution seems to have compensated the sessile fungi with the choice of thousand 'sexes' to readily find* *an appropriate complementary mating type.* This chapter is not an encyclopedic account on sexual reproduction in fungi but provides a concise picture of it.

10.1 Meiosis – Reproduction

Since manifestation of sex some 1.6–2.0 BYA (Butlin, 2002), the unicellular yeasts seem to have struggled to bring together the independent elements namely sex and meiosis. For example, *Candida glabrata*, *C. parapsilosis* and *C. tropicalis* have genes required for mating and meiosis. Yet, sex is not observed in them. In contrast, *C. lusitaniae* and *C. guilliermondii* have established sexual cycles including ascospore formation. However, meiosis is not yet described in them (Sherwood and Bennett, 2009). In filamentous fungi, sex is inseparably combined with meiosis. In them, for example, *Coprinopsis cinerea*, meiosis with prolonged prophase is synchronized in ~ 10 million cells within the mushroom cap (see Wallen and Perlin, 2018). Among Ascomycota, meiosis is immediately followed by mitosis to generate eight meiospores. In the unicellular yeast *Saccharomycodes ludwigii* (Fig. 1.35 C) and filamentous *Podospora anserina*, meiosis is not followed by mitosis; as their progenies are binucleates, they do not undergo clonal multiplication.

10.2 Thallism

The heterothallic fungi require two complementary mating types (or partners) arising from two different thalli to ensure mating and sexual reproduction, whereas homothallic fungi are self-fertile with complementary mating types hailing from a single thallus. Both these modes of sexual reproduction share common key features, i.e., ploidy changes, meiosis and recombinant progeny production but differ in others, i.e., complementary mating types arising from the same or different thalli. There are different types of homothallism, of which the following may be mentioned. In *Cochliobolus* spp, the two idiomorphs (for definition, see below) required for mating coexist in the same genome and are fused into one *MAT* locus on the same chromosome. But they are unlinked and located at different positions in the genome, as in *Aspergillus nidulans*. However, the heterothallic *A. fumigatus* and *A. oryzae* carry only one or other *MAT* idiomorph. Remarkably, some homothallic fungal species hold only one *MAT* idiomorph and yet exhibit a sexual cycle. Transition between one mode to the other is common. Both modes can occur concomitantly in different species of the same genus (e.g., homothallic *Aspergillus nidulans* and heterothallic *A. fumigatus*, see Ni et al., 2011) rarely even within the same species (e.g., *Candida albicans*, see Sherwood and Bennett, 2009).

10.3 Bipolar vs Tetrapolar

Within complementary mating types, sex determination is encoded by the MAting Type (MAT) locus. However, the MAT system can be within a single biallelic locus (bipolar) or two unlinked multiallelic sex loci (tetrapolar) (see Ni et al., 2011). In basidiomycotes, the bipolar and tetrapolar mating systems exist. Based on phylogenetic analysis, the common ancestor of all fungi is most likely the Bipolar Mating System (BMS), although Raper (1966) proposed many hypotheses for the emergence of BMS from tetrapolar ancestors of the Basidiomycota. The Tetrapolar Mating System (TMS), in which homeodomain (Ho) and pheromone/Pheromone Receptor (P/R) loci remain unlinked and the two mating types must differ at both loci to ensure mating, is peculiar to some basidiomycotes alone. In the bipolar *Cryptococcus neoformans*, the number of alleles of the *MAT* locus varies from two to thousands in a mating type. Still, > 40% basidiomycotes engage the BMS. Notably, *Ustilago hordei* engage BMS located in the same sex chromosome; in it, recombination is suppressed over the entire 500 kb region between Ho/PR loci. Thereby *U. hordei* maintain the BMS. Incidentally, for a list of genome size and chromosome number of Saccharomycotina, Dujon and Louis (2017) may be consulted.

10.4 Pheromones and Receptors

The mating types of fungi are indistinguishable except at a molecular level (Wallen and Perlin, 2018). Their outbreeding systems include up to thousands of alleles at one end of spectrum and self-fertile inbreeding system at the other. Both unicellular and filamentous fungi have evolved systems to detect mating partners through mating type-specific peptide Pheromones (P) and Receptors (R). In the chytrid *Allomyces macrogynus* female, for example, the pheromone is sirenin and that of male is parisin. Trisporic acid derivatives serve as mating pheromones in Zygomycota (e.g., *Phycomyces blakesleeanus*). In *Saccharomyces cerevisiae*, the pheromone is a lipid-derived peptide. The structurally simpler chemical compounds engaged by the chytrids and zygomycotes are completely unrelated among themselves as well as from those of basidiomycotes and ascomycotes (Ni et al., 2011).

In Ascomycota, pheromones and their receptors, that detect the appropriate mating partners, are not part of the mating type locus, whereas they are on the same locus in Basidiomycota (e.g., *Ustilago maydis*, Bolker et al., 1992). A pheromone precursor, *ccg-4* accumulates, when nitrogen is depleted, which triggers sexual reproduction (Bobrowicz et al., 2002). A better-known pheromone-precursor genes are *MF1-1*, which are expressed as *MAT1-1* in a strain but as *MF1-2* in another strain of the rice blast *Magnaporthe grisea* and *M. oryzae*, respectively (Shen et al., 1999).

10.5 Sex Determining Genes

Basidiomycota: In them, the dikaryotic state is established by clamp connections that help to sustain a pair of compatible nuclei in successive synchronous mitotic divisions. In free-living Agaricales, many thousands of 'sexes' are generated through variations among the multiple alleles within the mating type loci. There are two mating type genes namely *A* and *B*, which independently regulate different stages in dikaryon development and allow self-versus non-self recognition (Kues et al., 1992). As many as seven different *A* genes have been identified, of which four encode homeodomain proteins and thereby determine the mating type specificity. Nevertheless, for compatible mating, only one of these alleles must be different between the two partners but one heteroallelic combination in the mating pair is adequate to ensure *A*-regulated development.

The *B* mating locus encodes pheromones and pheromone receptors that also serve to identify potential mates and govern adjacent cell fusion during mating (Asante-Owusu et al., 1996). Owing to the multiallelic state of both the pheromone and pheromone receptor-containing locus, this system generates a potential for outcrossing in > 50% potential mating types (O'Shea et al., 1998). But in *Ustilago maydis*, the mating type with *a* locus encodes the pheromone and pheromone receptor system (PRS). In an *a* allele, *a1* is smaller (4.5 kb), while that of *a2* is larger (8 kb) (Bolker et al., 1992). Their sequences show no homology to each other and are absent in strains of complementary mating type, i.e., the allele *a1* is specific to *a1* strain. The *b* locus of *U. maydis* contains two unrelated homeodomain proteins that are different in the N terminal region. Thus, they are not true alleles. These two different forms are called idiomorphs.

Ascomycota: In them, the mating type locus contains genetic elements that facilitate recognition of a cell in the complementary mating types by the expression of mating type-specific genes. The genes also control the entry into the sexual phase and prevent the formation of mixed heterokaryons during the vegetative clonal phase (Glass et al., 1990). The following are selected as representative models for the mating type loci: 1. Sexual identity in ascomycotes is determined by a single locus *MATa* or *MATα*. In *Saccharomyces cerevisiae*, for example, the *MATa* region is shorter than that of *MATα*; it encodes regulatory proteins that control by α-specific or a-specific genes (Astell et al., 1981). The products of *MATα* locus activity determine the α or a phenotype, which encodes two genes α1 and α2. *MAT1α1* controls expression of α-specific genes, while *MAT-1α2* suppresses a-specific genes. 2. The fission yeast *Schizosaccharomyces pombe* is unstable as diploids. In it, mating type regions consist of three components, *mat1*, *mat2-P* and *mat3-M* (Kelly et al., 1988). Their cell type is determined by alternate alleles present at the *mat1* locus either as *P* in *h+* cells or *M* in *h–* cells. Both *Sa. cerevisiae* and haploid *Sc. pombe* cells secrete pheromones that are recognized by cells of complementary mating type. 3. In *Neurospora crassa*, the mating types are termed *A* and *a*. The *A* mating type locus contains sequences responsible for recognition of cells of the complementary mating type and *a* region encodes a high degree of amino acid similarity to *MATα1* of *Sa. cerevisiae*. The other region contains two functional segments, one of which is responsible for maturation of sexual organs and the other specifies mating identity and induction of mating (Staben and Yanofsky, 1990). 4. In *Candida albicans*, the mating type loci are named *MTLa* and *MTLα*. Both encode regulatory sequences similar to those of *Sa. cerevisiae* (Hull and Johnson, 1999). The structure of the mating type loci, their expression and formations are shown in Fig. 10.1.

With regard to heterothallism in *Aspergillus fumigatus*, *Neurospora crassa*, *Podospora anserina* and homothallism in *Sordaria macrospora*, the following may be noted: In heterothallic pezizomycotines, the *MAT1-1* idiomorph comprises the α domain gene, the PPF domain (containing the condensed proline and proline-phenylalanine residues) gene and an HMG (High Mobility Group) domain gene (Table 10.1). The α domain protein is obligately required for

TABLE 10.1

Constituents of MAT1 and MAT2 of homothallic and heterothallic pezizomycotines (based on Ni et al., 2011).

Species	MAT1				MAT2	
	HMG	PPF	α-domain	PPF	?	HMG
Heterothallics						
Aspergillus fumigatus			Present			Present
Neurospora crassa	Present	Present	Present		Present	Present
Podospora anserina	Present	Present	Present			Present
Homothallic						
Sordaria macrospora			Present	Present	Present	Present

MAT1-1 identity and sexual development. Deletion of PPF or HMG domain gene does not have an impact, whereas the deletion of both the genes dramatically decreases fertility (cf p 150). *MAT1-2* idiomorph includes two genes: an HMG domain gene, required to establish the *1-2MAT* identity and a small Open Reading Fame (ORF) of an unknown function. Notably, the domain PPF is not present in the homothallic *S. macrospora*.

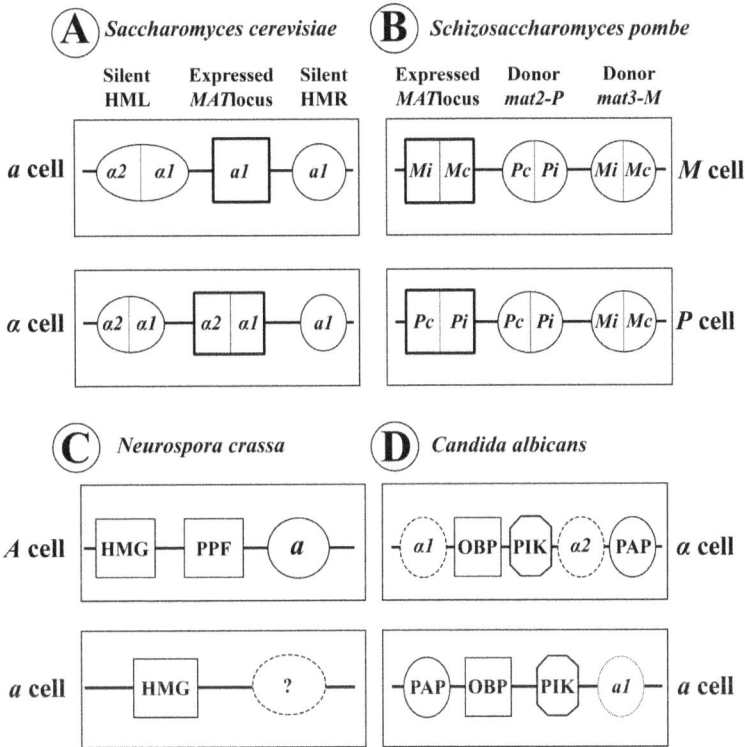

FIGURE 10.1

Representative mating type loci of the Ascomycota. (A) organization of the mating type loci of *Saccharomyces cerevisiae* containing homeodomain proteins that regulate the expression of mating type genes. (B) *Schizosaccharomyces pombe* containing homeodomain proteins at its mating type loci consisting of both expressed and donor regions. (C) *Neurospora crassa* existing as two haploid mating types, each of which contain genetic information at the mating type loci to govern expression of mating type-specific genes (see also Table 10.1). (D) *Candida albicans*, which does not have two different mating type loci, is found in diploid strains cultivated in the laboratory (modified and redrawn from Wallen and Perlin, 2018).

Interestingly, it is known that the haploid cells of one of the mating types in a few yeasts namely *Sa. cerevisiae*, *Sc. pombe* and *Kluyveromyces lactis* can switch to produce progeny of the complementary mating type during mitosis. The switching involves a 'cassette mechanism' (Fig. 10.1) with one active expression locus and two transcriptionally silent *MAT* allele copies residing in another location within the genome (Hicks et al., 1979). The *MAT* silent cassettes are an essential part of the system functioning as templates for switching of the mating type at various frequencies. *MAT* switching occurs through a DNA Double-Stranded Bread (DSB) repair mechanism. The Ho endonuclease is cell cycle-controlled and expressed only in mother cells during the late G1. Thus, only mother cells are licensed to switch.

10.6 Parasitic Fungi

Approximately, 40% fungal species are parasites (Table 3.9). For phytopathogenic basidiomycotes, meiosis and mating are essential, as dikaryotic teliospores alone can penetrate the host. But in ascomycotic parasites, the haploid ascospores are infective. In both, meiosis and fusion precede prior to infection, but the infective spores are dikaryotics in the former but haploids in the latter. In either case, mating is costlier for parasites, as they, especially obligate host-specific ones, undergo selective pressures from an array of biotic and abiotic factors both from inside and outside of their host plants. Phytopathogenic parasites must encounter reactive oxygen produced by host plants in response to fungal invasion, which can damage the fungal genome integrity. Animal parasites must overcome immune response arising from the host.

For parasitic **Basidiomycota**, the following four representative examples are described, as adequate information is available for them (e.g., Wallen and Perlin, 2018). Interestingly, *Microbotryum lychnidis-dioicae* replaces pollens on anthers of its host *Silene latifolia* with diploid teliospores, which are subsequently transmitted to new host plants by bees and other pollinators, i.e., only in the diploid state, the basidiomycotes can infect a new host. On the host, the spores germinate, their basidia undergo meiosis and produce haploid spores. When the spores of compatible mating type encounter each other, they conjugate to produce a dikaryotic hyphae that grow inside the second host tissue. The mating type loci of *M. lychnidis-dioicae* is the first evidence for the existence of sex chromosomes known from fungi (Hood, 2002). Their mating types namely *A1* and *A2* are defined by the two alleles at loci for homeodomain proteins and for pheromone/pheromone receptor system. *A2* has two genes, that encode pheromone precursors, while *A1* contains a single locus (Badouin et al., 2015). The germinating promycelium usually contains three cells but only two at low nutrient and temperature conditions. In the two-celled promycelium, the second septation fails to take place. Hence, it yields only a two-celled promycelium, with one single nucleated cell and another with two nuclei. Then, many of the same mating type each with two nuclei are formed as their locus in two compartments are linked (Hood and Antonovics, 1998). The production of two nuclei-holding mating type can infect a new host without having to encounter another compatible cell of a different mating type followed by mating and production of dikaryotic teliospores (see Wallen and Perlin, 2018).

Cryptococcus neoformans is a lung parasite of humans. In its bipolar mating system, the mating type, either *a* or *α* is defined by the genetic components of the mating type locus. Under external suitable conditions, fusion between cells of complementary mating type occurs to produce dikaryotic hyphae. Many pheromones precursor and pheromone receptor genes are dispersed throughout the mating type locus (Lengeler et al., 2002). Both of its mating types can infect independently. The pair of mating types undergo haploid filamentation and produce spores but the *α* cells are more common. As a result, almost all clinical isolates are of *α* mating types (Brefort et al., 2005). More surprisingly, Lin et al. (2005) reported the mating between two *α* mating type cells, followed by meiosis and recombination. So much so, sexual reproduction between the same mating type occurs within the host itself. Thus, *C. neoformans* has gained all the benefits of sexual reproduction to repair DNA damage caused by the host response, without the need to find a compatible mating partner. In contrast, *Armillaria mellea* persist only in a vegetative dikaryotic state in nature (Anderson and Ullrich, 1982).

Sporisorium reilianum exists in two varieties, which are host-specific; one of them SRZ infects maize and the other SRS sorghum (Zuther et al., 2012). Their mating type locus is composed of two unlinked genomic regions *a* and *b*, which code for a pheromone and pheromone receptor, and as homeodomain protein, respectively. The *b* locus exists in > five different alleles and encodes two subunits of a heterodimeric homeodomain transcription factor. It's *a* locus can be one of three alleles, each containing two pheromone precursors (Schirawski et al., 2005). These different versions are designated as *a1*, *a2* and *a3* idiomorphs. Each of these idiomorphs encodes different pheromones and pheromone receptors. Irrespective of these features and narrow host specificity of the two varieties of *S. reilianum*, there is a remarkable similarity between pheromones of the two (Radwan, 2013). Therefore, there is every possibility for inter-variety cross mating within *S. reilianum*. Notably, the mating system is bipolar in *Ustilago hordei,* but tetrapolar in *U. maydis*. Despite differences in the mating system and genetic constituents of the mating type loci, there is a high degree of similarity between the pheromone precursor and pheromone receptor genes in *U. maydis* and *U. hordei* (Bakkeren and Kronstad, 1994).

The parasitic **Ascomycota** seem to alter the course of reproductive (to reduce the pressure) cycle or depend more and more on clonal multiplication. In the rice blast *Magnaporthe* complex, infection begins, when a haploid spore forms an appressorium; then, it can penetrate, develop and establish infection inside the host plant. For it, mating is a costly process. Not surprisingly, the need for the formation of appressorium in *Magnaporthe* is not obligatory. Still, it has the two *MAT1-1* and *MAT1-2* idiomorphs within the locus of mating type (Arie et al., 2000). Nevertheless, the strains of complementary mating types of the blast collected from the field remain relatively infertile (cf p 147), albeit their isolates cultured in the laboratory are fertile (Li et al., 2016). In the field, the relatively infertile parasites are able to sense a potential host, due to the new pheromone precursor arising from *MAT1-2*, which encodes a protein with an HMG-box domain.

Another ascomycote parasite *Fusarium oxysporum* represent a complex, causing vascular wilt disease in a hundred and odd plant species. Despite its clonal life style, the parasite has still retained the mating type locus *MAT1-1* or *MAT1-2*. Though the function of these two idiomorphs remains largely unknown, *MAT1-2* idiomorph encodes a protein with an HMG-box domain. This novel protein pheromone precursor is able to sense and send chemical signals to and from the potential host plant – similar to that of *Magnaporthe* complex.

Among ascomycotous parasites, *Candida albicans*, *Pneumocystis carinii* and the facultative parasite *Aspergillus fumigatus* infect the lungs of men and mammals. The challenges for these pathogens are that they have to (i) overcome the onslaught of the host immune system and (ii) compete with a large number of microbes inhabiting within the same host's site. Holding a relatively smaller genome (Ma et al., 2016), *P. carinii* has three mating type genes namely *mat Mc*, *mat Mi* and *mat Pi*, which seem to be bound together within the genome. The absence of switching from one to other mating type indicates that *P. carinii* is a homothallic, a system, in which another mating partner is not required to complete sexual reproduction. As a single individual, it can produce cells of different mating types that can successfully mate. Hence, *P. carinii* reproduce clonally alone. No evidence has been thus far produced to show that *A. fumigatus* is capable of sexual reproduction. *C. albicans* exist as diploids but genes from the mating type (e.g., *MAT1-1* or *MAT1-2*) locus occur only as a single copy (Hull and Johnson, 1999). Cultured cells of *C. albicans* grow either as a white or opaque colony. Opaque cells are 10^6 times more efficient at mating than white cells. Interestingly, white cells alone are found in the mammalian host. Their ascospores are highly antigenic, which leads to a greater and more effective immune response from the host (Forche et al., 2008). As a result, the white cells of *C. albicans* have evolved the parasexual reproduction system, which allows some DNA repair but not mating and its products.

Regarding the alternations in sexual reproductive cycle of parasitic fungi, the situation seems to recollect that of unisexual fish (Pandian, 2011b, 2021b). Having arisen as obligate parasites on multiple occasions, each taxon/ group has chosen to explore its own uncommon pathways for reproduction. Accordingly, the basidiomycotous parasites, characterized by meiosis and external fusion, have opted for one of the following pathways: 1. Production of dikaryotic spores from two-celled promycelium and thereby avoiding meiosis and mating, as in *Microbotryum lychnidis-dioicae* (see p 148), 2. Internal sexual reproduction within the host by engaging mating between individuals belonging to the same mating type (e.g., *Cryptococcus neoformans*, see p 149) and 3. Persisting in the dikaryotic clonal state alone, as in *Armillaria mallea*. The pezizomycotine parasites, characterized by internal fusion and haploid infective spores, have retained their respective mating type loci, but have opted for one or another of the following options to reproduce: 4a. Reproduction by clonal multiplication alone, as in *Candida albicans*, *Fusarium oxysporum* and *Pneumococcus carinii* or 4b. Multiply parasexually, as in *Aspergillus fumigatus* (see p 150). 5. Induction of infertility among mating types in the field and engaging a new pheromone system for identification of a potential new host and communication with them, as in *Magnaporthe* complex.

10.7 Mitochondrial Inheritance

As in higher eukaryotes, the majority of fungi achieve uniparental mitochondrial inheritance, i.e., their offspring possess mitochondria from only one of the two mating parents. *Agaricus bisporus*, *Ustilago maydis*, *Aspergillus nidulans*, *Coprinopsis albicans*, *C. cinerea*, *C. neoformans* and *Neurospora crassa* are some examples for the uniparental inheritance. However, there are others like isogamic *Saccharomyces cerevisiae* and *Schizosaccharomyces pombe*, in which the progenies inherit equal contribution of mitochondria from two gametes into the heteroplasmic zygote. In them, homoplasmy is rapidly established so that each daughter cell possesses mitochondria of one parental gene type or recombinant of two parental gene types. The transient coexistence of two different mitochondrial genome within a cell may evoke conflict/competition between them. Hence, it is avoided (e.g., *A. biporus*) but when it occurs, the uniparental inheritance is reestablished (e.g., *Sa. cerevisiae*). In filamentous fungi, the inheritance of two different mitochondria is avoided, as mating between two compatible mycelia is achieved by mutual migration of two nuclei, while all the cytoplasm including their organelles, are left behind (see Ni et al., 2011, Wallen and Perlin, 2018).

11

Lichens

Introduction

Lichens are a unique paradigmatic group of fascinating symbiotic organisms. In them, mutually beneficial association exists between heterotrophic filamentous fungi, the mycobionts and populations of compatible autotrophic plants, the photobionts. The mycobionts absorb and provide water, minerals and micronutrients to the photobionts, which release photosynthates to the mycobionts. Hence, lichens represent a stable well-balanced self-supporting microecosystem. Lichenization is a very successful symbiosis with 20% of all known fungi forming lichens. Despite their poikilohydric nature, they occur in almost all terrestrial habitats, including cold and hot deserts (Garrido-Benavent and Perez-Ortega, 2017). According to Nash (2008), lichens exhibit intriguing morphological variations in miniature within the terrestrial autotrophs. They display an array of colors, which range from orange to yellow, red, green, gray, brown and black. They vary in size from < 1 mm to 2 m in the long pendulous epiphytes that hang from tree branches. Except for a few ephemerals (e.g., *Vezdaea*), almost all lichens are perennials, some of which are estimated to survive for over 8,600 years (e.g., *Rhizocarpon geographicum*, Miller and Andrew, 1972). In some ecosystems, the biomass of epiphytic lichens may exceed several hundred kg/ha (Coxson and Nadkarni, 1995). Lichens occur and grow, where water and carbon are available. They do not possess a mechanism for internal water regulation. Hence, they hydrate themselves along with diurnal and seasonal water availability and temperature conditions as well as extremes of weather and climate (Kranner et al., 2008). Being poikilohydrics, their mycobiont and photobiont cells lose water during desiccation but recover rapidly, when water becomes available. They survive over dramatic wetting (> 160% water/thallus dry weight) and drying (to < 20%) (see Kappen, 1988).

Regarding their biogeography, a striking feature is their bipolar distribution recognized since the days of eminent naturalists like Humboldt

(1767–1835). The very first estimate showed that 25% of Antarctic species are characterized by bipolar distribution. The incidence of 17 species (59%) of the genus *Cladonia* in the Falkland Islands at 51–52°S and 15 species (37%) in Navarino Island at 54–55°S is notable. Recent estimates indicate that 742 lichen species are bipolar, of which more specifically 171 species are shared between the Arctic and Antarctic and 169 species between the Arctic and New Zealand. For more details on their origin during the geological past and geographic range, Garrido-Benavent and Perez-Ortega (2017) may be consulted.

Some lichens are useful as food (e.g., *Parmelia* in India, *Umbilicaria* in Japan), fodder (e.g., *Cladonia rangifera* for reindeer and musk ox in Tundra and Siberia), medicine (e.g., *Usnea* to stop bleeding, incidentally, *Usnea* merits research to know about the presence of heparin), dyes (e.g., *Roccella montagnei* and *Lasallia pastulata*, as indicator dyes in litmus paper), tannin (e.g., *Lobaria pulmonaria* in leather industries) and cosmetics (e.g., *Evernia*, as an essential in the manufacture of cosmetic soaps). A few are harmful; penetrating lichens damage glass surface, buildings, statues and portraits. Their penetration depth ranges from 0.3 mm to 16 mm. In forests, pendant lichens like *Usnea* are inflammable and cause in spreading fire. For those interested in lichen taxonomy Awasthi (2013) and in *Lichens Biology* Nash (2008) may be consulted. The objective of this chapter is to complete the account on fungi, as ~ 20% of which constitute the mycobiont partner of lichens rather than to provide an exhaustive picture of lichens.

11.1 Mycobionts and Photobionts

Lichens are classified as fungi: they do not have independent scientific names (Rikkinen, 2002). The estimated number of described species increased from ~ 17,000 (Hale, 1974) to 19,409 (Lucking et al., 2017). Incidentally, many authors (e.g., Nash, 2008) do not include mycorrhizas as lichens. For, there are succinct differences between them. 1. The **septate mycelia** of the basidiomycotous and ascomycotous lichens hold **aquatic algae** as photobionts, but the **aseptate mycelia** of the Glomeromycota are held on (ectomyccorhiza) or inside (endomycorrhiza) the roots of the **terrestrial photobionts** namely bryophytes, ferns, gymnosperms and angiosperms. Whereas the latter multiply **clonally alone**, the former reproduce **clonally and/or sexually**. However, the glomeromycotous filaments serve as mycobionts and the plants as photobionts. Hence, this account considers them together as lichens.

Approximately, 20% of fungal species are lichenized, although Garrido-Benavent and Perez-Ortega (2017) claimed 29%. More than 85% of the aquatic mycobionts engage chlorobionts, 10% have cyanobionts and < 5% engage other microbial photobionts (see Zachariah and Varghese, 2018). For example, epiphytic diatoms are frequent photobiotic partners in relatively dry habitats. Of 13,500 lichenized ascomycotous species, 1,550 species (~ 12%) in 130 genera and 50 families are cyanolichens (Rikkinen, 2002). Molecular studies have shown that lichen type of symbioses have independently arisen at > 100 lichenization events and involved re- and de-lichenization during the diversification of extent fungi (Aptroot, 1998).

Mycobionts: According to Honegger (2008), there is no evidence for any fundamental difference between lichenized and non-lichenized fungi. Still, they differ from each other in their nutritional strategies. For example, the saprotrophic fungi acquire 45–50% fixed carbon from the substratum, i.e., dead organic matter but the symbiotic fungi only up to 30% (Lewis, 1973, Hawksworth et al., 1983). The following fungal taxonomic groups enter into symbiosis with photobionts (see Honegger, 2008).

Glomeromycota		230 species	1.7%
Basidiomycota		50 species	0.3%
Ascomycota		13,500 species	98.0%
	Subtotal	13,780 species	
Lichenicolous fungi*		1,300 species	
	Total	15,080 species	

*see Table 3.9. Different values are reported by different authors for the parasitic lichens.

Clearly, Pezizomycotina are pre-adapted to enter into symbiosis to form lichens. Within them, the Order Arthoniales, especially the family Roccellaceae and Helvellacea in Pezizomycetes (Pfister, 2015), Pleosporales in Dothideomycetes (Schoch and Grube, 2015) and 99% of the Lecanoromycetes (Gueidan et al., 2015) contribute the mycobionts of lichens (see also Honegger, 2008).

Photobionts: *Nostoc* is by far the most common cyanobiont in lichens, especially in lecanorales. Along with *Nostoc*, *Gunnera* occurs among angiosperms, *Cycads* in gymnosperms and *Anthoceris* in bryophytes and ferns. The others like *Gloeocapsa* and *Chroococcidiopsis* are important photobionts in lichens. Some cyanobionts like *Gloeocapsa* and *Myxoscarcina*, which clonally multiply by binary- and multiple-fission, respectively, rarely retain the reproductive mode of multiplication, when lichenized (Rikkinen, 2002). Rarely, the chlorobionts reproduce sexually in a symbiotic state (Scheidegger, 1985), although sexual reproduction occurs regularly in culture conditions (Skaloud et al., 2015). Not surprisingly, Kroken and Taylor (2000) went to the extent of suggesting that the mycobionts suppress sexual reproduction in chlorobionts to prevent generation of less suitable genotypes.

In Lecanoromycetes, pairwise associations in lichens are also reported between fungal partners, and Trebouxiales/Trentepohliales as algal partners. For phylogenetic relations of pairing in lecanoromycetes, Miadlikowska et al. (2006) may be consulted. The other pairwise groups are between: 1. Dothideomycetes – Trentepohliales and 2. *Prasiola* clade of Trebouxiophyceae – Eurotiomycetes (see Spribille et al., 2022). In a long list, Rikkinen (2002) assembled relevant information for mycobionts and cyanobionts of 108 living species in 70 families and 28 orders, as summarized below:

Phylum	Photobiont (no.)			Remarks
	Species	Family	Order	
Chytridiomycota	8	4	3	All are parasites
Zygomycota	1	–	–	Endocyanosis
Basidiomycota	6	4	3	4 lichenized fungi, 1 tripartite
Ascomycota	93	62	22	6 parasitic species
Total	108	70	28	

In all, *Peltigera* spp are associated as hosting fungi in 39 lichenized species, of which two are with Agaricales. On the other hand, *Nostoc* spp are associated with 31 lichenized fungal species, of which nine are tripartite pezizomycotines and basidiomycotes (see below). The cyanobiont *Scytonema* is associated with nine mycobionts, of which three are parasites, two tripartites and four bipartites. More pairwise associations are listed below. It is likely that *the pairwise association may have emerged independently multiple number of times.*

Diaporthales		
Pyrothicaceae	*Pyrenothrix*	Parasite
Pseudoperisoperiaceae	*Myxophora, Pycorella*	
Sphaerophoraceae	*Sphaerophorus*	Tripartite
Stereocaulaceae	*Muhria*	
Licheniales		
Licheniaceae	*Thermutis, Zahlbrucknerella*	
Peltuloceae	*Phyllopeltula*	Bipartite
Ostropales		
Strictidaceae	*Petractis*	

The term cyanolichens refers to those lichens with a cyanobiont either as the sole photobiont or as the second partner, in addition to the primary photobiont, the algae. About 95% cyanolichens are bipartite, in which cyanobiont serves as the primary photobiont. However, they are tripartite in the remaining 3–4% cyanolichens. They have an alga serving as photobiont and a cyanobiont as nitrogen fixer. In cyanobionts, direct atmospheric fixation of atmospheric nitrogen is catalyzed by an enzyme complex nitrogenease. Being sensitive to oxygen, nitrogenase and its nitrogen fixing activity are spatially separated and located in modified cells called heterocysts within filamentous cyanobionts (see Pandian, 2022). These heterotrophic heterocystic cells are confined to a structure called cephalodium within the lichens either on their upper (Fig. 11.1) or lower surface (Brodo et al., 2001). The biomass of heterocystic cells may account for 35.6% of cyanolichens, in comparison with 4–7% in free-living cyanobionts. In fact, the dependence of cephalodistic fungi like *Peltigera aphthosa* is so obligatory that it is not capable of an independent life.

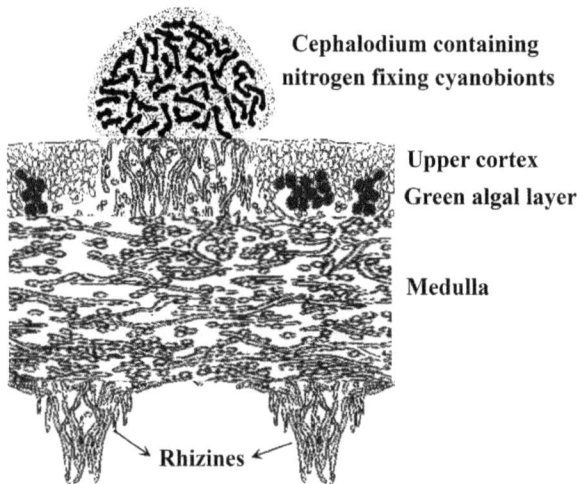

Cephalodium containing
nitrogen fixing cyanobionts

Upper cortex
Green algal layer

Medulla

Rhizines

FIGURE 11.1

Diagrammatic cross section through the thallus of *Peltigera* (free hand drawing based on Rikkinen, 2002).

Ribitol is a good marker for trebouxiophycean algae, as it is a motile compound translocated between the photobiont and mycobiont. Its concentration is 2.8% dry weight for chlorolichens, which is four (0.8%) times higher than that of cyanolichens. Similarly, chlorolichens have 1.1% trehalose, which is 20 times higher than that in the cyanolichens (Henskens et al., 2012).

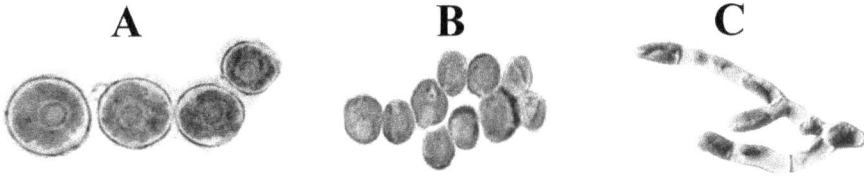

A   B   C

FIGURE 11.2

(A) *Trebouxia gigantea* within a thallus of *Xanthomaculina hottentotta*, (B) *Elliptochloris bilobata* and (C) *Dilabifilum arthropyreniae* (modified from Friedl and Budel, 2008).

About 40 genera of Chlorophyta and Cyanophyta are known as photobionts in lichens (Friedl and Budel, 2008). Three genera *Trebouxia, Trentepothlia* and *Nostoc* are the most common photobionts. Within Chlorophyta, the photobionts can be simple (e.g., *Trebouxia gigantea*, Fig. 11.2A), sarcinoid (e.g., *Elliptochloris bilobata* Fig. 11.2B) or filamentous (e.g., *Dilabifilum arthropyreniae* Fig. 11.2C). Figure 2.24 of Friedl and Budel (2008) shows that > 19 species in 11 genera are photobionts from the family Trebouxiophyceae. The following chlorophytes are some examples for the photobiont species and their occurrence in orders and families of mycobionts:

Photobionts	Mycobiont	Species
Trentepothlia	Arthoniales	*Roccella*
	Gyalectales	*Coenogenium*
Trentepothlia + Trebouxia	Ostropales	*Graphis*
	Caliciales	*Licidella*
	Lecanorales	*Trapetia*
Myrmecia	Verrucariales	*Catapyrenium*
Coccomyxa	Baeomycetaceae	
	Peltigeraceae	
Dictyochloropsis	Stictaceae	
	Peltigeraceae	

11.2 Mycorrhizas – The Root Fungi

Mycorrhiza, a characteristic symbiotic association between higher plants and fungus, is extended from the plant roots into the rhizosphere and the surrounding soil. The association confers a form of "biological fertilization" (Gianinazzi et al., 2010). Mycorrhizas consist of glomeromycotes with most species from the Helvelliceae in Pezizomycetes (Pfister, 2015) and 110 species in 60 genera of Dothideomycetes (Schoch and Grube, 2015). Due to the positive effects for plants, especially under conditions of abiotic and/or biotic stresses, it may not be a surprise that > 89% of all terrestrial plants are associated with mycorrhizas (see Dighton, 2009). The obligate higher plants depend heavily or entirely on mycorrhizas, whereas facultatives depend on them, only at times of need. Some fungal species are host-specific, for example, *Suillus grevillea* associate with only the European larch *Larix decidua*. However, most others may have broad host-specificity.

Estimates: For the first time, estimates are made to show the relations between the number of potential photobiont plants and the proportion of lichenized fungi (Fig. 11.3A) and mycorrhizal species number, as well (Fig. 11.3B). Values for photobionts were drawn from Pandian (2022). From values reported by Wang and Qui (2006), the actual numbers of lichenized fungal species were estimated (Table 11.1). Of lichenized fungi, 13,500 species are associated with aquatic photobionts; they comprise 12,000 species or 89% chlorobionts and 1,500 species cyanobionts. Mycorrhizas are engaged by > 89% angiosperms. The remaining mycorrhizas are hosted by bryophytes, tracheophytes or gymnosperms. Clonal cyanobiont spores survive for a day but for 60 d, when buried. But those of chlorobionts survive for > 60 d. The spores of bryophytes and tracheophytes are dispersed to a maximum distance of 1–5 mon (Pandian, 2022). For these reasons, their ability is limited to < 6%. In all, *the number of photobiotic plant species determines the proportion of lichenized fungi* (Fig. 11.3A). As mycorrhizas dominate with 89% contribution to lichenization, a quantitative estimate was also made; the contribution by different mycorrhizal types decreases in the following order: arbuscular > orchid > ecto > ericoid. Within angiosperms, 44,272 species remain non-mycorrhizals, albeit a value of 51,500 species is reported by van der Heijden et al. (2014). *Within lichenized fungi, mycorrhizas, especially the arbuscular play a major role* (Fig. 11.3B).

TABLE 11.1

Number of photobiont plant species distributed in freshwater and terrestrial habitats (condensed from Pandian, 2022). *Wang and Qui (2006, [†]as % of host species number of 281,701.

Taxon	Photobiont species (no.)	Host species (no.)	Lichenized species (%)[†]
Cyanophyta	–	1,500	0.53
Chlorophyta	–	12,000	4.26
Terrestrial			
Bryophyta	21,925	10,086*	3.58
Tracheophyta	11,850	6,162*	2.19
Gymnosperms	1,079	1,079*	0.38
Angiosperms	295,146	250,874*	89.06
Total	330,000	281,701	

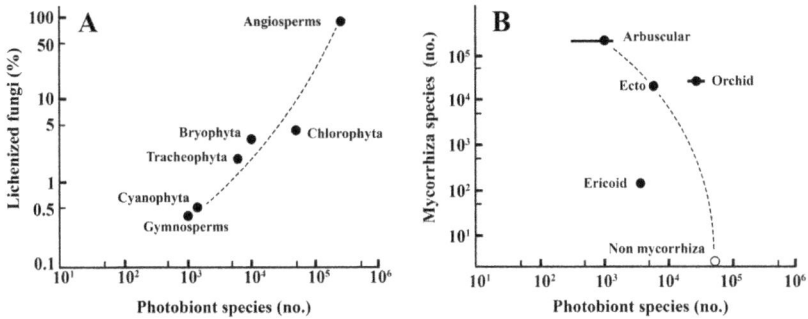

FIGURE 11.3

Relations between photobiont species number as functions of (A) proportion of lichenized fungi (based on Table 11.1) and (B) mycorrhizal species number (from data reported by van der Heijden et al., 2014).

Mycorrhizal groups: Based on the external or internal association with plants, mycorrhizas are broadly divided into three groups (Fig. 11.4): (i) Ectomycorrhizas (EcM), (ii) endomycorrhizas and (iii) ecto-endomycorrhizas (with mixed behavior between the first two). In the 6,000 speciose ectomycorrhizas (van der Heijden et al., 2014, however, see also Tedersoo et al., 2010), the obligate basidiomycotous and ascomycotous fungi form a mantle of hyphae around the root and completely envelop the root tip (Fig. 11.4A) of the ferns, cold-temperate pines, larches in boreals and subalpine forests as well as oaks and poplars in deciduous forest (Dighton, 2009). According to Ganugi et al. (2019), the hyphae of the endomycorrhizas grow inside the root but also penetrate the root cell walls and are enclosed in the cell membrane wall, as well, ensuring a more invasive symbiotic relationship between the mycorrhizas and the plant (Fig. 11.4B). In 86% of all vascular plants, the 300–1,600 speciose arbuscular mycorrhiza species

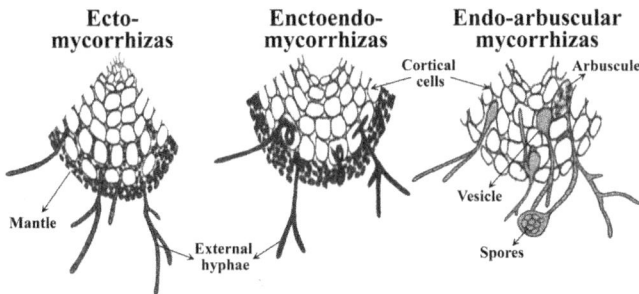

FIGURE 11.4

Different levels of association between mycorrhizal fungi and plant roots. Note arbuscular mycorrhizas penetrate the cortical cells of the root and form arbuscule and vesicle. Ectomycorrhizas cover the plant root system with a mantle of fungal tissue and the hyphae surround the plant cells within the root cortex (based on Ganugi et al., 2019).

are intimately associated (for species number, see van der Heijden et al., 2014, see Pandian, 2022). The mycorrhizas receive 10–40% of photosynthate carbohydrate but provides water, minerals and micronutrients.

Endomycorrhizas are further divided into five major subgroups: (i) arbuscular, (ii) ericoid (ErM), (iii) arbutoid, (iv) monotropoid and (v) archid mycorrhizas (OrM). But the Arbuscular Mycorrhizas (AM) are the most ubiquitous and have attracted many studies during the last 20 years, especially from the agronomic sector. Between 1997 and 2018, the number of publications increased from a few in 1998 to > 100/y in 2018 (Silva-Flores et al., 2021). These publications are reviewed from time to time (e.g., Dighton, 2009, Ganugi et al., 2019). Interestingly, there are also a dozen books on mycorrhizas. Of them, Fortin et al. (2009) on *Mycorrhizas: The New Green Revolution* and Phillips (2017) on *Mycorrhizal Planet* may be mentioned.

In contrast to AM, the others are characterized by restricted distribution, especially in the family Ericaceae. For example, the ericoid mycorrhizas (ErM) are restricted to the members of Ericaceae, Epacridaceae and Empetraceae. Their hyphae invade cortical cells of fine roots, in which they form coils. Arbutoid mycorrhizas (ArM) are extremely host-specific and restricted to members of two genera *Arbutus* and *Arctostaphylos* within the Ericaceae and some genera in the Pyrolaceae. Monotropoid mycorrhizas (MnM) are restricted to the Monotropoideae in the family Ericaceae. Their host plants have lost autotrophic ability. The MnM form bridges between their non-photosynthetic host plants and the neighboring photosynthetic plants, from which MnM draw photosynthetic nutrients (see Dighton, 2009).

The nutrient uptake-enhancing properties of the arbuscular mycorrhizas (AM) are: 1. The ability of AM to extend and explore a vastly greater area of the soil than the host plants roots can reach. Their narrow hyphal diameter (2–20 μm) allows them to access smaller pores that are unreachable to plants roots with implications in terms of water and micronutrient absorption. Evidences are available to show the ability of AM to adjust hyphal diameter depending on the soil porosity and thereby provide nutrients to the plant, independent of soil texture (e.g., Drew et al., 2003). 2. The ability of AM to absorb from soil containing –40.0 MPa water – a measure of water potential – is far greater than those bacteria (–10.0 MPa), Protozoa (–0.03 MPa) and germinating seed (–1.5 MPa) (see also Table 3.3). Hence, water absorption ability of mycorrhiza is phenomenal; as a result, the fungal hyphae can absorb water from the driest soil, whereas plant roots can do it only from moist soil. 3. Studies have also linked the ability of AM to enhance the acquisition of water, micronutrients and minerals, especially phosphate, the lowest soluble ion in soil water by AM. For example, the ability to acquire phosphate is increased by 1.5 times and zinc by 1–3 times in wheat plants (Ganugi et al., 2019). 4. Furthermore, many studies have confirmed that the

AM help plants to resist abiotic and biotic stress. For example, they synthesize trehalose, containing two molecules of glucose to help plants to better tolerate draught. In the presence of AM *Glomus mossae*, the peanut *Arachis hypogaea* significantly improves plant growth (28%), pod production (22%) and seed weight (12%). Despite the pathogenic effects of *Fusarium solani* and *Rhizoctonia solani*, the mycorrhizas negate the suppression of plant growth and increase its weight (26%), pod/plant (35%) and seed weight (39%). The ectomycorrhizals *Laccaria laccata* and *Paxillus involutus* significantly reduce the pathogenic effects of *Phytophthora cinnimomi* on the growth of chestnut tree seedlings of *Castanea* spp (Dighton, 2009). On the positive role played by mycorrhizas in agriculture, commercial forestry and pollution, Dighton also provides useful information.

Nutrient transfer: Efflux is known for glucose from *Nostoc* to the lichen *Peltigera polydactyla* but polyol and ribitol from *Trebouxia* to fungus (Drew and Smith, 1967, Richardson et al., 1967). Experiments performed on > 90 symbionts revealed that cyanolichens export glucose, whereas chlorolichens export polyols consisting of four (e.g., teritol), five (ribitol) or rarely six (e.g., sorbitol) carbon atoms (Smith, 1980). Utilizing these nutrients, the mycobionts build up mannitol for their respiration and growth (see Smith, 1963). This 'nutritional model' explains the lichen symbiosis (see Nazem-Bokaee et al., 2021, and also Spribille et al., 2022). The secondary metabolites, that are often deposited extracellularly by the fungal partner on the hyphal surface, are diverse and numerous with > 700 described compounds (Elix and Stocker-Worgottor, 2008). Serving as deterrents, the fungal derived secondary metabolites protect the lichens from herbivory (Solhaug et al., 2003). The deterrents also absorb the UV and provide photoprotection to both algal and fungal symbionts, especially during secondary anhydrobiosis (Beckett et al., 2021, see also Chapter 14).

   Genomic studies have provided interesting information on nutrient acquisition by two EcM in the agaricale *Laccaria bicolor* and peziziale black truffle *Tuber melanosporum* (Bonfante and Genre, 2010). The genome of *L. bicolor* lack genes encoding invertases, the enzymes that hydrolyze sucrose into glucose and fructose. Hence, the fungus has to depend only on the photosynthate released by its host as glucose. In contrast, *T. melanosporum* has one invertase gene, indicating its potential capacity to access the sucrose pool of its host plant. Incidentally, the biotrophic plant pathogenic fungus *Ustilago maydis* also possesses a sucrose transporter localized in its plasma membrane. Of the 47 putative transporters among the regulated genes of *L. japonicus*, 28 are important for nutrient acquisition. Analysis of EcM genomes has shown the capacity of free-living mycelium to absorb organic and inorganic nitrogenous substances in the form of ammonium, nitrate and peptides from the soil. The AM, which have so far proved to be unculturable in the absence of a host, are unable to absorb carbohydrate, except from their hosts. Root colonization is vital for propagation. An unusual feature of the clonal spores of AM is that when their hypha germinating from a spore is unable to find an appropriate host, the spore arrests the hyphal growth, retracts its cytoplasm back into its spores and becomes dormant. It may revive germination again and again, until it finds a suitable host. However, with a wide range of its hosts, it may find a suitable host at the first germination itself, as its nutrient lipid storage can sustain for a few days only (see Bonfante and Genre, 2010).

The arbuscular mycorrhiza *Coccidioides immitis* was once considered to multiply clonally alone. However, their population studies have shown successful existence of genetically defined populations. From their multinucleated clonal spores, the progenies may inherit different combinations of nuclei (see Fig. 1 of Taylor et al., 2015). This account proposes to extend this discovery to lower fungi, especially to the Glomeromycota. Within their tubular, non-septate (see Fig. 1.11B) coenocytic (syncytial) hypha, the nuclei and other organelles are continuously translocated from one part of the hypha to other by the streaming cytoplasm. *As the nuclei are scattered throughout the streaming cytoplasm, a fragmentation, as in Acaulospora (uwyo.edu), may produce asymmetric daughter fragments, each containing different numbers and possibly genetically different types of nuclei.* Hence, the scope for introducing genetic diversity does exist in Glomeromycota. It is also known that the spores of Glomeromycota are multinucleated (e.g., *Glomus etunicatum*, Jany and Powlowska, 2010). Hence, *it is likely that the different spores may carry different numbers and possibly genetically different types of nuclei to generate genetic diversity among the progenies.* This may prove to be a fruitful research area to understand the genetic diversity generated by the multinucleated hyphae and spores of Glomeromycota.

11.3 Structure and Growth

Structure: As 96% fungi are terrestrial (see p 65), lichens are also terrestrial except for a few found on the silicious rocks of freshwater (e.g., *Hymenelia lacustris*) or the sea shore (e.g., *Caloplaca marina*). The fungi constitute 90% of the lichen biomass (Honegger, 2012). In terrestrial habitats, they may be (i) corticolous growing on the bark of trees (e.g., *Graphis, Usnea*), lignicolous directly on the wood (e.g., *Calicium, Cyphelium*), saxicolous on the moist rock surface (e.g., *Poria, Xanthoria*) or terricolous growing directly on the ground (e.g., *Cladonia floerkeana, Collema tenax*). Based on external structures, they are divided into the following five groups: (i) **Foliose** lichens are flat with leaf-like lobed thallus attached to the substratum by rhizines (e.g., *Collema, Peltigera, Xanthoria*). Based on the distribution pattern of the algal component among the fungal component, two types of foliose lichens are recognized. In the homoiomerous type (Fig. 11.5A), the algal components are uniformly distributed among the inter-protective layer of fungal components within the thallus. But the heteromerous type (Fig. 11.5B) is more structurally organized with (a) the cortex consisting of a thick outer protective zone of the thallus composed of compactly interwoven fungal hyphae arranged at the right angle to the thallus surface, (b) algal layer composed of algal cells embedded between tangled network of loosely interwoven hyphae, (c) the central medulla with loosely interwoven fungal hyphae spaced between them, (d) lower cortex arranged as in the upper one but from which (e) the rhizines arise. (ii) **Crustose** lichens encrust the ones with a thin, flat thallus firm in texture. They are wholly or partially embedded in the substratum with visible fruiting bodies (e.g., *Buellia, Graphis* [Fig. 11.6A], *Lecanora, Lecidia*), (iii) **Fruticose** lichens are well-developed shrub-like thallus that may be cylindrical or branched. They are attached to the substratum by a basal mucilaginous disk (e.g., *Cladonia, Letharia, Usnea* (Fig. 11.6B)), (iv) **Filamentous** lichens are also well-developed and are ensheathed or covered by only a few hyphae (e.g., *Coenogonium, Racodium,* Fig. 11.6C) and (v) **Leprose** lichens are less common and found by a cluster of algal cells enveloped by fungal hyphae providing a simple thallus with a powdery appearance (e.g., *Lepraria incana* (Fig. 11.6B). According to Lisci et al. (2002), (vi) **Endolithic** lichens occur only on calcareous stone, in which the thallus is completely immersed. They are noted only when the fruiting bodies emerge from the stone. The ascomycetous lichens are characterized by the fruiting body perithecium (e.g., *Dematocarpon*) or a disk-shaped one called apothecium (e.g., *Parmelia*).

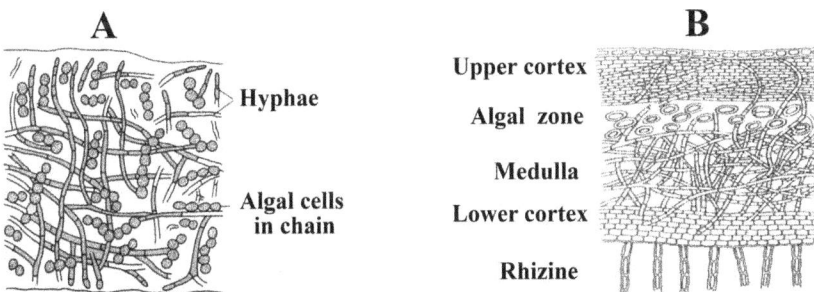

FIGURE 11.5

Examples for (A) Homoiomerous and (B) heteromerous types of foliose lichen. All others are free hand drawings.

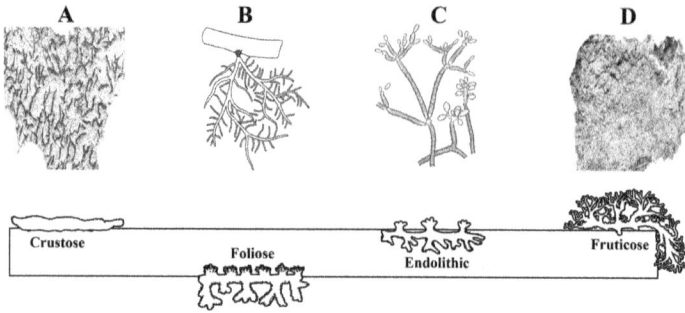

FIGURE 11.6

(A) Crustose lichen: *Graphis* (*inaturalist.org*). (B) Fruticose lichen: *Usnea*. (C) Filamentous lichen: *Racodium cellare*. (D) Leprose lichen: *Lepraria* (Wikipedia ikipedia). The lower row shows the lichens as growing on or in or within the rocky substratum (free hand drawing based on Lisci et al., 2002).

Growth: Lichen-forming fungi have a limited life span, after which they senesce and die. Some have the span of only a few months (Honegger, 1993), while the slow growing (0.006–1.0 mm/y) *Rhizocarpon geographicum* may live for 9,000 y. With average annual growth rate of 2 mm, *Parmelia centrifuga* may require 500 y to grow to a meter size in the Arctic (see Honegger, 1993). The radial growth in the foliose *Xanthoparmelia coloradoensis* on the placodioid *Lecanora novomexicana* is accelerated with increasing wedge width, become linear at the size of 2.01 mm and grows at 0.12 mm/y. The fastest growing 'lace lichen' *Ramalina menziesti* linearly increases to 43 cm/y (Sharnoff, 2014). Armstrong and Bradwell (2010) summarized available values for the annual radial growth rate of crustose lichens. They range from 0.03 mm/y in *Rh. geographicum* to 2.84–6.05 mm/y in *Lecanora muralis*. For *Rh. geographicum*, many values from different locations are reported. The trend for thallus size vs radial growth shows that the growth rate is fastest at ~ 0.6 mm/y in a thallus of 3.5 cm diameter. But it declines on either side of the thallus size (Fig. 11.7).

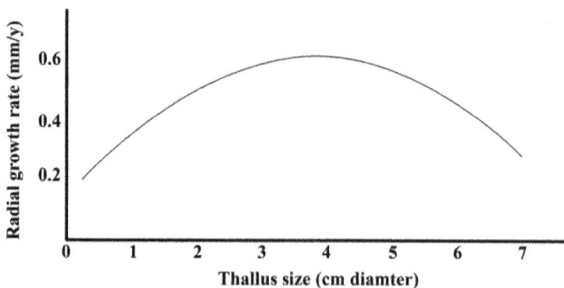

FIGURE 11.7

The relationship between radial growth rate and thallus size of *Rhizocarpon geographicum* growing on slate rock surfaces in Wales, UK (modified and redrawn from Armstrong and Bradwell, 2010).

During the hyphal growth, lichens are capable of breaking stones and rocks. For information on their role along with bryophytes in formation of the present landscape, Pandian (2022) may be consulted. Lichens are capable of causing damage to rocks and buildings mechanically and chemically. Their hyphae can penetrate up to the depth of 16 mm. They are also known for secreting carbonic acid, lichenic acid and oxalic acid (Lisci et al., 2002).

11.4 Propagation of Lichens

Most lichens are capable of clonal multiplication and/or sexual reproduction. As described earlier, glomeromycotous mycorrhizas are clonals. *In ascomycotous lichens, both mycobiont and photobiont propagate together in clonal multiplication, while they propagate as separate independent units in each generation of sexual reproduction.* This important difference between clonal and sexual propagation has three major consequences: 1. The clonal propagules facilitate successful co-dispersal, both the lichen partners, in contrast to the sexual spores, which require finding an appropriate photobiont on time to relichenization and establishment of thallus *de nova*. 2. Clonal multiplication facilitates a greater survival and recruitment so that the frequency and abundance of lichens is far greater than that of using sexual reproductive strategy. For example, Dal Grande et al. (2012) estimated that 70% of lichen thalli across worldwide populations of *Lobaria pulmonaria* are derived from clonal propagules. Among corticolous lichens in Wisconsin, Hale (1955) found that 72 of the most frequent species are clonal sorediates or isidiates, while 89% of the rarest species are neither sorediate nor isidiate. 3. In clonals, the same genomes of the symbiotic partners are passed on from the parental generation to progeny generation and thereby their 'genetic combination' is sustained. Apart from these, lichens develop a thallus ranging from the two-dimensional thin crustose structure to large structurally complex three-dimensional motifs called macrolichens. The former can ill-afford to develop clonal organs and spores. Consequently, they may have to necessarily depend solely on sexual reproduction and rarely an accidental fragmentation (see Bowler and Rundel, 1975).

Clonal multiplication: In ascomycotous lichens, clonal multiplication occurs frequently through clonal spores and less frequently by fragmentation. Vertical fragmentation is facilitated by erect bushy lichens like *Cladonia*. The hypha of some lichens breaks up (more like spores at the tip) into small bodies called **oidia**, which germinate into new lichens. **Isidia** and **soredia**, together called diaspores, are the most common haploid clonal spores produced by ascomycotous lichens. Isidia are small, grayish black or coral-like corticoid, cylindrical, detachable outgrowths developed on the thallus (e.g., Lecanoromycete: *Parmelina tiliacea*, Fig. 11.8A). Soredia are a grayish white bud like ecorticated thallus granules released via an ostiole in the cortex called **sorelia** (e.g., *Pertusaria flavicans*, see Gueidan et al., 2015, Fig. 11.8C). Besides, pycnidia = (conidiomata) also generate clonal spores called pycnidiospores (Fig. 11.8B). They are minute flask-shaped structures located on the surface of or embedded within the thalli. These spores may also serve as spermatia, as they are found attached to trichogyne in many lichens. Of the two diaspores, soredia are more common; for example, its incidence is known in 14 of 17 families within Lecanorales (Table 11.2) but for isidia, it is known only from 6 of 17 families. They co-occur in 5 families. Only in Umbilicariaceae and *Pseudovernia* within Parmeliaceae, isidia occur independently. Others like *Lobaria* and *Sticta* within the family Stictaceae, are sorediate or isidiate or both or bear none of them.

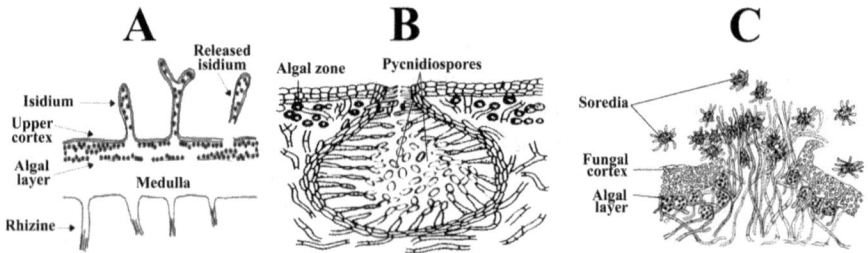

FIGURE 11.8

Enlarged view of cross sections of different lichens to show the clonal spores (A) isidia (*plantscience4u.com*), (B) pycnidia, (C) soredia (free hand drawings from Margulis et al., 2009).

Sexual reproduction: As mentioned earlier, sexual reproduction necessitates the reestablishment of relichenization between distinct units of myco- and photo-bionts. The haploid ascospores, released from the ascus, are sexual spores (Fig. 11.9C). The ascomycotous mycobiont arises from the 'internal' fusion between spermatia from the **spermagonium** (Fig. 11.9B) and trichogyne from the ascogonium (= carpogonium, Fig. 11.9A). The probability

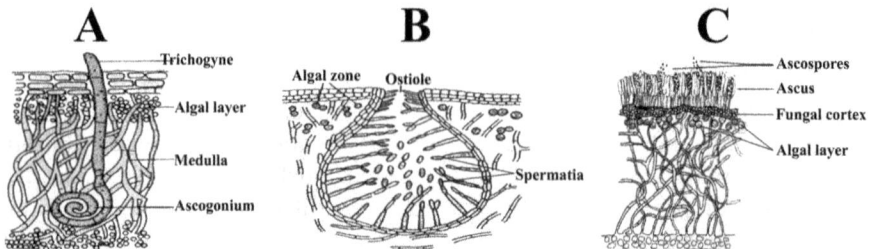

FIGURE 11.9

Enlarged view of cross sections of different lichens to show the clonal spores (A) female and (B) male sexual organs and (C) ascus and ascospores (free hand drawings from R. Parveen, Gautum Buddha Mahila College, Bodhgaya, India).

of an immediate contact of a potential mycobiont with a compatible photobiont is very low. However, this is neutralized by the following two features: (i) the wide low specificity of mycobiont and (ii) many mycobionts are able to form a temporary association with less compatible algae and remain in that 'symbiotic state' until the contact with more compatible photobiont is established (Gassmann and Ott, 2000, Beck et al., 2002). For example, *Chrysocephala* can part with the chlorobionts belonging to different orders, namely Trebouxiales (e.g., *Trebouxia*) or Prasiolaes (e.g., *Stichococcus*). The family Verrucariaceae (e.g., *Hydropunctana rheitrophia*) is known for its ability to associate with many unrelated photosynthetic partners until they find appropriate photobionts. This sort of 'promiscuity' seems to follow clear rules based on climatic factors and soil chemistry (see Skvorova et al., 2022). However, a high degree of specificity is displayed by *Phlyctis argena*, which associates only with one algal species *Dictyochloropsis splendida* (Beck et al., 2002). At this point, it is also necessary to distinguish 'selectivity' from 'specificity'. The former means selection of preferential interaction between symbionts and the latter refers to cell-cell interactions with absolute exclusivity (see Insarova and Blagoveshchenskaya, 2016). With regards to highly selective mycobionts, the selections begin with the recognition of the appropriate photobiont among the surrounding organisms. Besides photobionts, *Peltigera hymenina*, for example, is usually surrounded by 236 bacterial, 74 viral, eight eukatyotic and six Archean organisms (Zachariah and Varghese, 2018). Subsequently, the following cascade of reactions proceed: (i) intensive exchange of signaling at the 'precontact' stage, (ii) following wall-wall contact, the formation of appressorium and haustorium-like structure and polysaccharide cover, (iii) coating of photobiont cells by the mycobiont hyphae and formation of soredia-like structures, (iv) formation of dense matrix (v) coordinated development and (vi) differentiation of the thallus (see Insarova and Blagoveshchenskaya, 2016).

Lecanorales: Propagative strategies differ from species to species, and in some cases within a genus. For example, *Phlyctis distorta* reproduce sexually, whereas *P. grisea* clonally via soredia (see Insarova and Blagoveshchenskaya, 2016). Therefore, it may be difficult to quantify the number of lichen species that reproduce (i) clonally alone, (ii) sexually alone and (iii) both clonally and sexually. The data reported by Bowler and Rundel (1975) also confirmed it. They assembled relevant information for 70 genera in 23 families of Lecanorales. 1. According to Gueidan et al. (2015), the Lecanorales consist of 5,695 species in 250 genera in 19 families. Hence, the data assembled by Bowler and Rundel may represent only 28% of Lecanorales. 2. Table 11.2 shows that 14 families are sordiates and sexuals, while only 6 families are isidiates and sexuals as well as 5 of them are both sorediates and isidiates.

TABLE 11.2

Estimations on the proportion of clonal and sexual reproduction in Lecanorales (modified and condensed from Bowler and Rundel, 1975). *but in Europe 8% clonals.

Family/genus	Clonals (%)				Sexuals (%)
	Soredia	Isidia	Both	Neither	
Buelliaceae	2				98
Pannariaceae	17				83
Nephromataceae	17				83
Anziaceae	50				50
Candelariaceae	50				50
Coccocarpiaceae	66				34
Lecideaceae	8				92
*Lecidea**					100
*Rhizocarpon**					100
Lecanoraceae					
Haematomma	29				71
Lecanora	1				99
Ochrolechia	20		5	5	75
Ramalinaceae					
Ramalina	30				70
Desmaxieria	17			17	83
Umbilicariaceae		9			91
Peltigeraceae					100
Peltigera	17	24			59
Parmeliaceae					
Asahinea	50			50	50
Cetrelia	75				25
Parmelia	16	32			52
Pseudevernia		33		67	67
Cetraria	13	5		18	83
Collemataceae			33	6	61
Heppiaceae					
Heppia					100
Peltula				17	83

Table 11.2 contd. ...

...Table 11.2 contd.

Family/genus	Clonals (%)				Sexuals (%)
	Soredia	Isidia	Both	Neither	
Physciaceae					
Anaptychia	16	6			78
Dirinaria	40		20		40
Physcia	44	3	10	3	42
Pyxine	33	15			52
Cladoiaceae					100
Cladonia	38	2	2		58
Dactylina/Thamnolia				100	100
Evernia	40			60	60
Letharia	50				50
Usnea	17	1	34		48
Alectoria	36	4	4	44	56
Stictaceae	18	9	9	27	63

3. None of the crustose genera in the families Acarosporaceae (4 genera), Baeomycetaceae (1 genus), Gyalectaceae (2 genera), Lichinaceae (17 genera) and Placynthiaceae (2 genera) are either sorediate or isidiate. They all reproduce sexually alone. However, only 8–9% (see Table 11.2) of 235 species are sorediates within the large crustose genus *Lecidea*. On the other hand, of 236 foliose lichen species from several families world over, only 3% bear the clonal diaspores. Further, only 1% of 118 foliose *Lecanora* species are sorediates. Hence, *it is difficult to generalize that all the crustose lichens reproduce sexually alone and all the folicolous lichens multiply clonally alone.*

From the data assembled in their long Table 1, Bowler and Rundel (1975) proposed that from the ancestral sexual lichens, chemical diversification led to the emergence of A, B and C races. From the race A, sub-races, sexual A, and sorediate A arose. The sub-races, sexual B, sorediate B and isidiate B appeared from the race B. From the race C solely sexual C or sorediate C or isidiate C sub-races as well as with both the diasporous sub-races were diverged (Fig. 11.10). Though data from their Table 1 support this proposed course of racial diversification, no genomic or molecular evidence is yet provided to support the proposed racial diversifications.

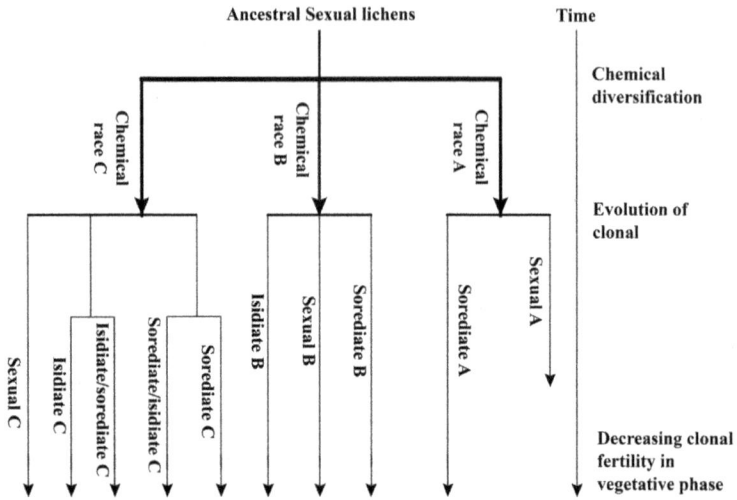

FIGURE 11.10

The proposed hypothetic model of evolutionary diversification in lichens (based on Table 12.1, Bowler and Rundel, 1975).

Spore dispersal is a process of fundamental importance for the spatial dynamics, evolution and migration of the population. Species with poor dispersal ability may become extinct from the entire landscape, as isolation decreases the chances of recolonization in habitats (see Ronnas et al., 2017). Lichens have developed alternate strategies for recolonization either by clonal multiplication or sexual reproduction. The number of publications available for lichenized clonal spores are relatively less than those of ascospores among ascomycotes. Being large (50 mm) and consisting of both mycobiont and photobiont cells, the number of *Lecanora conizaeoides* sorediates drastically decreases from 1,614 on a 2 cm branch and 705 (not shown in Figure) on a 5 cm branch and to < 20 at 20 cm distance, and to almost none beyond 90 cm distance (Fig. 11.11A). Greater water content seems to hasten the faster settlement of the spores. Hence, sorediates from wet thalli settle faster and disperse a shorter distance than its counterpart, the dry thallus (Fig. 11.11B). In epiphytic lichens like *Lobaria pulmonaria*, clonal propagules are dispersed to a longer distance of 10 m (Ronnas et al., 2017). In all, *the sorediates may disperse a maximum distance of < 10 m*. In contrast, the conidia (e.g., 3 to 8 μm, Pfister, 2015) of non-lichenized free-living ascomycotes (e.g., *Aspergillus, Cladosporium*) can be dispersed over distance of hundreds of kilometers (Lecey, 1996).

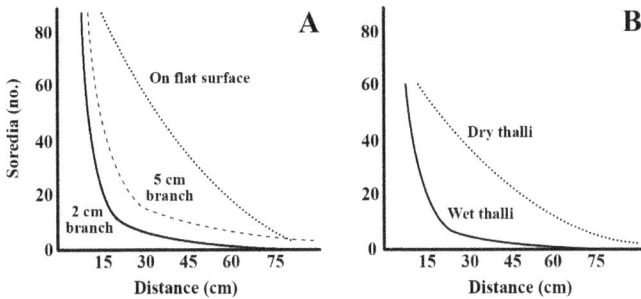

FIGURE 11.11

Dispersal of *Lecanoro conizaeoides* sorediates on (A) flat surface, 2 and 5 cm tree branches and (B) from wet and dry thalli (simplified and redrawn from Bailey, 1966).

In contrast to short distance dispersal of sorediates, the ascospores of *L. pulmonaria* disperse several kilometers (Ronnas et al., 2017). The sexual spores of epiphytic lichens are lighter than clonal diaspores (Brunialti et al., 2021). Hence, they are able to travel longer distances, reaching more remote sites. However, they incur heavy loss of the spores, as only a few of them could find an appropriate photobiont partner. Consequently, they avoid sibling competition and have a lesser tendency to form clusters.

For the sexually reproductive propagules, more data are available for *Lobaria*. Merinero et al. (2017) collected 3,048 sexually reproductive *L. serobicularia* from 18 populations along the climatic gradients of the Iberian Peninsula. They found that the probability of the lichen thallus commencing sexual reproduction increased from ~ 50 mm size to 280 mm size and reproductive allocation also grew with increasing thallus size. Water being the essential requirement, precipitation was related with the size at sexual maturity. With increasing level of rainfall, the threshold size at thallus maturity increased from 35% in 70 cm raining zone to 70% in the 180 cm raining area. *In this lichen, sexual reproduction excludes clonal multiplication. Briefly, lichen growing in the rain-rich climatic zone delayed sexual maturity as far as possible to provide a longer period for clonal multiplication.* However, on commencing sexual reproduction, lichen excluded clonal multiplication.

12

Past: Fungal Evolution

Introduction

Evolution is an ongoing process, and speciation and extinction are its unavoidable by-products. Extinct organisms have left their remnants in the form of fossil fuel by microbes, coal by plants, imprint by soft-bodied fungi, plants and animals, amber by plant gum secretions over the insect inclusive fungi therein and fossil by skeletonized/shelled animals. Due to their small size and fragile structure, the remnants of fungi are not readily fossilized. Yet, an array of fungal residues in the form of spores, hyphae, mycelia and reproductive structures can be traced as remnants throughout much of the geological strata. These remnants include their 1. impressions (e.g., the ascomycotous *Xylomites asteriforonis* on *Podozamites* leaf of Jurassic) and 2. compressions, i.e., coalificationized preservation on plants (e.g., *Sphaerites suessi* scattered on the leaf of Cenozoic) yield valuable information. More information can be traced from the thin band of 3. carboniferous coal (e.g., ascal spores of *Clavascina orlovensis*, *Discascina perforata*), 4. charcoal (charcoalified pieces of *Prototaxites*), 5. fungal hyphae in mummified conifer wood (from Miocene – Pliocene in Alaska and Canada), which are yet to be mineralized and 6. ambers serving as an avenue for fungal fossil research (spores and other residues from the gut and feces of the preserved insects from Cretaceous–Cenozoic ambers). Thanks to the contributions of hundreds of dedicated paleomycologists, it is now possible to have a glimpse over the vanished world from *Fossil Fungi* of T.N. Taylor et al. (2015).

12.1 Geological Time Table

Using the carbon dating method, geologists have developed a procedure to estimate the age of Earth during the geological past. Accordingly, the age of the Earth is considered a little longer than 4 billion y. The period between those times and the present is divided into Paleozoic, Mesozoic and Cenozoic eras (Table 12.1). The Paleozoic era is further divided into four epochs, and each of the Mesozoic and Cenozoic eras into six epochs. Organisms, especially anaerobic bacteria began to appear some 2.5 BYA. Initially, the oxygenic photosynthetic cyanophytes and subsequently chlorophytes began to appear and caused the Great Oxygenation event by 2 BYA. Their abundance led to the rapid transformation of the Earth's atmosphere from essentially anoxic to its present state (Bekker et al., 2004). Around 1.6–2.0 BYA, organisms also discovered sex, which was successfully manifested in them (Butlin, 2002). So much so, representative fungal species of almost all fungal phyla are found during Cambrian some 600 MYA (see Fig. 12.2). Table 12.1 shows that (1) The Earth began to warm up during Ordovician and it recurred during Mississippian and Oligocene (at least once during each era). Hence, climate change and global warming, that are encountered these days, are not new to Earth. (2) The extinction of organisms is also not new. In fact, it was widespread among organisms during the Permian and trilobites even during the Cambrian. (3) Speciation is not new too. During the chequered history of evolution, speciation and extinction have recurred, in response to the continuous change in the Earth's climate and habitats.

Apart from these well-established geological time tables, the pre-Cambrian period is now considered under three eons, namely Archean from 4.5 BYA to ~ 3.5 BYA, Proterozoic from 3.5 BYA to 1.2 BYA and Phanerozoic from 1.2 BYA to 600 MYA (Fig. 12.1A). Today's Earth surface was certainly different from, when it was condensed out by the Sun's dusty rotating nebula. Older rocks were obliterated by heavy meteorite bombardments that

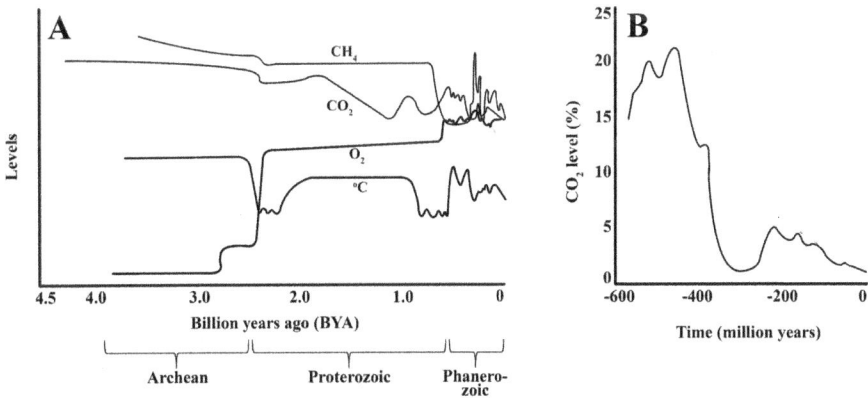

FIGURE 12.1

(A) Earth's changing atmosphere and climate through geological time, as proposed by Hessler (2011) (simplified and redrawn from Hessler, 2011). (B) Changes in atmospheric CO_2 during the last 600 million years (modified and redrawn from Berner, 1997).

TABLE 12.1

Geological time table with remarks on origin, evolution and extinction of organisms as well as climate change (based on Wallace, 1991, Kenrick et al., 2012, *kremp.com*).

Epoch	Million years ago (MYA)	Remarks
Paleozoic era (600 to 360 MYA)		
Cambrian	600	Abundant cyanophytes and chlorophytes. Appearance of representative fungi of most major phyla
Ordovician	500	Appearance of Charales. Occurrence of spores and sporangia of land fungi. **Warming climate**
Silurian	425	Appearance of first land plants, Bryophyta and animals, especially burrowing arthropods. **Continents became increasingly drier**
Devonian	405	Appearance of vascular plants, lycopodiales and ferns. Ascendance of bryophytes. Fungus-plant interaction in Rhynie chert. **Frequent glaciations. Atmospheric CO_2 reduced**
Mesozoic era (359 to 160 MYA)		
Mississippian	359	Appearance of gymnosperms and trees. **Warm and humid climate**
Pennsylvanian	310	Dominance of ferns and gymnosperms as well as amphibians and insects
Permian	280	Widespread extinction of plants and animals. *Reduviasporoites* phytopathogenic fungi appear. **Cooler/drier atmosphere. Widespread glaciations**
Triassic	220	Dominance of gymnosperms. Fungal fossils. Extinction of ferns. Appearance of dinosaurs. **Deserts appear**
Jurassic	181	Appearance of flowering plants. Dominance of gymnosperms. Rapid evolution of dinosaurs
Cretaceous	160	Appearance of monocots, oak and maple forests, modern grasses and cereals. Coevolution between flowering plants and insects. Amber preserved fungi appear. Massive extinction of dinosaurs
Cenozoic era (159 to 1 MYA)		
Paleocene	159	Appearance of mammals
Eocene	54	Appearance of hoofed mammals. Coevolution between Neocallimastigomycota and ruminants. **Erosion of mountains**
Oligocene	36	Modern monocotyledons appear. **Warmer climate**
Miocene	25	Rapid evolution of angiosperms, appearance of unicellular fungal spores
Pliocene	11	Declining of forests – spreading of grasslands. Appearance of man. **Lot of volcanic activities**
Pleistocene	1	Age of man. Large scale extinction of plant and animal species. **Repeated glaciations – End of Ice age – Warmer climate**

also created the Moon. The then much dimmer Sun was ~ 30% lower in its luminosity than today. Volcanic outgassing contributed to much of the atmospheric nitrogen. Concentration of greenhouse gases, carbon dioxide (CO_2, ~ 22%, Berner, 1997), methane and others were high, leaving very little space for oxygen (Hessler, 2011). The earliest organisms may have been prolific CH_4 producers, prior to the advent of oxygenic photosynthesizers (Fig. 12.1B). Our knowledge of the primeval atmosphere is based on some geophysical and geochemical (e.g., preservation of ancient atmosphere as bubbles in glacial ice, see Hessler, 2011) experiments and other considerations. According to the widely accepted hypothesis, the primordial atmospheric composition changed later to a redox-neutral state (see Fay, 1992).

Since its origin, the Earth has witnessed two mega-scale changes (i) weathering of rocks, and (ii) atmospheric composition. The seemingly structurally simpler but chemically active, sessile fungi and plants have contributed at mega-level to these changes. In contrast, the contribution by structurally more organized motile animals is limited to a supplementary role in weathering by burrowing ancestral arthropods (see Kenrick et al., 2012).

12.2 Weathering and Landscape

Since the geological past, rocks are broken by a process called weathering, which led to the formation of the present landscape. Weathering can be caused by abiotic and biotic factors. The former can be divided into physical/mechanical and chemical factors. Mechanical weathering involves breaking of rocks by physical force without changing the chemical nature. It is caused by extreme cold and hot temperatures. When the rocks shrink due to the cold, water seeps into the cracks, freezes and expands, and thereby causes further breakdown. By blowing sand and small rock pieces and hitting them against hard rocks, wind, as a mechanical force, may also serve as an abiotic factor for the formation of the landscape. Chemical weathering is caused by rainwater, especially acidic rain, which erodes rocks, particularly limestone and creates pits and holes. Carbon dioxide (CO_2) readily combines chemically with water to form carbonic acid ($CO_2 + H_2O = HCO^{3-}$) that erodes the surface of rocks and soils, especially involving calcium and magnesium silicates (see Kenrick et al., 2012). Hence, weathering was an ongoing slow process ever since Earth came into existence from 4.5 BYA to ~ 2.5 BYA, i.e., prior to the origin of life. Biological weathering is accomplished by fungi and bryophytes through the production of organic acids. Estimates of global weathering (km^3 rock/y) suggest that weathering by biotic factors is nearly three times faster than by abiotic factors. Incidentally, the withdrawal of CO_2 by plants and water for the formation of carbonic acid may reduce atmospheric CO_2 level and also cool the climate (Porada et al., 2016). However, this account is limited to the description of biological weathering alone.

In land rocks, the biotic weathering was commenced by bryophytes and lichens from some 460 MYA (Kenrick and Crane, 1997). Lichens with rhizines and bryophytes bearing single celled rhizoids with relatively larger surface areas release larger quantities of an array of organic acids (e.g., carbonic acid), alkalinolytes and chelating agents (see Stretch and Viles, 2002, Porada et al., 2016). Bryophytes and fungi possess fewer robust tissues than vascular plants; further, none of their tissue is lignified. Therefore, they hold lower fossilization potential and are likely to be under-represented in fossil records (Kenrick et al., 2012). Nevertheless, their impressions and adpressions comprising a carbonized film indicate that bryophytes and fungi especially the crustose ones and lichens contributed a lot to weathering. The estimate for global weathering by these non-vascular land plants suggests a potential flux of 2.8 km^3 rock/yr. This level is ~ three-times faster than that of today's global chemical weathering flux.

12.3 Fossil Records

Evidences indicate the occurrence of fossil lichens and fungi in the form of hyphae and/or spores almost throughout the geological strata in different geographical zones (Table 12.2). They are known from the Cambrian (e.g., *Farghera robusta*) to Pleistocene (e.g., *Anzia electra*) for lichens and from the Triassic (e.g., *Glomus fasciculatum*) to Eocene (e.g., *Metasequoia milleri*) for the fungal hyphae. Interestingly, evidences are also available for the occurrence of pharganospores of *Reduviasporonites* from Permian and conidiophore tufts of *Paleopyrenomycites devonicus* from Devonian. In fact, *Paleorhiphidium* with anastomosing filaments (not hyphae) is perhaps the first Cambrian fossil evidence, but its fungal status remains to be confirmed. The terrestrial fungi in the form of sporangia and their spores are known from Ordovician. The Rhynie chert from the Aberdeenshire, Scotland has brought definitive evidence for the terrestrial fungal–plant interaction since ~ 410 MYA. Hence, successful colonization of the terrestrial realm involved a symbiotic association between plants and fungi. The coiled hyphae of *Gigasporites*

TABLE 12.2

Reported fossils of lichens and fungi (compiled from Taylor et al., 2015).

Species	Era/location	Remarks
Lichens		
Thuchomyces lichenoides, Diskagma buttoni	Precambrian - South Africa	Septate branched hyphae terminating in conidiophores
Farghera robusta	Cambrian, Australia	Crustose on rocks
Cyanolichenomycites devonicus, Chlorolichenomycites salopensis	Devonian, UK	Charred lichen thalli
Winfrenatia reticulata	Devonian, Rhynie chert	Liverwort/algal photobionts
Daohugouthallus ciliiferus	Triassic, Germany	Highly branched thallus
Honeggeriella complexa	Cretaceous, Canada	Heteromerous foliose thallus
Anzia electra (Lecanorales)	Eocene, Baltic amber	Chlorolichen – heteromerous lobed anastomosing hyphae
Fungi		
Glomus fasciculatum	Triassic, sediments, USA	AB fungi
Glomeromycota	Triassic, Antarctica	Silicified specimen similar to modern *Sclerocystis*
Notophytum krauselii	Triassic, Antarctica	From nodular AB fungi
Gigasporites myriamyces	Triassic, Antarctica	Coiled hypha in cycad fossil *Antarcticycas schopfi*
Glomites cycestris	Triassic	Arbuscular in *A. schopfi*
Metasequoia milleri	Eocene, Canada	Hyphae in root cortex of fossil conifer
Paleopyrenomycites devonicus	Devonian, Rhynie chert, UK	On stems and rhizomes of lycophyte *Asterohylon mackie*
Reduviasporonites	Permian	Phragmospores, similar to phytopathogenic *Rhizoctonia*
Rossellinia congregata	Oligocene	Wood
Onakawananus varitas	Cretaceous sediment, Canada	Ascomycotes associated with conifer mycelia
Diplodites sweetie	Cretaceous	On pre-mineralized angiosperm fruits
Glomorphites intercalaris	Permian, silicified Peat of Antarctica	Spores
Monoporisporites minutus	Cretaceous – Paleocene	Their spores are related to teliospores of the Uredinales
Inaperitisporites	Cretaceous – Eocene	Basidiomycote-like fossils but with conidia
Granodiporites Dyadosporites cannanorensis	Cretaceous Miocene	2 celled spores

Table 12.2 contd. ...

...Table 12.2 contd.

Species	Era/location	Remarks
Hypoxylonites	Cretaceous – recent	Unicellate spores similar to Xylariaceae
Basidiosporites	Paleocene – Eocene	Unicellate spores – from sediments of Arabian Sea
Diporisporites	Miocene – Pleistocene	Unicellate spores
Exesisporites annulatus	Pleistocene	Unicellate spores

myriamyces and arbuscules of *Glomites cycestris* are well preserved since the Triassic in the fossilized cycad *Antarcticycas schopfi* (Table 12.2). Remarkably, the highly branched foliose lichen *Daohugouthallus ciliiferus* of the Jurassic is the most impressive compressed evidence; in it, each branch terminates in one to several wart-like lobes that look like sorelia.

12.4 Divergence and Species Diversity

The timings of divergence are known for all phyla and some classes of Ascomycota since the origin of fungi around 1.2 BYA (T.N. Taylor et al., 2015). For other classes, relevant information was collected from computer searches. Whereas Gueidan et al. (2011) indicated the divergence of Lecanoromycetes from Eurotiomycetes during the late Devonian between 371 MYA and 380 MYA, only hints for the divergence are indicated for Dothideomycetes, Laboulbeniomycetes and Coelomycetes. In fact, the timings could not be traced for Hyphomycetes. To them, species numbers were added to each taxonomic group from Table 1.1. So much so, Fig. 12.2 shows a more or less complete picture on the geological timings of diverging phyla and classes along with species number. The following may be inferred from Fig. 12.2: 1a. Clonality, unicellularity and environmental stability have decelerated species diversity in fungi. 1a. Restriction of propagation by clonal multiplication alone has reduced diversity to 230 and 260 species in Glomeromycota and Neocallimastigomycota. 1b. Unicellularity limits the diversity to 1,309 species in Saccharomycotina and 1c. Stability of aquatic habitat also limits it to 1,840 species in Chytridiomycota/Oomycota. 2. In the labile terrestrial habitats, fungal diversity, as symbionts/biotrophs or saprotrophs, had to depend on plant diversity. On land, the diversity was overtaken that of oceans by 125 MYA. Angiosperms consisting 79% of photosynthetic plants increased terrestrial productivity equal to that in the oceans, covering 71% of the Earth's surface (Costello and Chaudhary, 2017). Thus, the evolution of angiosperms has coincided both in terms of diversity and productivity since 160 MYA. The thin vertical line in Fig. 12.2 shows that *Cretaceous period between*

160 and 125 MY witnessed the explosive diversity of all terrestrial taxonomic groups of fungi along with flowering plants.

From Fig. 12.2, it is also possible to draw the phylogenetic tree for the Kingdom of Fungi. Being the very first attempt, no claim is made for its precision. Accordingly, Neocallimastigomycota seems to have originated from Chytridiomycota. Among the higher fungi, the first seems to be the emergence of Saccharomycotina, from which Zygomycota and Pezizomycotina may have emerged.

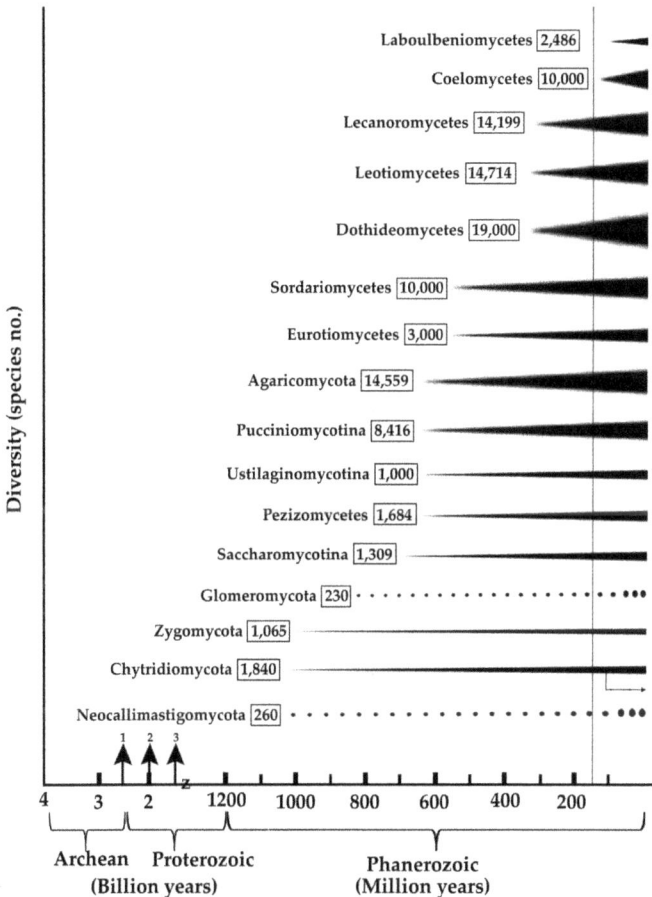

FIGURE 12.2

Divergence in geological times of different phyla/classes of fungi (compiled from Taylor et al., 2015, Gueidan et al., 2011, Poinar, 2016, Hongsanan et al., 2016, Samarakoon et al., 2016, Liu et al., 2018, Lucking and Nelsen, 2018). Vertical arrows bearing 1, 2 and 3 indicate the origin of life, atmospheric oxygenation by cyanobionts and discovery of sex, respectively. Bent arrow indicates the emergence of terrestrial amphibian parasites. Numerals in boxes indicate the species number.

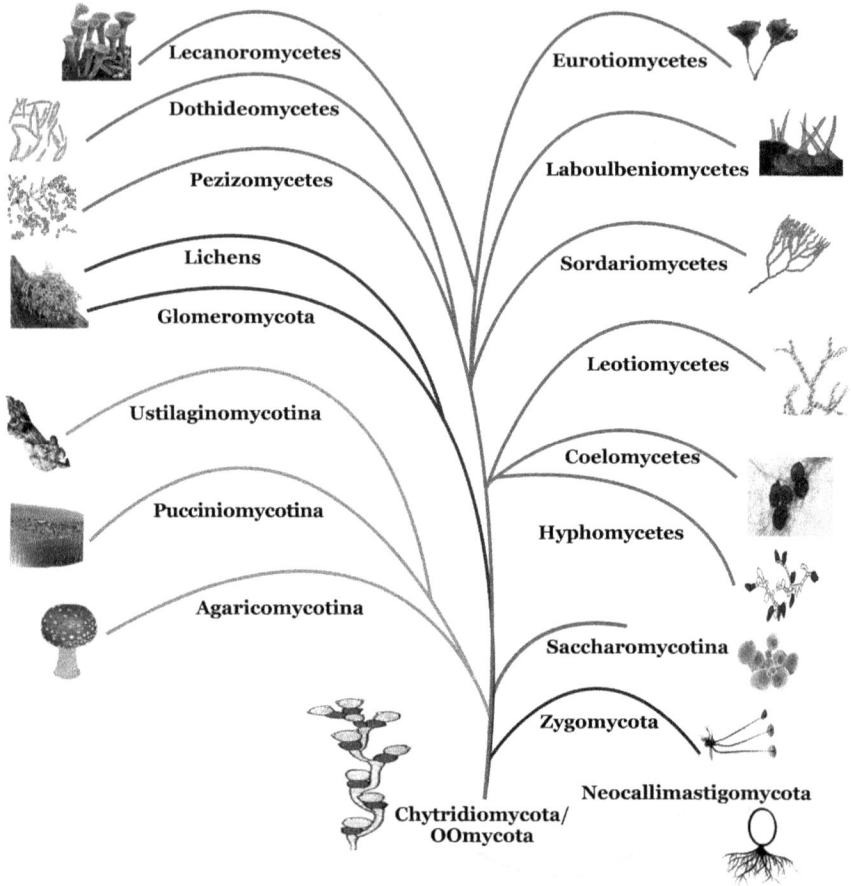

FIGURE 12.3

The proposed course of evolution in Kingdom of Fungi (based on Margulis et al., 2009, Taylor et al., 2015 and others). Figures for different taxa are drawn from the following non copyrighted sources: Chytridiomycota: *Allomyces* (*nathanvoshell.weebly.com*), Zygomycota: *Rhizopus nigricans* (*labotienda.com*), Pucciniomycota: *Puccinia sorghi* (*ipcm.wisc.edu*), Ustilaginomycotina: *Ustilago maydis* (*creativecommons.org*), Agaricomycotina: *Amanita muscaria* (*creativecommons.org*), Glomeromycota, Pezizomycetes: *Cladosporium* (*adelaide.edu.au*), Lecanoromycota: *Cladonia* (*alchetron.com*), Dothideomycetes: *Mycosphaerella* (*alchetron.com*), Eurotiomycetes: *Penicillium* (*microbiologynotes.com*), Laboulbeniomycetes: *Pyrixiophora* (*website.nbm-mnb.ca*), Sordariomycetes: *Neurospora crassa* (E. Kalkman, Wegeningen University), Leotiomycetes: *Monilia* (*atrium.lib.uogeulph.ca*), Coelomycetes: *Phoma* (*inspq.qc.ca*), Hypomycetes: *Pyrigemmula* (Magyar et al., 2011), Saccharomycotina: yeast (rearranged from *genengnews.com*), Lichens: fruticose *Letharia vulpina* (*creativecommons.org*), Neocallimastigomycotina: (modified from Gruninger et al., 2014).

The number for the taxonomic hierarchy from class to species has also been assembled for different taxonomic groups of fungi. For some groups, precise data could not be collected. Benny et al. (2001) reported 10 orders,

18 families and 122 genera for Zygomycota but the species number is not given. The values reported by Maharachchikubura et al. (2016) for family and genus of Sordariomycetes are also not compatible with species number. Considering the reliable values for Ascomycota alone, the Lecanoromycetes and Dothideomycetes seem to be the most diversified group in terms of orders, families, genera and species (Table 12.3). However, the values are 13, 40 and 249 species/genus in Dothideomycetes, Pucciniomycotina and Leotiomycetes, respectively, i.e., the Dothideomycetes have not diversified at species/genus level, as in Leotiomycetes. Similarly, it increases also from 10 to 227 and 545 species/family in Pezizomycetes, Pucciniomycotina and Leotiomycetes, respectively. At the order level, it increases from 280 species in Pezizomycetes to 1,092 in Lecanoromycetes and 2,102 in Leotiomycetes. In all, Leotiomycetes with 14,714 species have passed through the highest diversity at the order, family and genus levels, whereas the most (19,000) speciose Dothideomycetes have gone through the lower diversity at all these levels (Table 12.3).

TABLE 12.3

Number of classes/subclasses, orders, families, genera and species in some fungal taxonomic groups. Values in bracket indicate the number of species/order/family/genus. *eol.org*.

Taxa	Class/subclass	Order	Family	Genus	Species	Reference
Neocallimastigomycota	–	–	–	6 (43)	260	Gruninger et al. (2014)
Glomeromycota	3	5 (46)	14 (16)	29 (8)	230	Oehl et al. (2011)
Ustilaginomycotina	2	10 (170)	26 (65)	115 (15)	1,700	Begerow et al. (2014)
Pucciniomycotina	9	20 (421)	37 (227)	210 (40)	8,416	Aime et al. (2014)
Saccharomycotina	6	7 (187)	17 (77)	93 (14)	1,309	Kurtzman and Sugayama (2015)
Lecanoromycetes	2	13 (1,092)	72 (197)	–	14,199	Gueidan et al. (2015)
Eurotiomycetes	3	7 (428)	24 (125)	282* (11)	3,000	Geiser et al. (2015)
Arthoniomycetes	1	3 (500)	11 (136)	74 (20)	1,500	Schoch and Grube (2015)
Dothideomycetes	2	32 (594)	191 (99)	1,495 (13)	19,000	Pem et al. (2021)
Pezizomycetes	1	6 (280)	10 (168)	41 (41)	1,684	Pfister (2015)
Sordariomycetes	5	16 (625)	–	–	10,000	Eriksson (2005)
Leotiomycetes	–	7 (2,102)	27 (545)	59 (249)	14,714	Zhang and Wang (2015)

13

Present: Desiccation and Dormancy

Introduction

The negative anthropogenic activities have driven many eukaryotes to a point of near extinction (e.g., fungus *Erioderma pedicella*, IUCN Red List). Presently, efforts are being made to conserve them. *In situ* conservation demands legislation and its implementation by governments but it may be of local and temporal importance. The engraved stone edifices of the Indian Emperor Asoka (270-232 BC) reveal that he was the first to ban inland fishing operations during the spawning season. Employing scientific indices, IUCN, an independent organization under the umbrella of UNESCO, has recognized some as red listed endangered fungi (e.g., *Cladonia perforata*, IUCN Red List) and demanded urgent and strict implementation of laws, requiring different measures of conservation. On the other hand, *ex situ* conservation of fungi requires scientific techniques, which may be of global importance. Aside from these, organisms have also developed their own strategies like diapause and dormancy to conserve themselves for a short or longer duration. Whereas animals have developed strategies like diapause and dormancy, plants and fungi have opted for desiccation tolerance and dormancy.

13.1 Desiccation Tolerance

In terrestrial habitats, plants (e.g., Abscisic Acid [ABA] regulate stomatal closing and opening, see Pandian, 2022); aquatic animals including protozoans (e.g., contractile vacuole, see Pandian, 2023) have one or another active mechanism to control and regulate their water content. In contrast, fungi and lichens are poikilohydric eukaryotes, as they do not have active mechanisms to regulate their water content. Consequently, their water content fluctuates widely depending on the availability of water from the substratum, on which they grow. For example, it ranges widely from 52 to 88% in mycelia of *Collybia* sp on leaf litter, from 56 to 66% in its mycelia on twigs and from 80% in the basidiomycote *Phlebia chrysocreas* to 88% in *Mutinus bombasinus* growing on rotten logs (Lodge, 1987). However, both fungi and lichens possess a high capacity to tolerate desiccation, which has sustained their survival and to overcome drastic changes in water content. In fact, some bryophytes, a few ferns and 350 angiosperm species (Gasulla et al., 2021) (but not in gymnosperms, Alpert, 2000) do have this desiccation tolerance ability. Desiccation tolerance is the ability of an organism to equilibrate its internal water potential with that of moderately dry air and resume normal function after rehydration. This requires remarkable tolerance of water deficit. Theoretically, fungi and lichens can survive infinite cycles of desiccation and rehydration, albeit the desiccation duration may be limited by resource availability (Gasulla et al., 2021). The duration is estimated to last for 3 to 8 months (Kappen, 1973).

When dried, fungi and lichens enter a latent stage called anhydrobiosis and reduce metabolism to 50% until rehydration. In the hydrated lichens, photosynthesis is blocked but photobionts may continue to absorb light energy and mycobionts may also continue to absorb water and minerals. Strikingly, the dried thallus of fungi and lichens can tolerate temperature up to 60°C, whereas the hydrated thallus dies even at 35°C. The thallus, especially that of lichens can also withstand extreme low temperatures (–196°C) but can revive, when rehydrated slowly. These abilities have allowed them to conquer stressful terrestrial habits, including deserts and high mountains. The key mechanisms facilitating the desiccation tolerance of fungi especially lichens involve storage and utilization of polyols, heat shock proteins, a powerful antioxidant protection system, thylakoid and oligogalactolipids.

13.2 Spores – Dormancy and Germination

The eukaryotic fungi, algae, plants, protozoa and prokaryotic bacteria form spores, which are hardy and ubiquitous propagules (see Sephton-Clark and Voelz, 2018) that ensure survival, spatial and temporal dispersal through successive generations. In fungi, spore dormancy is a common phenomenon. Dormancy is considered as the incapacity of a viable spore to germinate under unfavorable conditions (Nara, 2009). Tommerup (1983) defined the dormant spore as "the one that fails to germinate, although it is exposed to physical and chemical conditions that support germination and hyphal growth of apparently identical but non-dormant spores of the same species".

Understandably, the fungi produce two types of spores namely subitaneous and dormant spores. For example, the arbuscular (AB) fungus *Gigaspora gigantea* produce mostly dormant spores during the late summer and autumn that can germinate with the onset of spring in the next year (Juge et al., 2002). Deacon (2005) noted that the oospores of *Phytophthora* spp require a 'maturation' period before they can germinate. During this period, the thick (2.0 μm diameter) wall is progressively thinned (0.5 μm) by digestion of the inner layers of the wall. By exposing the spores of AB mycorrhizal fungus to 4°C for a period of 0, 7, 14, 90 or 120 d, Juge et al. (2002) found that the minimum maturation duration for the spores of *Glomus intraradices* is 14 d prior to successful germination at 25°C. Tommerup (1983) reported that this period can be longer, i.e., 6 wk for *G. caleodonium*, 12 wk for *Gigaspora calospora*, 6 mo for *Acaulospora visleavis* and more specifically, 5 mo for *Entomophaga maimaiga* (Hajek et al., 2001). Teliospores of *Urocystis cepulae* and resting sporangia of *Synchytrium endobioticum* are thought to survive over 25 years in soil (e.g., Putnam and Sindermann, 1994).

It is a common laboratory practice to store freshly isolated fungal spores at 4°C prior to estimation of their dormancy duration and germination success (see Juge et al., 2002). However, many authors have not reported whether the storage at 4°C was under complete darkness or illuminated conditions. For, it is the burial or exposure of spores to complete darkness that prolongs the dormancy duration and subsequent successful germination. Spores, seeds and even fruit tissues of plants exposed to complete darkness in caves, within deep sediments or permafrost survived over hundreds and thousands of years and facilitated their revival (see Pandian, 2022). Still, due to the metabolic expense of resources during dormancy, only a fraction of these spores could be revived. For example, germination success of the ectomycorrhiza (EcM) *Rhizopogon roseolus* spores stored in water at 4°C dropped from 89 to 2% (Torres and Honrubia, 1994). Ishida et al. (2008) demonstrated a significant decrease in spore germination of *Laccaria, Inocybe, Hebeloma* and *Russula* EcM after one year of preservation. Of 10 EcM tested, only 50% spores of *Rhizopogon, Laccaria* and *Inocybe* were germinated after a period of 1 y preservation but those of *Pisolithus* and *Cortinarius* failed to germinate (Nara, 2009). Notably, the available data on decreasing viability and germination success are all related to EcM. It is not known whether the spores of other taxonomic fungal groups also suffer from it.

Bruns et al. (2009) brought an answer to the observed decreasing germination success of mycorrhizas with extended duration of dormancy. Some 400–500 MYA, the mycorrhizas emerged on land, which was already well colonized by bryophytes and vascular plants. Unable to compete with the bryophytes and vascular plants, the mycorrhizas opted to act as symbionts in them (see Gasulla et al., 2021). The chances for the basidiospores of *Rhizopogon* EcM in the pine forest decrease due to (i) the loss of individual spore viability with age and (ii) reduction in total number of spores (per unit area) through fungivory in soil. Therefore, Bruns et al. buried 2.5×10^8 spores of *Rhizopogon* in retrievable containers to inoculate different plots in selected areas of the pine forest. They found that (1) with increasing spore concentration, all the four species namely *R. occidentalis*, *R. olivaceotinctus*, *R. salebrosus* and *R. vulgaris* broke dormancy at a statistically indistinguishable rate, (2) a minimum of 1–8% of the spores were initially receptive to pine roots, (3) the proportion of seedling colonizing the pine root system increased in all 4 species as a function of spore density and (4) the minimum number of spores required to colonize 50% pines decreased from 11 to 70-fold during the fourth year of dormancy than that at the first year of dormancy. These findings led Bruns et al. to suggest that the true longevity of the spores in their banks can be 99 years. For, *R. occidentalis* and *R. olivaceotinctus* are common in spore banks of the same region, despite their rarity as mycorrhizas or fruiting bodies in forests.

Feofilava et al. (2012) distinguished the deep, constitutive endogenous dormancy of sexual spores (zygotes of zygomycotes, basidio- and asco-spores) from the exogenous dormancy of clonal spores. (i) Highly viscous cytoplasm, due to accumulation of trehalose and (ii) moderate dehydration are characteristics of endogenous dormant spores. In contrast, exogenously dormant spores (a) retain up to 60% water (in comparison to 80% in growing vegetative cells) and (b) are relatively less viscous. Whereas cyst metabolism depresses one by 500,000 times in *Artemia* embryo (see Pandian, 1994), that in the hypometabolism in dormant fungal spores does not exceed 50%. The fungal spore walls are fortified by amino polysaccharides, glycans, sporopollenin, mucoran and others. As a cryo-protector, arabit serves in the agaricale *Lentinula edodes*, but mannite in ascomycetes. Almost all fungal spores possess a biochemical mechanism associated with special compounds called autoinhibition that prevents germination. For example, the autoinhibitory gloiospores in *Colletotrichum gleosporioides* is the most studied and is also synthesized. Germination in *Penicillium paneum* is controlled by the volatile inhibitor 1-octen-3-ol. Rust fungi have their own specific autoinhibitor *cis*-ferulic acid methyl ester (Feofilova et al., 2012).

According to Feofilova et al. (2012), plant hormones like gibberellins, abscisic acid and fusicoccin can also be obtained from *Fusarium moniliforme*, *F. amygdally* and others. More interestingly, lycopene, a pigment useful in containing cancer, cardiovascular disease, asthma and the like, which was earlier produced from tomatoes, is now obtained from mucoraceous fungi at a much cheaper cost. For more information on the use of fungal spore and of biotechnological application, Feofilova et al. (2012) may be consulted.

There are international agencies to collect, preserve and supply breeding lines of agriculturally important plants like paddy (International Institute of Rice Research, Manila). Hitherto, no effort has been made to collect and preserve useful breeding lines of beneficial fungi. It is likely that some international companies possess very useful breeding lines of economically important fungi. But that may lead to a sort of gene imperialism. Institutions like Food and Agriculture Organization (FAO) should come forward to establish an international institute to collect, preserve and supply breeding lines of beneficial fungi.

13.3 Long Lasting Dormancy

Available evidences have shown that the dormancy duration of fungal spores can be 4 y or up to 99 y, when stored in the spore bank (see Bruns et al., 2009). However, the duration can be longer than hundreds and thousands of years, when the dormant spores/vegetative mycelia are buried deep amidst museum specimens or within the deep marine sediments at, say, 4°C or deep within the permafrost at subzero temperatures. The limited information available on this aspect calls for more research in this area.

Desert fungal spores: A few years ago, Faculty of Science, University of Copenhagen collected > 200 spores, whose age varied from 2 to 250 years. From them, spores of the desert coprinus fungus *Podaxis* was germinated in the Department of Biology by Benjamin Schantz-Conolon. This unique achievement demonstrated that spores of the desert fungus *Podaxis* remain dormant for centuries before germinating, when conditions allow.

Sedimented fungal spores: Buried within the deep marine sediments below 1,000 m depth, the fungi and their spores are exposed to complete darkness in aquatic medium at +4°C. Roth et al. (1964) isolated marine fungi from a depth of 4,450 m. Subsequently, researchers have found evidence of fungi thriving at greater depths in 'nutrient-starved' sediments of > 100 million years old. This discovery has the potential to turn the sediments into gold for biologists and pharmaceutical companies seeking antibiotics to combat the growing problem of drug-resistant bacteria (Monastersky, 2012). For the revival of sedimented fungal spore, the first report was published by Dr. Chandralata Raghukumar, National Institute of Oceanography, Goa, India. From the calcareous sediments of Bay of Bengal at a depth of 965 m, her team isolated spores of *Aspergillus ustus* and successfully germinated them under simulated deep-sea conditions (Raghukumar et al., 1992). Employing polyclonal antibodies, a calcofluor and fluorescent optical brightner, they isolated conidial spores of *A. terreus* from sediments collected at 5,700 m depth from Chago Trench off Sri Lanka in the Indian Ocean. As a climax of their persistent endeavor, they successfully germinated – under the simulated deep-sea conditions – the 180,000* years old spores of *A. terreus* (Raghukumar et al., 2004).

Permafrost fungi: The solid icy permafrost is located below the seasonally thawing layer (Graham et al., 2012). Indigenous fungi are likely to penetrate permafrost layers either from the surface or from other layers due to potential vertical movement of water. Once trapped in permafrost, the vegetative fungal mycelia – as in amber – remain 'amberized'. Admirable efforts have been made to collect these 'ambers' from permafrost at different depths; the age of these amberized fungi are also dated. To enable the identification, efforts have also been made to identify them through culture and germination, and/or molecular techniques using DNA sequencing, cloning and so on. Kochkina et al. (2012) published rare information on the filamentous fungi in 36 samples from the Antarctic permafrost sediments. The following are reported from this publication: 1. The fungal population density is rather low. It ranges from two in *Cladosporium herbarum*/g permafrost at ~ 18 cm depth to 91 *Aureobasidium pullulans*/g at ~ 118 cm depth in Russkaya Station. In general, the density increased with expanded depth of permafrost sediment (Table 13.1). 2. Nevertheless, there was an appreciable level of diversity, i.e., 26 fungal genera in 36 samples. 3. *Penicillium* was the most common genus and was represented by 19 species. Next to it, *Cladosporium* had 4 species. However, the other 24 genera were represented by a single species alone. 4. The number of samples, from which a single species was collected, ranges from 1 in *P. olsonii* to 2 in *P. citrinum*, 3 in *P. expansum*, 4 in *P. variabilis* and 8 in *P. chrysogenum*.

Table 13.1 lists the viable mycelial fungi collected at different depths of permafrost sediment from six stations of the Antarctic. Remarkably, the age of fungal taxa increased from 7,485 y at Bellingshausen to 23,705 y at Banger Oasis and to > 50,000 y at Novolazarevskaya. Strikingly, the temperature was also depressed from –1.0°C at Bellingshausen to –8.3° and –10.8°C at Novolazarevskaya and Russkaya Stations, respectively. Hence, *the permafrost temperature plays a key role in determining the (i) dormant duration and (ii) viability of the fungal mycelia. However, no fungal mycelium could be collected from the > 3 million years old permafrost of the Beacon Valley at –22.5°C. Clearly, permafrost can preserve fungi like P. expansum and Mycelia sterilia only at –8.3°C between the depths of 60 and 70 cm, as in Novolazarevskaya, but at –1.0°C at 970 cm depth, as in Bellingshausen, where the fungal diversity increases to 3 and 4 species at 325 and 620 cm depth, respectively.*

---

* wrongly stated as 1.8 million y in Pandian (2021b, 2022).

TABLE 13.1

Viable mycelial fungi in Antarctic permafrost sediments (condensed from Kochkina et al., 2012).

Station	Age (yr)	Temperature (°C)	Depth (cm)	Density (no./g)	Fungal taxa
Bellingshausen 62°S, 59°W, 15 m altitude	7485	−1.0	325	18	*Aspergillus sydowii, Chryosporium europae, Penicillium chrysogenum*
			620	22	*Cladosporium sphaeromum, P. chrysogenum, P. variabilis, P. walksonanii*
			970	53	*P. chrysogenum, P. olsonii*
Progress 69°S, 76°W 96 m altitude	–	−8.5	235	4	*P. chrysogenum, Cryonyces sp, Coelomycetes*
Banger Oasis 66°S, 100°E 7 m altitude	23,705	−7.8	120	11	*Cl. herbarum, Cl. antarcticum, Chaetophoma, Coelomocytes, Phoma eupyrena*
			323	6	*Cl. cladosporioides*
Russkaya 74°S, 136°W 64 m altitude	–	−10.8	118	91	*Gilocladium sp, Aureobasidium pullulans, Exophiala sp, Ascochyta sp*
			23	5	*Cl. cladosporioides*
			18	2	*Cl. herbarum*
Novolazare-vskaya 70°S, 11°E	More than 50,000	−8.3	65	17	*P. expansum,* Mycelia sterilia
Beacon Valley 77°S, 160°E 1,000 m altitude	More than 3 million	−22.5	390–600	10	None

Information is also available for the density of fungi from permafrost sediments of both the Arctic and Antarctic. Figure 13.1 shows a comparative trend reported for fungal density as a function of permafrost depth. Firstly, the fungal density is several times greater in the Arctic than in the Antarctic. Secondly, Antarctic permafrost is unable to preserve fungal mycelia beyond 1.7 m depth, while that of the Arctic can do it up to 6 m depth. Reasons for this important discovery remain to be known.

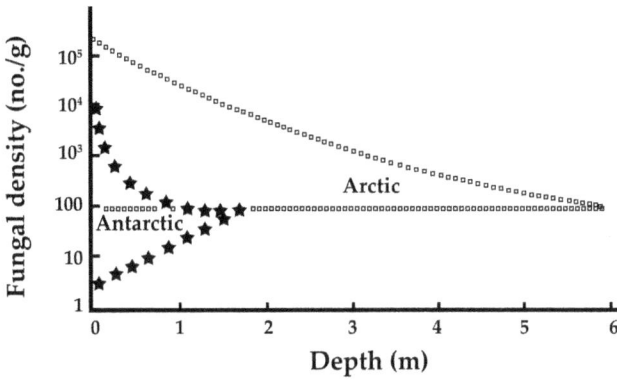

FIGURE 13.1

Fungal density as a function of depth in the Antarctic and Arctic permafrost sediments (modified and redrawn from Kochkina et al., 2012).

Notably, fungi have left intact spores and mycelia for revival and germination after long lasting dormancy, whereas animals and plants have left only their cysts and spores/seeds for revival and germination after a long dormancy.

14

Future: Climate Change

Introduction

Due to anthropogenic activity, the Earth and its organisms are being more frequently threatened by bacterial epidemics and viral (e.g., COVID-19) pandemics, the serotonin-induced swarming of desert locusts and environmental factors like violent earthquakes, tsunamis, cyclones, storms and pollutants. But they are transient localized episodes. There is no historical evidence to show that any of them either singly or in combination have wiped out an eukaryotic species, *albeit* they may drastically reduce population size of the affected taxa. Contrastingly, climate change is a long-lasting phenomenon that covers the entire Earth. Not surprisingly, researches on the unprecedented increase in atmospheric carbon dioxide (CO_2) concentration and consequent global warming, ocean acidification and rise in sea level during recent years have become the hottest research area. One consequence is the shift of organisms towards the poles. An estimate for the shift of non-motile plants along with mycorrhizal fungi is in the range of 65 cm/y (Kelly and Goulden, 2008), which may be compared with 0.6 km/y for motile animals (Parmesan and Yohe, 2003). In the aquatic system, due to increase in surface water temperature at $0.004°C/y$, fish like *Sardinella longiceps* migrated northwardly from the coast of Kerala at rate of 88 km/y (Vivekanandan, 2011). This chapter intends to present an emerging scenario in the context of species diversity.

14.1 Air – Water Interaction

The Earth is surrounded by the atmosphere consisting of ~ 78% nitrogen, ~ 21% oxygen, 0.03% carbon dioxide (CO_2) and others. With the advent of the industrial era in 1750 and the accompanied ever increasing energy extraction from fossil fuels has increased greenhouse gas (GHG) emission and concentrated the atmospheric GHG. In turn, the concentration has led to global warming and ocean acidification, which are collectively known as climate change. Global warming has also begun to melt the polar caps, which leads to the rise in sea level and submergence of coastal areas in seawater. Since 1750, the level of atmospheric CO_2 has risen from 280 ppm to 410 ppm in 2019 and is predicted to go up to 550 ppm by 2050 (Table 14.1). During the last 250 years, the levels of other greenhouse gases have also increased from 715 ppb to 1,866 ppb for methane and from 270 ppb to 332 ppb for nitrous oxide (IPCC, 2021). As a consequence, the global mean temperature increased at the rate 0.2°C/decade over the last 30 years. Most of the added energy is absorbed by waters of the oceans (up to 700 m depth), where temperature increased by ~ 0.6°C over the last 100 y and is continuing to increase (see Pandian, 2015). The available database (Emergency Events Database: *emdat.be*) indicates that the frequency of events like drought, storm and flood increased from ~ 100/y during 1990 to > 200 in 2016; of them, the frequency of drought alone increased from ~ 50 time/y in 1990s to > 120 time/y in 2016.

TABLE 14.1

Changes in climate features during the last 10–30 years and predicted changes by 2050 (from Pandian, 2015), *by 2080s.

Climate features	Last 10–30 years	By 2050
Atmospheric CO_2 (ppm)	385	550
pH of oceanic waters (unit)	–0.1	–0.1 to –0.3
Sea surface temperature (°C)	+0.4	+1.5
Coral bleaching (time/y)	+2	+15–25
Sea level rise (mm/y)	1	8*
Hypoxic aquatic system (no.)	400	680
Wind speed %/1°C increase	3.5 %/1°C	Increases

Besides absorbing atmospheric temperature, oceans also absorb CO_2, as it combines with water chemically. Covering 70% of the Earth's surface and holding 97% of its water, they serve as a buffer to CO_2 concentration. Consequently, the daily uptake of atmospheric CO_2 by the oceans is 22 million metric ton (mmt). Since the advent of the industrial era, oceans have absorbed 127 billion metric ton (bmt) carbon as CO_2 from atmosphere. The CO_2 absorbed by the oceans range between 25 and 40%, i.e., a third of atmospheric carbon emission. Without this 'ocean sink', the atmospheric CO_2 concentration would have by now increased to 450 ppm and a consequent increase in temperature on land.

Hydrolysis of CO_2 increases the hydrogen ion (H^+) concentration with concomitant reduction in pH and carbonate (CO_3^{-2}) concentration. This process of reducing sea water pH and concentration of carbonate ion is called 'ocean acidification'. Consequent to the acidification process, the mean pH level of the world's oceans has declined by –0.1 unit and –0.1 to –0.3 unit reduction is expected by 2050. The decrease in sea water pH and carbonate ion concentration is one of the most persuasive environmental changes in the oceans and poses one of the most threatening challenges to marine organisms. The progressive reduction in the availability of carbonate ion (CO_3^{-2}) renders the acquisition of biogenic calcium carbonate ($CaCO_3$) by calcifying organisms energetically costlier, but may not totally inhibit the acquisition. In fact, the reduction in pH is more critical for the calcifying poikilothermic organisms than the increase in sea water temperature (see Pandian, 2015).

With climate change and the consequent melting of polar caps, rise in sea level threatens to engulf most of our coastal cities. It is in this context, studies on the estimation of sea level rise have become important. Interestingly, some protozoans serve as sensitive indicators of the sea level rise. Foraminifers and testate amoebae are the key constituent of microfossils and are useful in reconstruction of data on past sea levels, especially in the Holocene since the 1880s. These microfossils are valuable sea level indicators, as their modern counterparts are distributed in narrowly defined niches in the intertidal zone at specific level. The sea level raise can be estimated, when an indicator fossil species is dated and its occurrence level (height) is measured and compared with those of its living counterpart.

14.2 pH and Tolerance

Prior to describing the biokinetic pH range of fungi, a preamble is required. A fact known but not adequately recognized is that along with cycling of water, the pH also undergoes a similar cycle from an alkaline state in waters to an acidic gaseous state following evaporation and returning to the alkaline state. The acidic rain water returns to the alkaline state, when it directly reaches the soil or gradually passes from the acidic to alkaline state as water flows through streams and rivers. For example, the pH of the Kodayar River (8°3'N) in Tamil Nadu is characterized by pH 6.7 (Murugavel and Pandian, 2000). In fact, many montane lakes are acidic (e.g., Kodaikanal Lake at an altitude of 2,285 m in Tamil Nadu). These acidic montane waters are known for low diversity (5 phytoplankton species), abundance (159 indi/l) and productivity (0.11 g $C/m^3/d$), in comparison to the alkaline ground water (> 14/l, 412/l, 1.97 g $C/m^3/l$) (see Fig. 15.2 B_2). Conversely, species diversity, their abundance and community composition of fungi remain almost equal in both acidic and alkaline waters in the aquatic system as well as soil.

Available information on soil pH indicates that typically, most fungi taxa thrive over a wide pH range from 4.0 to 8.3, in which their density and diversity remain unaffected. To demonstrate it, Rousk et al. (2010) made a comparative study in fungal and bacterial communities across a pH gradient in an aerobic soil. Their study for fungal communities covers a total of 4,700 classifiable sequences among 16 soil samples used for 290 fungal sequence/soil sample. The fungal copy number remained at about 2×10^3 at pH range from 4.0 to 8.3 (Fig. 14.1A). The number of Operational Taxonomic Unit (OTU)/ sample also showed a similar trend for Fungal OTUs – pH relationship, albeit a weak, non-significant increase with increasing pH (Fig. 14.1B). The composition of fungal communities was also related to soil pH, although the relationship was weaker. Mantel tests corroborated this pattern, indicating a significant but a weaker relationship between community composition and soil pH (Fig. 14.1C). When the fungal communities were examined at more detailed levels of taxonomic resolution, different patterns of relationship became apparent. Accordingly, the relative abundance of the ascomycete group Hypocreales increased from zero at pH 5.0 to 0.05 level at pH 8.3 (Fig. 14.1D). Conversely, the abundance decreased from pH 5.0 to 8.5 for another ascomycete group Helotiales (Fig. 14.1E) and basidiomycote group (Fig. 14.1F). In all, almost all fungal groups grow and flourish between pH 4.0 and 8.3, and their abundance, diversity and community composition remain almost unaffected within this wide range of pH.

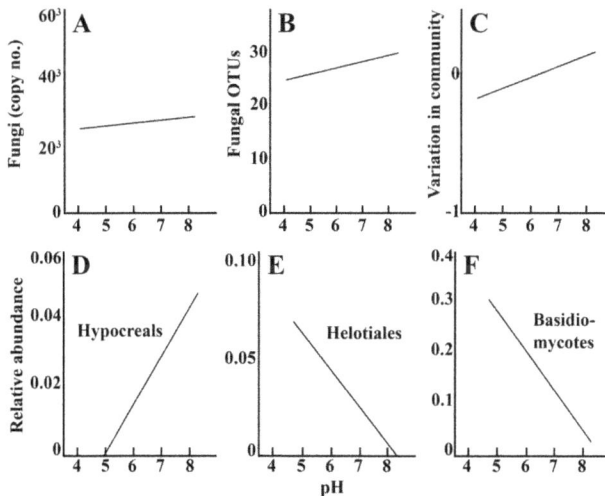

FIGURE 14.1

Relation between pH and (A) fungal sequence copy number, (B) fungal OTUs (diversity) and (C) variation in fungal community composition. Relative abundance as function of pH in (D) Hypocreals, (E) Helotiales and (F) Basidiomycotes (compiled, modified and simplified from Rousk et al., 2010).

14.3 Dispersal and Distribution

In terrestrial habitats, > 96% fungi are propagated by their clonal and/ or sexual spores through wind over the air or animal vectors. Hence, the mechanisms of spore dispersal may play a crucial role on recruitment and distribution of fungi. As rust fungi generate the maximum number of five spore types, this account has chosen to consider their dispersal. From information summarized in Table 14.2, the following may be generalized: 1. Spermatia, aeciospores and teliospores are large and thick walled. Hence, they are resistant to desiccation and UV radiation. However, their dispersal range is limited to < 10 m distance. Their chances of being affected by climate change is remote. Spermatia are an exception to it. They are dispersed by insects and other animals, which, especially dipterans and birds, can disperse them to longer distances. They are likely to suffer due to warming and consequent latitudinal and altitudinal shift of their vectors. 2. Basidiospores and urediniospores are small (see Fig. 9.4). Survival of the former depends on temperature and consequent desiccation. Despite their small size, the urediniospores are thick walled and ornamented to reduce settling speed. They are capable of dispersing over longer distances at higher altitudes (see Table 9.4).

TABLE 14.2

Size, structure, desiccation and UV resisting character and dispersing ability of rust fungal spores (drafted from Helfer, 2014).

Spore type	Remarks
Basidiospore	Small (10–20 µm in diameter) as the size of a water droplet. High humidity and narrow temperature range required for survival. Dispersal range is ~ 300 m (Kinloch, 2003).
Spermatium	Dikaryotized spores produced from spermogonia in the aecial host plant. Small hyaline and vulnerable to desiccation and UV radiation damage. Dispersed by mostly insect vectors (Roy, 1994). Diptera (11 families, Naef et al., 2002) and Hymenoptera are the main vectors. With latitudinal and altitudinal shift of vectors, the outcrossing mechanism may suffer.
Aeciospore	Produced from aecial host plants. Larger (> 25 µm), often thick walled and more resistant to environmental changes. These features limit dispersal to < 10 m distance (Barnes et al., 2005), although viable aeciospores are intercepted at 300–2,000 m altitude, suggesting their ability to disperse over 720 km (e.g., *Conaritum ribicola*).
Urediniospore	Produced in vast numbers, small (15–30 µm) but often ornamented with a thick wall and fine spines (to reduce settling speed) promote long distance travel by air (Wang et al., 2010), e.g., soybean rust dispersed from Asia to Africa and finally to South and North Americas (Isard et al., 2005).
Teliospore	Largest, long-living, thick walled, dark colored spores. Resistant to desiccation and UV radiation (Brown, 1997).

14.4 Mycorrhizal Fungi

Mycorrhizal fungi: Being symbionts in 89% terrestrial plants, mycorrhizal fungi, i.e., arbuscular (AM), ecto- (EcM) and ericoid- (ErM) mycorrhizas play key roles in plant distribution and productivity as well as dispersal and community interactions. Table 14.3 summarizes some traits of AB, EcM and ErM mycorrhizas. To elevated atmospheric eCO_2 concentration, warming and precipitation shift, the responses of mycorrhizas have been studied intensively. Not surprisingly, there are overwhelming, observational and experimental evidences to show that the mycorrhizal fungi impact an array of traits like (i) plant productivity, distribution and community composition (e.g., Jiang et al., 2017), (ii) decomposition and soil nutrient recycling (e.g., Langley et al., 2006), (iii) soil microbial community composition (e.g., Kyaschenko et al., 2017) and (iv) soil carbon stability (e.g., Jackson et al., 2017). The traits listed in Table 14.3 are responsive to climate change. For example, mycorrhizas vary in melanin content (Wright et al., 1996), which has been linked to water availability (Deveautour et al., 2019). Increase in melanin content may reduce carbon storage (e.g., Clemmensen et al., 2015). Variations in helicase activity can promote cold tolerance (e.g., Branco et al., 2017).

TABLE 14.3

Varying traits among the three mycorrhizal fungi (modified from Bennet and Classen, 2020).

Trait	AM fungi	EcM fungi	ErM fungi
Distribution (%)	~ 100	~ 30	~ 5
Primary nutrient delivered	Phosphorus	Nitrogen	Nitrogen
External hyphal exploration type	Species-specific variations and distance	Short, medium, long-distance	Unknown
Melanization	Spores, hyphae	Spores, hyphae	Hyphae
Helicase activity	Yes	Yes	Unknown
Host abiotic tolerance promoted	Yes, greatly	Yes	Unknown

Bennett and Classen (2020) conducted four Web of Science literature searches that yielded 458 relevant publications, of which 271 were related to warming, 38 to precipitation and 27 to extremes. After deleting duplicate publications, they processed 150 publications; meta-analysis of these more relevant publications led them to reach the following major conclusions: 1. Over 92% of these publications were related to the northern hemisphere, i.e., the USA and Europe. Indirectly, they call for studies from tropical countries like India, where the impact of climate change will be more intense. 2. Of eCO_2, temperature and precipitation shift, temperature and rainfall variabilities have greater but variable impacts on mycorrhizas than that by eCO_2. 3. The three mycorrhizal types, namely AM, EcM and ErM vary in their responses to climate change.

Host plants, that encounter the negative effects of climate change, can either adapt or disperse to more suitable habitats or go extinct. The buffering effects of mycorrhizas provide the required time for their host plants to progressively adapt to climate change or disperse to more suitable habitats and thereby avoid extinction. Mycorrhizas can increase host plant tolerance to other abiotic stresses associated with temperature and precipitation shift, which should decrease plant extinction risk. For example, AM fungi may alleviate the negative effects of variations in precipitation (Mohan et al., 2014) through either improved water flow (Delavaux et al., 2017) or increase water use efficiency (Andreo-Jimenez et al., 2015). Mycorrhizas can also produce protective compounds inducing the synthesis of temperature tolerant substances like trehalose (e.g., Lenoir et al., 2016) in the host plants. Experimental studies on the field and greenhouses have revealed variable responses for the impact of increased eCO_2 on mycorrhizas. This is also true for flowering and seeding by plants (see Pandian, 2022). Manipulation studies by Bennett and Classen (2020) have shown that 25 and 40% positive influences against the increased eCO_2 in the field and greenhouses, respectively. This sort of variable responses is also true for change in Relative Humidity (RH). They found an overall relatively stronger positive effect of temperature on colonization of AM fungi. AM fungi facilitate plant dispersal (Reinhart and Callaway, 2006, van Grunsven et al., 2014). However, EcM fungi limit plant dispersal (e.g., Nunez et al., 2009).

14.5 Rusts – Pathogenic Fungi

Rust fungi are constituted by 7,800 species in 166 genera of Pucciniales and 14 families of Urediniales. They are obligate biotrophs and exclusively plant pathogenic fungi on vascular plants. The heteroecious life cycle of Pucciniales involves two exclusive and unrelated host plant taxa to complete their life cycle. These cycles can comprise up to five spore stages; the spermatia and telia of many Pucciniales are incapable of infecting the alternative host. This sort of pronounced host specificity is therefore a limiting factor, especially considering the dependence of heteroecious rusts on (i) the presence of two unrelated and susceptible host species within spatial proximity, (ii) the successful spore dispersal within an appropriate time and (iii) the prevalence of favorable environmental conditions during both infective periods.

Notwithstanding these ecological constraints, they are globally distributed and cause > 25% losses on harvest of commercial crops (Table 3.6). With a combination of short generation times spanning from weeks to months and a vast number of propagules, the adaptive evolutionary potential of rust is relatively great, when faced with rapid environmental change (Pringle and Taylor, 2002). Many of their epidemics are attributable to unintentional introductions through the transfer of infected plants by man rather than to extension of the range due to climate change. The unintentional introduction of infected chestnut *Cryphonectria parasitica* (Ringling and Prospero, 2018, see also Table 9.4) into a hitherto disease-free chestnut range is a well-known example. Yet, agrochemical drift and elevated concentrations of eCO_2 and ozone (O_3) have enormously changed the chemistry of water, air and soil (Helfer, 2014). Whether these climatic changes have increased the frequency and range of rusts in their host plants are explored here.

TABLE 14.4

Impact of climate change on rust fungi (modified from Helfer, 2014).

Climate change	Potential change	Affected rust taxa
Climate warming	Increased survival of pathogens	*Puccinia graminis* on cereals and grasses. *Phakospora pachyrhizi* on soybean crop
	Restricted survival of pathogens	*Pu. striiformis* on cereals. *Uropyxis petalostemonis* on white clover
Higher humidity	Increased disease incidence	*Hemileia vastatrix* on coffee
Extreme weather	Enhanced air borne spore dispersal	Many species
	Water dispersal	*Chrysomyxa weirii*
Spread of alien plants or rusts	Increased geographical range	*Pu. hieracia, Phragmidium ivesiae*
	Increased host range	*Pu. psidii, Cronartium ribicola*
	Genetic recombination and somatic hybridization	*Melampsora, Cronartium* and *Puccinia*
Increased CO_2	Increased/restricted pathogen fitness	*Graminicolous* rusts
	Unchanged pathogen fitness	Poplar rusts
Increased O_3	Increased pathogen fitness	Poplar rusts
	Decreased rust fitness	Wheat rusts

Firstly, the reported threats to plant health have been increasing; the number of new disease outbreaks has increased from 17/y between 1961 and 1990 to 76/y between 1991 and 2002, i.e., the outbreaks have increased by nine times mostly due to climate change (Fletcher et al., 2010, Black, 2013). Secondly, dramatic host range expansion has been reported for alfalfa rust

Uromyces striatus (Skinner and Stuteville, 1995) and in the rust of South American Myrtaceae *Puccinia psidii* (e.g., Kriticos et al., 2013). In *U. striatus*, 141 out of 345 taxa from 11 genera on 27 legume host plants (i.e., 41%) were susceptible to single uredinial isolates (clones) of the rust on first encounter, albeit these plants have not co-evolved with the pathogen. Similarly, 129 species in 33 genera of Myrtacea in Australia became susceptible to the Brazilian *P. psidii*. In Brazil, the host suffer mildly but in Australia more aggressively (Coutinho et al., 1998). Thereby, relatively harmless endophytes or mildly pathogenic fungi have become aggressive pests, when host plants suffer from other biotic and/or abiotic stresses of climate change (Manter et al., 2005). More examples are listed in Table 14.4 for the impact of climate change in pathogenic fungi.

14.6 Mycotoxic Fungi

On the basis of available information, eCO$_2$ concentration is predicted to increase from 385 ppm during the last 30 years to 550 ppm by 2050. Consequently, global temperature has risen at the rate of 0.02°C/y during the last 30 years (Table 14.1) and is likely to increase ~ 2°C by 2050 (see Medina et al., 2015). Research on mycogenic fungi and their contamination due to climate change is a nascent area with very limited numbers of publications. While fungi are tolerant to elevated eCO$_2$ alone, they become less tolerant, when other environmental factors like rise in temperature and/or drought are combined (Magan and Aldred, 2007). To examine the effects of elevated eCO$_2$, temperature and/or drought on mycogenic fungi and mycotoxin contamination, the impacts of three factors may have to be considered together. For example, *Aspergillus flavus* grow and produce much faster aflatoxin B$_1$ under elevated temperature and drought conditions. Due to prolonged hot and dry conditions, myxotoxin contamination increases to 69% in Serbian maize (Kos et al., 2013). The mycogenic fungi *Fusarium graminearum* and *F. verticillioides* are reported to grow much faster, when exposed to different combinations of temperature increase from 25 to 35°C at different levels (350 and 1,000 ppm) of CO$_2$ (Medina et al., 2015, see also Vaughan et al., 2014). Hence, researches on climate change on mycogenic fungi shall increasingly become more and more important in the years to come.

14.7 Green Shoots and New Hopes

Ocean acidification leading to reduction in pH is critical to sustain normal functioning of organisms, especially for calcifying poikilothermic ones than elevation in temperature. No fungi are known to fortify its body by calcification. In the aeroplankton, the majority of fungal spores are exposed to acidic moistures. The pH range for tolerance (4.0–8.3) and optima (5.0–8.3) are wider than those for plants and animals. For example, those for fishes range from 4.0 for the mudminnow *Umbra pymaea* in the acidic waters of the Netherlands to 9.5 for the endemic carp *Chalcaburnus tarchi* in Lake Van, Turkey (see Pandian, 2015). Hence, it is unlikely that fungi, at any stage of their life in aquatic or terrestrial habitat, may suffer on exposure to acidic pH.

Fungi grow and thrive, where water and carbon are available. Most of them grow between 10° and 40°C with optima between 25° and 35°C. However, *Chaetomium* sp grow optimally at 50°C and survive even at 60°C (see Evert and Eichorn, 2013). Culture studies have revealed that the optimum for external release of digestive enzymes is 40°C for *Penicillium citrinum* (Chrzanowsa et al., 1995), 55°C for *Aspergillus oryzae* (Vishwanathan et al., 2009) and 60°C for *Trichoderma rosei* (Savitha et al., 2011). In the prolonged dry Arabian Deserts characterized by 55°C and 10 cm precipitation/y, 302 fungal species exist, of which *Aspergillus* comprises 32 species, *Alternaria* 10 species, *Chaetomium* 122 species, *Cladosporium* 18 species, *Fusarium* 27 species, *Glomus* 23 species, *Penicillium* 93 species and others including arbuscular mycorrhizal and lichenized fungi (Ameen et al. 2021a, b). These are new hopes and green shoots. Hence, with the appearance of more of them, the mother Earth shall continue to witness burgeoning life forms and species diversity, hopefully inclusive of man.

15

Eukaryotes: Species Diversity

Introduction

The eukaryotes include four kingdoms: (i) Protista (32,950 species), (ii) Fungi (106,761), (iii) Plantae (374,000) and (iv) Metazoa (1,543,196). In them, *species number increases with increasing structural complexity*. In prokaryotes, comprising bacteria and others (Table 15.1), the ubiquitous clonality limits the number to ~ 10,000 species. During the checkered history of evolution, the eukaryotes were successful and witnessed the maximal diversity. Predictions for the existing number of species in the Earth range from 3 to 100 million. During the last 300 y, 2.1 million eukaryotic species were described. Mora et al. (2011) estimated that to describe the remaining 7.5 million species, it may require 303,000 taxonomists, 1,200 years and ~ 364 billion US$ at the rate of 48,500 US$ to erect a species. During the last 20 y, the description rate of eukaryotic species has remained at 6,200 species/y. Hence, only ~ 1 million species remain to be described, which may require 40,400 taxonomists, 160 years and ~ 50 billion US$. The growing concern for conservation of biodiversity culminated at signing of the Convention on Biological Diversity in 1992. In view of its importance, Levin (2001–2013) brought out 5,484 paged seven volumes on Encyclopedia of Biodiversity including many chapters authored by leading experts on different aspects of biodiversity ranging from habitat to ecosystem, extinction to existing, invasion to economics and so on. Viewing it from a holistic angle, the series on Evolution and Speciation in Animals, Plants, Protozoa (Pandian 2021b, 2022, 2023) and Fungi has identified environmental factors: 1. Space, 2. Light-Temperature, 3. Precipitation-Liquid Water, as well as biological attributes: 4. Cellularity, 5. Symmetry, 6. Clonality, 7. Sexuality, 8. Modality and 9. Motility that either accelerate or decelerate species diversity. This chapter briefly summarizes how these factors and attributes govern biological diversity in eukaryotes.

15.1 Collection and Cataloguing

In the process of collection of specimens for species description, the size and motility are important considerations. Eukaryotes that are sessile and visible to the naked eye have attracted and are readily amenable for collection. Motile organisms demand efforts to collect them. So are burrowing animals. The collection and cataloguing of unicellular animals (e.g., Protozoa), plants (unicellular algae) and fungi (e.g., yeasts) had to wait for the discovery and improvements in microscopy between 1609 and 1750. The description for even macroscopic plants and vertebrates had to wait for Rundle et al. (1991) in the Atacama-Sechura Desert, and Newton and Newton (1997) and Mallon (2011) in the Saudi Arabian Desert. In the peculiar Thermal Vents and Cold Seeps at 1,000–2,000 m depths of oceans, it had to wait until 1990s (see Pandian, 2017). Of course, the complete absence of light beneath 400 m depth eliminates the existence of plants. Enormous efforts in terms of funds, team work and technologies including dredging and other facility are required to discover the existence of organisms in the harsh abyssal (3,500–6,500 m depth) and hadal (6,500–10,200 m) habitats. They have brought to light the existence of a few radiolarian protozoans up to 8,000 m depth (see Finlay and Esteban, 2018), unicellular yeasts (7 genera) and multicellular filamentous (10 genera) fungi (Vargas-Gastelum and Riquelme, 2020) and animals (162 species, see Pandian, 2021a) from the Mariana/Philippine Trenches (below 10,000 m depth). Strikingly, the animal species number decreases drastically from 41,350 at the surface to 162 between 6,500 and 10,210 m depth (Fig. 15.1).

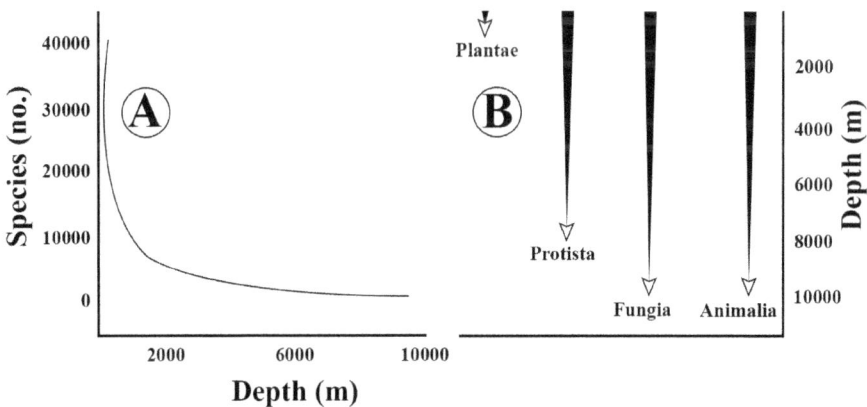

FIGURE 15.1

Vertical distribution of (A) animal species and (B) eukaryotic kingdoms in the oceans.

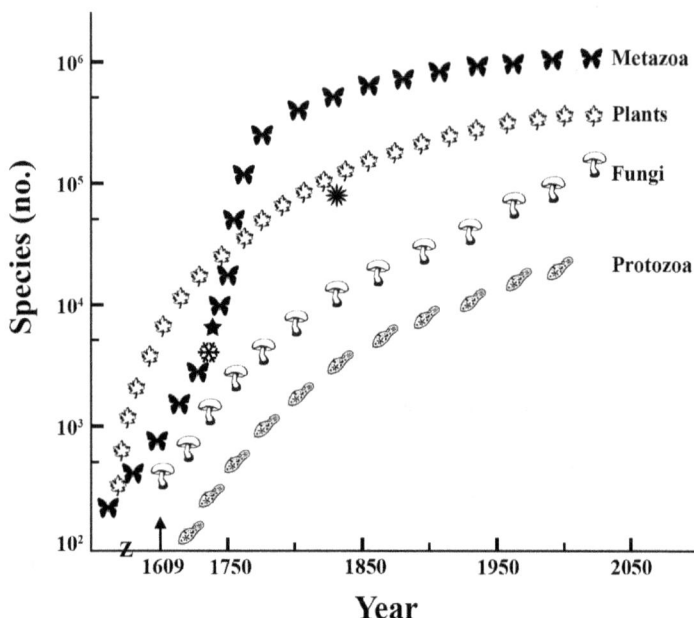

FIGURE 15.2

Number of described species as function of year in motile heterotrophic Metazoa (✹, Mora et al., 2011), sessile autotrophic plants (✿, Pandian, 2022), multicellular sessile heterotrophic fungi (♄, Hawksworth and Lucking, 2018) and unicellular slow motile heterotrophic protozoa (✺, Pandian, 2023). Symbols marking the number [5,900 species] for plants (★) and [4,326 species] for animals (✿) are drawn from *Systema Naturae* (Linnaeus, 1735) and 97,205 flowering species (✳) from *Genera Plantarum* (Bentham and Hooker, 1862–1881).

Direct extrapolation of the species number – accumulation curves, which usually show asymptotic trajectories over time – as shown for Metazoa, Plants, Fungi and Protozoa in Fig. 15.2 – have been used as a method to predict the number of species for one or another taxonomic group. Employing different procedures, Mora et al. (2011) estimated the currently catalogued and predicted species number for the four major eukaryotic kingdoms and two prokaryotic major kingdoms (Table 15.1). According to them, there are 1.2 million catalogued eukaryotic species. But their predicted number is 7.77 million species for Metazoa alone. This conclusion is misleading and not correct (see also Hawksworth and Lucking, 2018). Their currently catalogued values are far lower than the presently estimated and accepted values. Firstly, their value 215,644 is lower than the currently accepted number of 374,000 species for Plantae (Christenhusz and Byng, 2016). So are their values 953,434 for Metazoa, 43,271 for Fungi and 8,118 for Protista, instead of the currently estimated values of 1,543,196, 106,761 and 32,950, respectively. Secondly, the value of 7.77 million species predicted for Metazoa is unacceptably too high. For, (i) the species-accumulation curve for Metazoa shows asymptotic trajectory over time (Fig. 15.2). (ii) Except for Arachnida, Mollusca, Annelida, Echinodermata and Hemichordata, the trend for the actual and predicted number of species for almost all other taxonomic groups including Insecta shows only a small difference between them (Fig. 15.3). Subtracting 1,400 fungal species/y (Hawksworth and Lucking, 2018), 2,000 vascular plant species/y (Christenhusz and Byng, 2016), a few for algae, bryophytes and protozoa from the current description rate of 6,200 eukaryotic species/y (Mora

TABLE 15.1

Catalogued and predicted species number of prokaryotes and eukaryotes, as compiled by Mora et al. (2011). For comparison, the presently estimated and expected numbers are also included. *Pandian (2021b), [†]Christenhusz and Byng (2016), [‡]this volume, [‡]Pandian (2023).

Kingdom	Catalogued	Predicted	Estimated	Expected
Eukaryotes				
Animalia	953,434	7,770,000	1,543,196*	1,800,000
Plantae	215,644	298,000	374,000[†]	584,000
Fungia	43,271	611,000	106,761[‡]	260,000
Protista	8,118	36,400	32,950[‡]	36,400
Others	13,033	27,500	–	-
Subtotal	1,233,500	8,740,000	2,056,907	2,680,400
Prokaryotes				
Archaea	502	455	–	
Bacteria	10,358	9,680	–	
Subtotal	10,860	10,100	–	
Grand total	1,244,360	8,750,000	2,067,007	

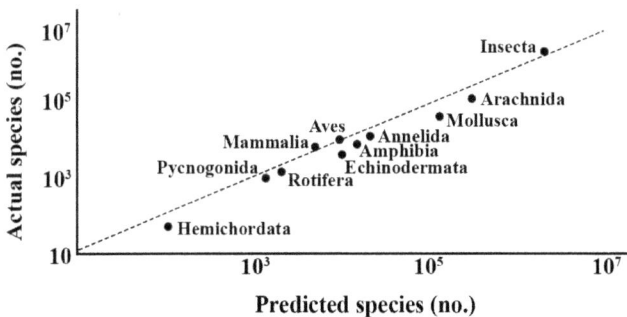

FIGURE 15.3

Estimations of actual and predicted species number for selected taxonomic groups of Metazoa (simplified from Mora et al., 2011).

et al., 2011), the scope for erection of Metazoa was reached at ~ 2,500 species/y. Considering this rate, the number of species to be described for Metazoa may be around 250,000 in the next 100 years. In all, the number for animals may not exceed 1.8 million (Table 15.1). The number of new species erected has remained around 2,000 species for vascular plants between 1995 and 2015. In 2016, the number decreased to < 100 new species (Christenhusz and Byng, 2016). With the discovery of new culture techniques, new species of chlorophtyic algae like Coccolithophores are being described (see Pandian, 2022). Even assuming 2,100 species are described for plant/y, not more than 210,000 new plant species may be described in the next 100 y. Hence, the species number for plants may not exceed 584,000 species. The predicted number of 36,400 species for Protozoa is in the acceptable range.

According to the Dictionary of Fungi (1943–2008), the currently accepted number for fungi is around 120,000 species (Fig. 15.4A). However, *catalogueoflife.org* and *Species Fungorum* Database indicate the description of 146,154 and 150,600 fungal species, respectively. Regarding the trend, the database recognizes the following three phases: (i) an ascending phase between 1750s and 1860s, (ii) a steep phase between the 1870s and 1880s, during which microscopy came into general use and intensive collections were also made from barely explored areas and (iii) a declining but constant phase from the 1890s to the present (Fig. 15.4B). During the last 40 years, the average rate of description has also remained around 1,400 species/y (Fig. 15.4C). The resurgence in species description is largely attributed to the increasing use of molecular techniques. Nevertheless, the average rate has remained at 1,400 species/y during both the Premolecular and Molecular eras. Assuming the rate of 1,400 species/y, the newly described fungal species may only make up another 140,000 during the next 100 years. In all, the number (currently accepted number 120,000 + estimated 140,000) may not exceed 260,000 species for fungi.

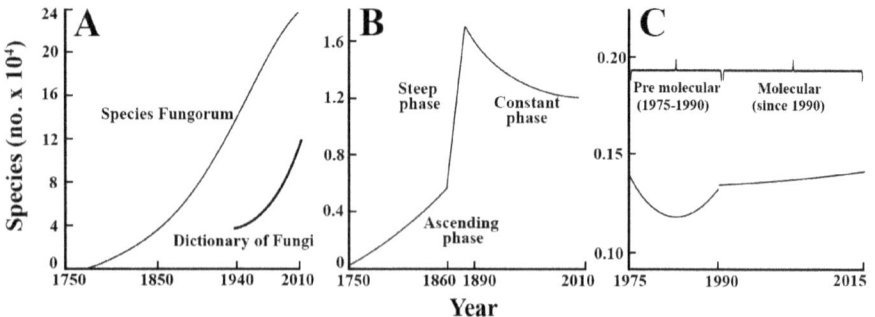

FIGURE 15.4

(A) Increasing catalogued species number of fungi from 1750 to 2010. (B and C) The trends for the relation between species number and year (simplified from Hawksworth and Lucking, 2018).

Nevertheless, a series of hypothetic estimates was made that ranges from 500,000 to 10 million fungal species, although 1.5 to 5 million species is more acceptable to many mycologists (e.g., Blackwell, 2011). These values are based on fungus : plant ratios. Depending on different regions and ecosystems, a wide range of ratios from 7.5: 1, 10.6: 1, 12.9: 1, 13.1: 1 to 19.1: 1 (see O'Brien et al., 2005, Taylor et al., 2010, Hawksworth and Lucking, 2018) have been proposed. From the ratios, Hawksworth and Lucking (2018) suggested that there can be 3.8 million fungal species. However, these predictions may not hold true. Firstly, the ratios are so widely scattered that there is no universally acceptable ratio for all fungus : plant ratio. Secondly, vast majority of phytopathogenic Pucciniomycotina (8,416 species) are species-specific. With the involvement of two host plants, the ratio for these fungi can be 2:1. Thirdly, ~ 46% fungi are decomposers. No information is yet available on the minimum and maximum numbers of fungal species required to decompose a single plant species. Hence, it is difficult to accept the fungus : plant ratio and the hypothetical predictions of 500,000 to 5 million fungal species. Recently, from an analysis based on species per genus, Hawksworth and Lucking (2018) arrived at 1 million species for fungi. Adding 94,059 species-level OTUs remaining in the GenBank with no names, they suggested ~ 2 million species for fungi.

However, *a fact that none of these mycologists have ever considered is that clonality and homothallism decelerate species diversity*. Except for the 10,911 speciose Agaricale complex, the remaining 90% fungi reproduce clonally and sexually (see Table 7.2). Of them, 16,889 species or 15.8% of fungi propagate only through clonal multiplication (see Table 7.1). With < 1% clonal species, Metazoa are enriched with 1,543,196 species. In plants, the existence of 39% clonal species (see Table 15.10) has limited the number to 374,000 species. With 100% clonality, the number is limited to 32,950 species in Protozoa. Clearly, clonality decelerates diversity and species number in organisms. *With 90% clonality, the chances for increase in fungal species number is limited*. Incidentally, the clonal cyanobionts require 0.76 (million years (MYs) to generate a species, whereas sexually reproducing chlorophytes need only 0.24 MYs to do it (Pandian, 2022). A second fact that decelerates species diversity is monoecy/hermaphroditism. For example, 95% of the 1,543,196 speciose Metazoa are gonochoric or dioecious. The existence of only 16% dioecy in plants has limited the number to 374,000 species. Similarly, the existence of 23% dioecy (see Table 15.14) has also limited the species number to 32,950 in Protozoa (Pandian, 2023). *The ~ 30% homothallism in fungi (see Table 7.2) may also limit species diversity*. Hence, this account retains the number for fungi as 260,000 species (see Table 15.1).

15.2 Habitats and Distribution

As mentioned earlier, oceans cover 70% of the Earth's surface with 97% of its water. Freshwater systems, however, cover only 1% and hold as small as 0.01% of its water. The remaining 29% of the Earth is covered by land (see Pandian, 2011a). Holding water mass of 1.3 billion km³ (*jbutler@uh.edu*), the more stable oceans provide 900 times more livable volume of space than that of land. However, the more labile land with widely varied niches provides a number of livable habitats and a better scope for diversity and speciation, and are able to support and sustain 77% of metazoan, 88% of plant and 96% of fungal species. In land, species richness seems to have overtaken that in the oceans ~ 125 MYA. It coincided with diversification of flowering plants, which constitute 79% of all photosynthetic eukaryotic plants (see Pandian, 2022). Flowering plants increased terrestrial productivity (52% global productivity, Field et al., 1998) equal to that in oceans and led to herbivory three times more than in oceans (Costello and Chaudhary, 2017). Besides, the aquatic habitats are environmentally stable but that of terrestrials are labile. Consequently, speciation requires a longer time in aquatic than in a terrestrial habitat. For example, the aquatic Nymphaeaceae require 1.3 million years to generate a species, compared to only 5,250 years in the terrestrial Poaceae (see Pandian, 2022).

In aquatic systems, light penetration is limited to a maximum of 400 m depth. Its direct and indirect effects on distribution of organisms have been elaborated earlier. Amazingly, it is the pelagic euphotic zone, constituting ~ 8% of the oceanic area, is diversly occupied by 86% marine organisms and the remaining 92% of the oceanic zones are thinly occupied by ~ 14% faunal and other organisms. Interestingly, density of water peaks at 4°C. Due to geothermal energy input, water at the deepest oceans remains in its liquid form. No doubt, geothermal energy also sustains life of many microbial species, which, in turn support limited fauna in isolated geothermal vents (see Pandian, 2017). But, *the origin, evolution and speciation fostered by solar energy input is several orders of magnitude greater than that of geothermal energy input* (see also Pandian, 1975).

To perform efficient photosynthesis, autotrophic plants require water and CO_2 and energy input from solar light. Photosynthesis can occur at extremes of –7°C in algae of the Antarctic soil and +75°C in blue green algae (Davison, 1991). However, it occurs at optimal efficiency only between 10°C and 30–35°C (*sciencing.com*, Ribeiro et al., 2006). It takes place in a cycle named after the discoverer Calvin in C3 plants. To suit the high intensities of light, and temperature prevailing in arid zones, it can also occur through C4 or CAM cycle in xerophytic plants that are a little more efficient and adapted to conserve water. The proportion of C3, C4 and CAM plant species is > 93.5, 2.2 and 4.3%, respectively, i.e., *despite the enhanced photosynthetic economy and efficiency, the dry and arid zones limit species diversity* (see Pandian, 2022).

On land, temperature and precipitation are decisively important factors that provide the highest scope for diversity and speciation. The average annual precipitation on the globe is 5.77×10^{14} m^3, of which 79% fall on oceans and 21% on land (Agrawal, 2013). Organisms can acquire water only in its liquid form. Water is available in a solid frozen form in the Arctic (14.0 million km^2) and the Antarctic (14.5 million km^2), in which biological activity is at its least. Deserts are characterized by evaporation (> 25 cm) exceeding precipitation (< 20 cm). Being a harsh habitat, water is scarce in liquid forms in deserts. Ward (2009) recognized as many as 23 deserts, which span over 28.5 million km^2 (*edu.seattlepi.com*) and cover 9% of land area, leaving only 20% area with relatively more precipitation, and hence more productivity and biodiversity. Of available productive land of 13.2 billion ha (b ha), 12% (1.6 b ha) is currently in use for cultivation of agricultural crops, 28% (3.7 b ha) under forests and 35% (4.6 b ha) comprise grassland and woodland ecosystems (*fao.org*).

Montane or alpine elevated ecosystems, located between 60°N and 60°S, cover 4.5×106 km^2 or 30% of the land surface (Korner, 1995). Hence, they compensate the productive land area more than that lost as least-productive deserts. In general, species diversity is particularly high in the montane ecosystems, some of which are known as biodiversity hotspots (Admassu et al., 2016). The highest altitude, at which plants and animals are known to inhabit is 6,480 m for the sandwort *Arenaria bryophylla*, 4,572 m for the flowering plant *Abeis squamata* and 3,300–5,000 m for the Tibetan yak *Bos grunniens* (see Pandian, 2022).

Apart from this sprinkling, there are quantitative estimates for the distribution of eukaryotes in Marine (M), FreshWater (FW) and Terrestrial (T) habitats (Table 15.2). Prior to an explanation for spatial distribution of eukaryotes, a preamble is required. Life has originated and existed in the sea longer than on land. Fossils reveal the existence of bacteria over 3.7 and 3.1 billion years ago (BYA) in oceans and land, respectively. Whereas all other eukaryotes are largely (~ 87%) known from terrestrial habitats (Table 15.2), the distribution of Protozoa alone is restricted to 6%. Of 3,300 speciose Cyanophyceae, some 689 species or 21% inhabit the moist rocks and damp

TABLE 15.2

Habitat distribution of eukaryotes (compiled from Pandian, 2021b, 2022, 2023 and this volume).

Kingdom	Habitats (%)		
	Marine	Freshwater	Terrestrial
Protozoa	74.0	20.0	6.0
Fungia	1.0	2.7	96.3
Plantae	4.8	7.2	88.0
Metazoa	15.1	7.8	77.7

soils in land. So are the 21,925 speciose bryophytes. Irrespective of structural simplicity, autotrophism seems to have let them to occupy land, albeit relegated to moist zones alone. However, *the combination of the structural simplicity and heterotrophism seems to have impeded the colonization of terrestrial habitats by Protozoa* (see Pandian, 2023).

Regarding habitat distribution, the unicellular Protozoa and multicellular Fungi present a contrasting picture. This contrast can be attributed to features of water and air. Water is 800 times denser than air and drastically reduces the dispersal rate of reproductive elements like the spores of fungi and lower plants. Most fungi are propagated by clonal and/or sexual spores. The chytrid zoospores, for example, are small and measure 1 to 6–7 μm in length. Hence, their dispersing ability is limited to shorter distances through aquatic *milieu*. So are the zoospores of algae. The reproductively active *Ascophyllum nodosum* attached to wood or plastic objects has travelled from the east coast of South America to the west coast of Africa at the speed of 13 km/d over a period of 430 d for its passage of 5,500 km through water (see Pandian, 2022). In contrast, the urediniospores of the coffee rust *Hemileia vastatrix* have been dispersed across the Atlantic from African Angola to South American Brazil at altitudes of 1,500–2,000 m within 6 d (see p 139). Notably, the differences in the *rate of reproductive element dispersal ranges from a few micrometers in aquatic fungi and 13 km/d for the aquatic alga A. nodosum to > 1,000 km/d for the terrestrial fungi.* Secondly, water is a universal solvent and many substances like enzymes dissolve rapidly in it. Being externally digesting eukaryotes, all fungi apply their digestive enzymes on food/substratum, digest and absorb micronutrients through the body surface. *The scope for dilution of these externally released digestive enzymes of fungi is greater in the aquatic milieu than in terrestrial habitats. These two features have eliminated the distribution of the multicellular fungi in waters, although the aquatic habitats abound with unicellular fungi.*

A major reason that limits vertical distribution of plants in aquatic systems is that light can penetrate to a maximum of 200–500 m depth. Whereas violet and blue radiations penetrate up to the 200–500 m depth, the red one carrying the highest quantum of solar energy is already diminished to zero level at 34 m depth. It is in this 'red light zone', most cyanophytes and chlorophytes flourish. A point to be noted is that the chlorophytes and cyanophytes are 'at home', as 68 and 76% of them are freshwater inhabitants. The phaeophytes dominate up to the depth of 150 m (Fig. 15.5A). Notably, Phaeophyta and most Rhodophyta are exclusively marine inhabitants.

The Kingdom of Metazoa consists of 1,543,196 species. Of them, 97.0 and 3.0% belong to the nine major and 26 minor phyla, respectively. Minor phyletic species are more 'aquatics' and are occupants of marine (41.3%) and freshwater (15.4%) habitats, in comparison to the major phyla that are more 'terrestrial' (78.1%) inhabitants. Among major phyla, Echinodermata and 14 minor phyla are all exclusively marine inhabitants. *Exclusive restriction to the more stabilized marine habitat limits species number to 7,000 in Echinodermata, whereas the number goes to 186,193/phylum among the remaining major phyla, which are distributed in marine, freshwater and terrestrial habitats. This is also true of minor animal phyla and plants.* Briefly, the ability of phyletic members to access and colonize the sea, freshwater and terrestrial habitats is decisively important for horizontal distribution and thereby species diversification (see Pandian, 2021b).

In fact, no plant exists below 400 m depth (Fig. 15.5A). The levels of temperature, oxygen and nutrient availability are drastically reduced below 400 m depth. Utilizing the sinking organic debris, heterotrophs, however, survive up to 10,200 m depth, albeit with decreasing diversity. For example, 41,500, 11,592 and 8,459 animal species are found at the surface, 100 and 200 m depths, respectively. The diversity is drastically reduced to 422 and 162 species at the abyssal (3,500–6,500 m) and hadal (6,500–10,210 m) zones. No detailed information is yet available for the vertical distribution of Protozoa and fungi. However, radiolarians are found at 8,000 m depth, and many yeasts and filamentous fungi are recorded from depths up to 10,000 m (Fig. 15.5A). Briefly, *depth decelerates diversity in aquatic habitats.*

On the other hand, with increasing altitude (above sea level, asl) in the montane zone, temperature decreases at $1°C/150$ m asl and atmospheric pressure at 1.2 kPa/100 m asl, and consequent partial pressure of the oxygen level. Another important factor is the soil depth and its moisture (precipitation) content, which are yet to be adequately recognized. To support the sustenance of herbs and trees, ~ 200 and ~ 100 cm soil depths are required. First and unexpectedly, *the diversity level in animals is less or equal to that of plants in montane zones* (Fig. 15.5B–C, see also Fig. 15.8A). Second, *the diversity level of poikilothermic flying insects* (Fig. $15.5C_2$) *is far less than that of homeothermic birds and bats* (Fig. $15.5C_1$). Third, the distribution of birds and bats are extended up to 4 km asl, whereas that of insects is limited to < 3 km asl except for orthopterans with limited flying ability. Other than temperature, *plant diversity depends on soil depth and its moisture content, but that of animals on their ability to regulate body temperature.* In the montane aquatic systems too, temperature is the prime factor that determines the diversity in plankton (12–14°C) and benthos (7–8°C). Fourth, *the level of diversity in aquatic insects is less than that in terrestrial insects in the montane zone.* The aquatic arthropods are limited to a maximum of ~ 2,000 m asl, except for chironomids, in which > 10 species are found up to 3,000 m asl. In Tamil Nadu, India, rare data are available for the diversity, density and productivity of planktonic algal species in the Kodayar River hailing from 1,312 m asl. In it, these values increase from 5 species/l, 159 no./l and 0.11 mg $C/m^3/d$ at 1,312 m asl to > 14 species/l, 412 no./l and 2 mg $C/m^3/d$ at the ground level (Fig. $15.5B_2$).

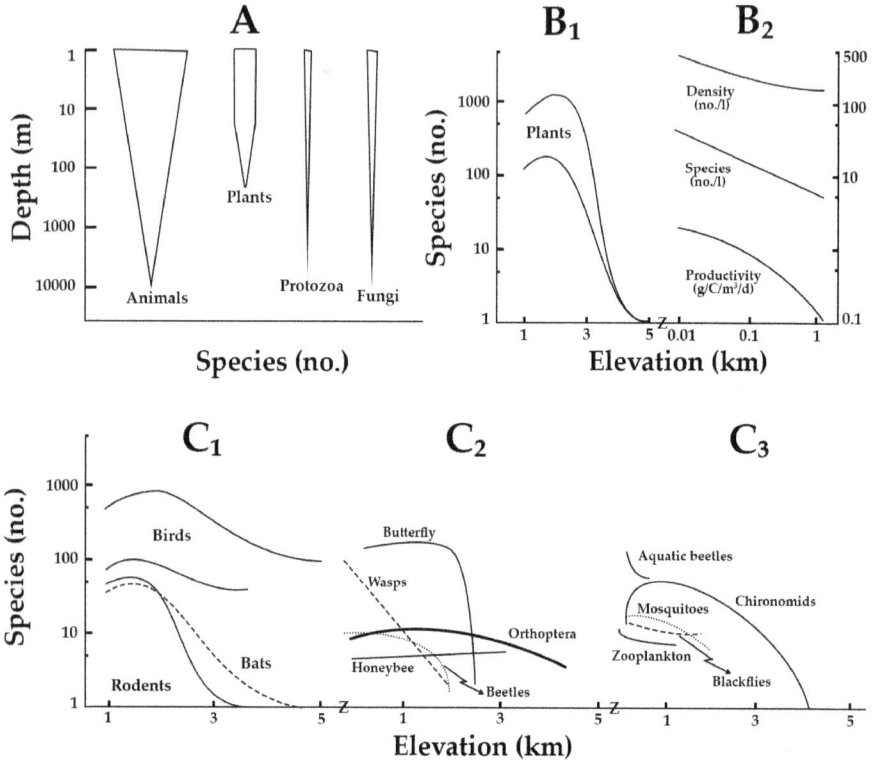

FIGURE 15.5

(A) Species diversity of Metazoa, plants, Protozoa and fungi as function of depth in marine habitats. (B–C) Relation between species diversity and elevation in montane habitats: (B_1) terrestrial plants, (B_2) aquatic algae, (C_1) homeothermic vertebrates, (C_2) poikilothermic terrestrial insects and (C_3) aquatic insects (based on Pandian, 2021b, 2022, 2023, this volume, Murugavel and Pandian, 2000).

Returning to aquatic systems, there has been a persistent tendency among terrestrial eukaryotes to return to aquatic habitats mostly to freshwaters (Table 15.3). Of terrestrial fungi constituting 102,123 species, 878 or 0.85% have returned to mostly freshwater and marine habitats. In plants, 255 fern species belonging to the families Salviniaceae and Marsileaceae have returned to freshwater. Within monocots, 48 species belonging to Zosteraceae have returned to the sea and another 222 to freshwater. Of 750 aquatic angiosperms, only 123 species undergo hydrophilous pollination. The remaining 627 are entomophilous. Being an inhibiting process, the hydrophilous pollination has drastically reduced the scope for angiosperms returning to aquatic habitats. In all, < 0.4% plants have secondarily returned to aquatic habitats. Among animals, most insects spend larval stages in water but adult stages in terrestrial habitats. In contrast, the adults of 29 freshwater and 7 marine turtle species go to land for laying their eggs. Among birds, 222 species are completely marine but 72 are partially marine. In all, 263 bird species are aquatics. Fifty five marine snakes and 130 marine mammals are viviparous and have no need to return to land. This also holds true for 124 freshwater mammals. In all, 4.9% of terrestrial metazoans have returned to aquatic habitats. Of 1,836,582 terrestrial eukaryotic species, 72,489 or 3.9% have secondarily returned to aquatic habitats.

TABLE 15.3

Estimated number of terrestrial species that have secondarily returned to aquatic habitats. *Dodds (2002), †*iucn.org*, ‡Veron et al. (2008).

Taxonomic group	Terrestrial species (no.)	Species returned to aquatic habitats	
		(no.)	(%)
Fungi (Grossart et al., 2021)			
Zygomycota	1,065	198	18.60
Basidiomycota	23,975	120	0.50
Ascomycota	77,083	560	0.73
Subtotal	102,123	878	0.85
Plantae (see Pandian, 2022)			
Ferns	10,560	255	2.40
Monocots	85,379	270	0.30
Dicots	210,004	750	0.40
Subtotal	305,943	1,275	0.40
Animalia (see Pandian, 2021a, b)			
Insecta	903,420	64,600	7.20
Mites	500,000†	5000*	1.00
Reptilia	9,545	219	2.30
Aves	10,038	263	2.60
Mammalia	5,513	254‡	4.60
Subtotal	1,428,516	70,336	4.90
Grand total	1,836,582	72,489	3.90

15.3 Light – Temperature

Based on climatic conditions, the Earth (510.1 million km²) is divided into (i) tropical (located between 0° to 23°N and 0 to 23°S with land area of 184 million km² or 36% landmass of the Earth), (ii) temperate (23°N to 66°N and 23°N to 66°S spanning over 295.4 km² area or 57.9% landmass of the Earth) and (iii) polar (above 66°N and 66°S, covering 16.5 and 14.2 million km² for the Arctic and Antarctic zones, respectively) zones (*nationalgeographic. com*). These zones are characterized by hot and humid weather (20 to 35°C, 100–450 cm precipitation/y, *fossweb.com*) in tropics, cool climate in temperate (0°C to 20°C, 75–150 cm precipitation/y, *fossweb.com*) and cold polar (0°C to –40°C, < 20 cm precipitation/y) zones (*ucmp.berkeley.edu*). Plants grow and thrive, where light, water and CO_2 are available. In the absence of autotrophic plants, the heterotrophic animals and fungi cannot be sustained for long. The duration of mean daytime is 4,320, 4,365 and 3,960 h/y for tropical, temperate and the Arctic zones, respectively. However, the intensity of solar radiation amounts to 380, 240 and 100 watt/m² in these zones. Hence, *it is the intensity of light rather than the diurnal duration that determines temperature range in these zones.* With regards to precipitation, tropical Asian countries are characterized by monsoon. Unlike spring, summer, autumn and winter prevailing in temperate zones, the tropical Asian countries are characterized by pre-monsoon (February to May), monsoon (June to September) and post-monsoon (October to January) seasons. Interestingly, India experiences the southwest monsoon between June and August but southeast India enjoys the northwest monsoon between September and December. With a narrow strip of land, countries like Myanmar and Malaysia have monsoons over a longer period of eight months. Not surprisingly, India and Malaysia are recognized as megabiodiversity countries (see Pandian, 2002).

Species diversity progressively decreases from the equator toward the poles. This aspect is elaborated as Latitudinal Gradients of Biodiversity (LGB). The gradients represent biogeographic patterns. They are defined as components of taxonomic/phylogenetic, functional, genetic and phenotypic dimensions changing with the latitudinal position on the Earth's surface (Willig and Presley, 2013). However, most investigations have considered species richness alone. For details on mechanisms and hypotheses, Willig and Presley (2013) may be consulted. For specific details on LGB in fungi, marine benthic communities and sexual dimorphism, Shi et al. (2013), Witman et al. (2004) and Estlander et al. (2017) may be consulted. Both species richness and diversity decrease as function of latitude (Fig. 15.6A). The mean numbers of species and genera also decrease from the tropics toward either pole. Notably, the decreases are more pronounced towards the southern pole (Fig. 15.6B). However, the reverse holds true for distribution of mites (Eggleton, 2000).

Aside from these plausible climatic features, vivid evidence is provided for the effect of temperature on generation time, life span and fecundity of parthenogenic daphnids and sexually reproducing amphipods. Both these

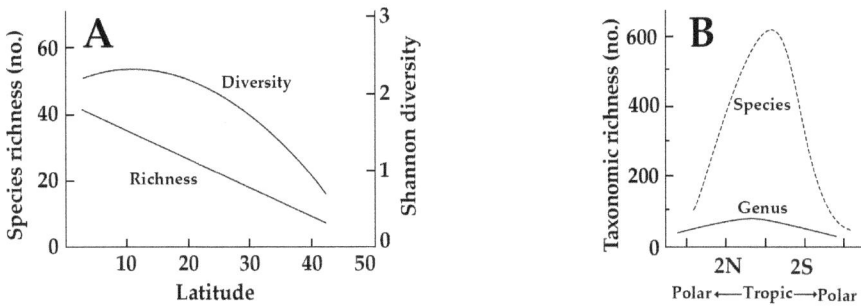

FIGURE 15.6

(A) Species richness and diversity as a function of latitude, (B) Taxonomic richness as a function of latitude. 2N and 2S represent the northern and southern temperate zones, respectively (compiled/ modified from Willig and Presley, 2013).

crustaceans brood their eggs and release completely developed young ones called neonates. Data assembled for parthenogenetic daphnids and sexual amphipods are limited to tropical and temperate zones. Repeated computer searches have not yielded corresponding data for the Arctic crustaceans. Absolute fecundity drastically decreases from 161 neonates in tropical daphnids to 16 neonates in temperate ones (Table 15.4). Similarly, the value also decreases from 164 neonates to 16 neonates in tropical and temperate amphipods, respectively. In general, more the number of egg/progeny, greater shall be the genetic diversity, especially among sexually reproducing animals like amphipods. Clearly, the tropical animal generates nearly 10 times more genetic diversity in their progenies than those of temperate ones.

TABLE 15.4

Estimated production of neonates by tropical, temperate and arctic daphnids and amphipods (compiled from Pandian, 1994, 2021b, Nair and Anger, 1979).

Characters	Parthenogenic daphnids		Sexually reproducing amphipods	
	Tropical	Temperate	Tropical	Temperate
Generation time (as % of life span)	20	50		
Reproductive life span (%)	80	50		
Molt frequency (no.)	15.0	5.5		
Berried molt (no.)	12.0	4.4	11	4
Neonates (no./brood)	13.4	3.6	15	4
Absolute fecundity (no./♀)	161	15.8	164	16

15.4 Precipitation – Liquid Water

Temperature and precipitation on land are decisively important factors that provide scope for speciation. As mentioned earlier (p 207), the average annual precipitation on the globe is 5.77×10^{14} m^3, of which 79% fall on oceans and the remaining 21% on land (Agrawal, 2013). First, the precipitation level is the highest in the tropics and decreases toward the polar regions (Fig. 15.6A). Second, annual Net Primary Productivity (NPP) also peaks in the tropics and tapers in either direction toward the polar regions (Fig. 15.7B). Third, species richness peaks also in the tropics and decreases toward the polar regions (Fig. 15.7A). Therefore, *it is tempting to conclude that precipitation is one of the causative factors in regulation of speciation. As if to confirm this conclusion, the number of vascular plant species increases with increasing NPP* (Fig. 15.7C). With increasing number of flowering plant species, that of herbivorous animal species is also known to increase since the Cretaceous epoch (cf Costello and Chaudhary, 2017). If availability of liquid water through a higher level of precipitation is a causative factor for species richness in eukaryotes, its scarcity may also cause scarcity in species number in deserts and polar zones. In support of this conclusion, which is known but not adequately recognized, evidences are brought for the first time.

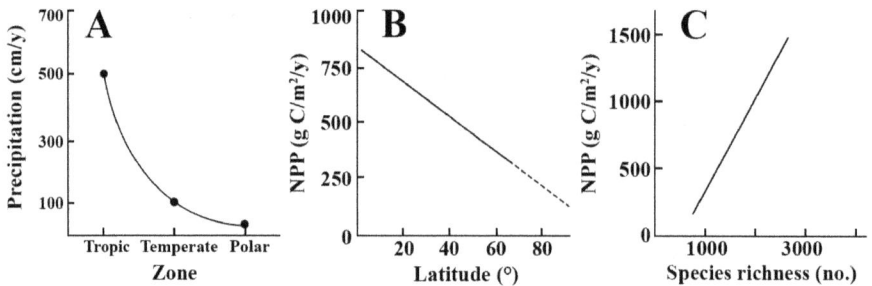

FIGURE 15.7

Precipitation and net primary productivity (NPP) as function of latitude (A) indicated as tropic, temperate and polar zones, (B) latitude (dotted lines indicate the extension to polar zone), and (C) vascular plant richness (based on Gillman et al., 2015 and others).

Water occurs as liquid or gaseous moisture within the soil and fog in air or solid ice (inclusive of permafrost). Most aquatic eukaryotes acquire water in its liquid form through the body surface. In terrestrial habitats, plants acquire water in the liquid form through roots or rhizoids. Some plants are able to absorb it from the atmospheric moisture/fog. Fungi absorb liquid water through the body surface and also via rhizines by lichens. Notably, water availability in its liquid form is extremely limited in harsh habitats like deserts and polar ice sheets. There are 23 deserts, which span over 28.5 million km² or covers 9% of land area. In them, evaporation exceeds precipitation. This account has chosen the following three representative deserts to describe their climatic features:

Sonoran Desert (North America) cover 223,000 km² area, up to 40°C, 24.5 cm precipitation/y. Aridity increases eastward from the Californian mountains.

Arabian Desert (Asia) cover 2,300,000 km², 55°C, < 10 cm precipitation/y.

Atacama-Sechura Desert (South America) covers a narrow strip of 10,500 km² extending from 10°S to 30°S, 17°C, 1.5 to 4.5 cm precipitation/y.

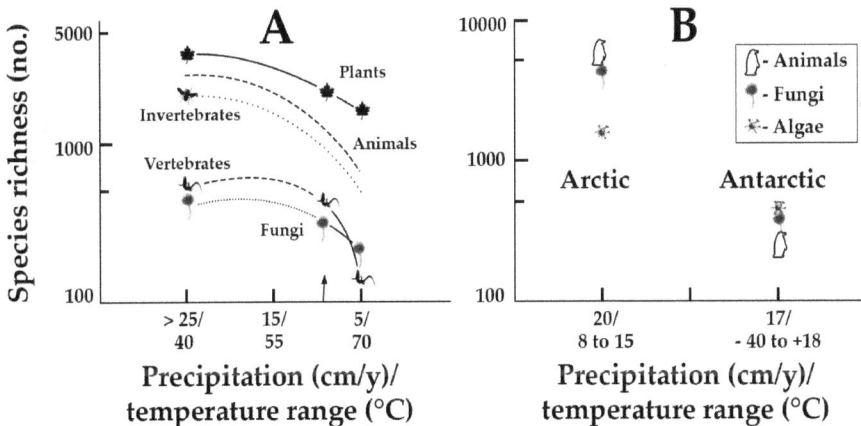

FIGURE 15.8

(A) Deserts and (B) Polar zones: Distribution of species richness in different eukaryotes as functions of precipitation and temperature (values are drawn from Table 15.5).

For these deserts, available information is limited to invertebrates in the Sonoran Desert, and fungi in the Saudi Arabian and Atacama-Sechura Desert (Table 15.5). For the Sonoran Desert, 229 fungal species (e.g., Ranzoni, 1968) and molecular based sequence numbers (Terminel et al., 2010) are reported. 1. Strikingly, precipitation plays a more decisive role in determination of species richness than temperature. The species number for plants, animals and fungi are the least in cooler (17°C) and the least rainy Atacama-Sechura Desert than those in the hotter (55°C) and relatively rainier Saudi Arabian Desert (Table 15.5). 2. Unexpectedly, the values available for species number are consistently higher for plants than animals (Fig. 15.8A). A reason for it may be that the C4 and xerophytic plants are more capable of tolerating scarcity of water than animals. 3. In all, there are only a few animals (2,689 species), plants (7,897) and fungi (931), which are equivalent to 0.0011, 0.0031 and 0.0004 animal, plant and fungal species per km² area (Table 15.5). *On the whole, there are 11,517 eukaryotic species inhabiting the 2.5 million km² desert area, i.e., 0.0045 species/km².*

TABLE 15.5

Species richness and distribution of animals, plants and fungi in selected deserts as well as the Arctic and Antarctic (for deserts, values are drawn from Pandian, 2021b, 2022, this volume and other sources indicated in text).

Habitat	Area (km²)	Animals		Plants		Fungi	
		(no.)	(no./100 km²)	(no.)	(no./100 km²)	(no.)	(no./100 km²)
Deserts							
Sonoran	223,000	2067	0.97	3900	1.79	400	0.18
Arabian	2,300,000	476	0.02	2067	0.09	302	0.01
Atacama	10,500	146	1.39	1930	18.38	229	2.18
Subtotal	2,533,500	2689	0.11	7897	0.31	931	0.04
Eukaryotes total	2,533,500	11517 (0.45%)					
Tropical evergreen forest (e.g., Milliken et al., 2010)							
Evergreen forest	13,500,000	2504727*	539	170000‡	126		
Amazon	6,700,000	102205†	6555	80000‡	119		
Polar							
Arctic	16,500,000	5500	0.030	1700	0.010	4350	0.028
Antarctic	14,200,000	250	0.001	450	0.003	401	0.003
Subtotal	30,700,000	5750	0.018	2150	0.007	4751	0.015
Eukaryotes total	30,700,000	12651 (0.041%)					

‡*rainforests.mongabay.com*, ‡*wwf.panda.org*, *amazonconservation.org*, †*nationalgeographic.org*.

In the Arctic and Antarctic, water is available in a solid frozen form, i.e., icy sheet/mounts. They are characterized by the following climate and eukaryotic species:

The Arctic spans over 14.0 million km² area, –8 to –15°C (*auroraexpeditions. com*), 20 cm precipitation/y (Smithsonian Magazine), enriched by 1,700 plant species (*twinkl.co.in*), 5,500 animal species (*tourradar.com*) and 4,350 fungal species (*arcticbiodiversity.is*).

The Antarctic spans over 14.5 million km^2 area with more icy mountains in the eastern zone, 17 cm precipitation/y, with 250 animal species (*oceanwideexpeditions.com*), 450 plant species (*brittanica.com*) and 401 fungal species, of which 350 are lichens (de Menezes et al., 2019).

First, the limited available values for animal, plant and fungal species are from less authentic sources (Fig. 15.8B). Second, there are no animals except for some mud-dwelling worms and snails in the Antarctic. Most vertebrates like penguins and marine mammals are visitors to the Antarctic. Fungi are reported from six samples collected from the Antarctic Peninsula and represent 51 species in 6 genera. With these constraints and limitations, the following may be noted: 1. The Arctic is warmer and rainier than the Antarctic. Consequently, liquid water becomes available in greater quantities to support 11,550 species in Arctics, in comparison to 1,101 species in the Antarctic. There are 12,651 species per 100 km^2 in the polar zones, which may be compared with 11,517 species per 100 km^2 in deserts. Hence, *liquid water scarcity decelerates species diversity in the desert and polar zones.* Available values for tropical evergreen forests (Table 15.5) are 539–6,555 species/100 km^2 and are greater than those of deserts and polar eukaryotes.

15.5 Cellularity

Most eukaryotes are multicellular. However, the Protozoa thrive as unicellular eukaryotes. Their 'omnipotent' cells engage subcellular organelles to perform almost all functions of metazoans. However, 1,230 species or 3.7% protozoans are multicellular colonials (see Pandian, 2023). A few fungi, are unicellular (for details, see p 4–5, Table 15.6). Some planktonic algae are also unicellular. A compromise had to be made to assess their approximate number as 12,000 species or 3.2% plants (see Tables 2.3, 4.1 of Pandian, 2022). *In all, of 2,056,907 described eukaryotic species, 47,494 species or 2.3% of them are unicellular.* At the kingdom level, the unicellularity drastically decreases from 96.3% in Protozoa to 3.2% in Fungi, 3.3% in plants and 0% in Metazoa (Table 15.6). In them, species diversity, however, increases from 32,950 to 1,543,195. Hence, *unicellularity decelerates the diversity but multicellularity accelerates it.* Further, it also impedes their access to terrestrial habitat.

TABLE 15.6

Estimation on the number of unicellular eukaryotic species. *see Table 1.4, †see Table 1.1.

Kingdom		Total species (no.)	Unicellular species	
			(no.)	(%)
Protozoa		32,950	31,720	96.3
Fungi	3504	106,761	3,774	3.5
Chytridiomycota-Myxomycota	1840			
Neocallimastigomycota	260			
Microbotryomycetes	200*			
Cystobasidiomycetes	25*			
Taphrinomycotina	140			
Saccharomycotina	1309			
Plantae		374,000	12,000	3.2
Metazoa		1,543,196	0	0
	Total	2,056,907	47,494	2.3

15.6 Symmetry

Prior to explaning symmetry, a preamble on sessility may have to be introduced.

Sessiles: With the emergence of multicellularity, eukaryotes began to organize their structure in the form of cylindrical (radial) filaments, in which polarity was first established, or flats in two dimensional thallus with not only antero-posterity but also dorso-ventrality. For example, the polarity is exhibited with location of anterior sterile zooids and posterior fertile zooids in the volvocid protozoan multicellular colony *Pleodorina*, which moves only in the antero-posterior direction (see Hyman, 1940). With their further expansion, eukaryotes had to go for layered structures. As locomotion in these layered eukaryotes became increasingly costlier, they had to opt for sessility. At this point of evolution, the sessile plants chose autotrophy but fungi and animals heterotrophy. In both aquatic and terrestrial habitats, the option of fungi to external digestion and absorption of micronutrients through the body surface has limited the multicellular expansion towards the surface area and also restricted their structural organization at the tissue-level. Being floated and drifted by waves or swayed by currents, the autotrophic algae have also remained mostly at the tissue-grade organization. However, the need for acquisition of water and nutrients from the soil/rock on land, the rhizines in lichens, rhizoids in bryophytes and roots in higher plants necessiated the development of multi-layered three dimensional structures namely organs and vascular system. Strikingly, sessility imposes homothallism/monoecy in fungi and plants. Hence, their evolution on land had to wait for the emergence of terrestrial motile animals to accomplish pollination/fertilization (e.g., Pucciniales) and dispersal of spores/seeds. The multicellular expansion led animals characterized by the combination of heterotrophy and internal digestion toward diploplastic sessility and triploplastic motility.

Nevertheless, the incidence of sessility is not uncommon among a wide range of animal phyla, albeit restricted to aquatic habitats alone. It occurs in diploplastic sponges and cnidarians as well as triploplastic polychaetes, molluscs, crustaceans and crinoids among major phyla, placozoans, myxozoans, rotifers, entoprocts, bryozoans, pterobranchid hemichordates and urochordates. In all, 44,580 species or 2.88% metazoans are sessile (Table 15.7). Within the 32,950 speciose Protozoa, 1,772 species (volvocales: 393 species, choanoflagellates: 175 species and ciliates: 1,254 species) or 5.3% are sessiles. Plants and fungi are also sessiles. Among 106,761 speciose fungi, the chytrids, myxomycotids (1,840 species) are motiles, and the 260 speciose neocallimastigomycotes are also motiles but within the vertebrate gut. Hence, 104,661 species or 98.6% fungi are sessiles. Among the 374,000 speciose plants, the immotile phytoplankton are drifters. The filamentous *Spirogyra*, the salviniacean and marsileacean ferns and others like *Eichornia* are aquatic floaters. On the whole, only ~ 25% eukaryotes are sessiles.

TABLE 15.7

Estimated sessility in eukaryotic kingdoms (compiled from Pandian, 2021b, 2022, 2023a, this volume).

Kingdom	Total species (no.)	Sessile species	
		(no.)	(%)
Protozoa	32,950	1,772	5.3
Fungi	106,761	104,661	98.6
Plantae	374,000	374,000	100.0
Metazoa	1,543,196	44,580	2.9
Total	2,056,907	525,013	25.55

Symmetry is the beautiful pattern that arises from structural organization in a balanced proportion. Bilaterality and radiality are recognized as the two major symmetrical patterns. In radial symmetry, body parts are developed radially around the central axis. But in bilateral symmetry, they are arranged on opposite sides of a median axis so that it can be divided into essentially similar halves only through a single plane. Sessility imposes radial symmetry in most eukaryotes, albeit with following exceptions: 1. Even at an individual level, the sessile protozoans exhibit bilateral symmetry. 2. On the other hand, the 391 speciose brachiopods are sessiles but are not radiales (see Pandian, 2021a). 3. In contrast, the 6,300 speciose non-crinoid echinoderms are radials but are motiles. 4. The 200 speciose Scyphozoa are also radiales but motiles.

With the expansion of multicellularity, sessility has left the tissue-grade fungi in varied asymmetric forms, including radial symmetry in mushrooms and radiobilaterality in a few fungi like *Coprinus sterquilinus* (Fig. 1.1B). Also in tissue-grade algae, the expansion has led to either cylindrical (radial) or flat symmetry. There are others, in which asymmetry fluctuates; for more information on it, Graham et al. (2010) may be consulted. With further expansion of multicellularity, the development of three dimensional structures became a necessity in terrestrial plants. Yet, sessility has strongly imposed radial symmetry in all the higher plants. For example, the fronds are arranged radially in all ferns. The distribution of vascular bundles, considered as an index, is radial in roots and stems of gymnosperms, monocots and dicots (Li et al., 2020, *berkshirecc.edu*, *floraldesignsbyalka.com*). The trimerous, pentamerous and multimerous petals are arranged radially in flowers. So are their stamens. Only in some plants, tetramerously petalled flowers as well as opposite and whorled phyllotaxies may indicate a type of radiobilaterality. Estimation reveals the incidence of radial symmetry in 24.6% eukaryotes (Table 15.8), compared to 25.5% sessility. The 1.0% decrease in radial symmetry is due to the inclusion Ostracoda + Cirripedia (11,204 species), Crepidularia (155), Bivalvia (459), Rotifera (1,950), Entoprocta (200), Brachiopoda (391), Hemichordata (25) and Urochordata (2,855), which are bilaterally symmetric but sessiles. In colonies like Bryozoa and Urochordata, individuals retain bilaterality although their arrangment may be radial. Clearly, *sessility imposes radial symmetry in most eukaryotes.*

TABLE 15.8

Estimatation for radial symmetry in eukaryotic kingdoms (compiled from Table 2.8 of Pandian, 2021b, 2022, 2023, this volume).

Kingdom	Total species (no.)	Radial species	
		(no.)	(%)
Protozoa	32,950	0	0
Fungi	106,761	106,661	98.6
Plantae	374,000	374,000	100.0
Metazoa	1,543,196	26,575	1.7
Total	2,056,907	505,236	24.6

In animals, symmetry assumes a greater significance and relationship to species diversity. Remarkably, Cuvier (1769–1832) was the first to recognize symmetry as an important characteristic for classification of animals. He divided Eumetazoa into two major groups namely (i) Radiata and (ii) Bilatterea. Hyman (1940) considered Porifera as 'asymmetric or radially symmetric Eumetazoa of cellular grade' organization, Cnidaria as 'primarily radial or radiobilaterally symmetric Eumetazoa of tissue grade' organization and Acnidaria as 'bilaterally symmetric Radiata'. To these, the secondary radially symmetric Echinodermata of organ/system grade organization is to be added. Hence, the sub-classification like the Enterocoela and Coelomata may not be true. Still, two points should be noted. The echinoderms are unique in that their embryos and larvae display bilateral symmetry but the adults pentamerous radial symmetry. Holothurians exhibit radiobilateral antero-posterior polarity. The existence of radiobilateral symmetry and secondary radial symmetry, i.e., bilateral larval symmetry and adult pentamerous radial symmetery indicates the inability of radial symmetry to assume total coverage of a taxonomic group through its entire ontogeny.

The primitive, structurally simpler (< 9 tissue types) Porifera, Cnidaria and Acnidaria are characterized by radial symmetry. Surprisingly, the structurally more complex primitive deuterostome Echinodermata are also radially symmetric, albeit secondarily. Equally surprising is the manifestation of bilateral symmetry in the unicellular Protozoa. In the 1,543,196 speciose Metazoa, 1.7 % (or 26,575 species) and 98.3% (or 1,502,884 species) are characterized by radial (Table 15.9) and bilateral symmetry, respectively. The former includes 6,644 species/phylum, in comparison to 250,481 species/phylum among the bilaterally symmetric six phyla. In other words, the number of bilateral species/phylum is 37.7 times more than that of radials. Hence, *manifestation of bilateral symmetry has greatly accelerated species diversity. But the radials have decelerated it. The imposition of sessility need not necessarily be limited to radials alone in Metazoa. However, sessility imposes radial symmetry in most eukaryotes.*

TABLE 15.9

Effect of selected characteristics on species number in radial and bilateral phyla. *See Pandian (2018, 2019, 2020, 2021b).

Phylum	Species (no.)	Symmetry	Clonality* (%)	Gonochorism (%)	Motility
Porifera	8,553	Radial	100.0	38.10	Sessile
Cnidaria	10,856	Radial	100.0	100.00	Sessile
Acnidaria	166	Radial	100.0		
Echinodermata	7,000	Radial	1.9	99.90	2 mm/s
Subtotal	26,575				

Table 15.9 contd. ...

...Table 15.9 contd.

Phylum	Species (no.)	Symmetry	Clonality* (%)	Gonochorism (%)	Motility
Protozoa	32,950	Bilateral	100.0	24.80	~ 4 mm/s
Platyhelminthes	27,700	Bilateral	0.7	0.16	~ 2 mm/s
Annelida	16,911	Bilateral	1.1	76.00	85 mm/s
Mollusca	118,451	Bilateral	0.0	77.10	3 mm/s
Arthropoda	1,242,040	Bilateral	0.0	100.00	11 m/s
Vertebrata	64,832	Bilateral	0.0	99.80	~ 111 m/s
Subtotal	1,502,884				
Total	1,529,459				

Radial symmetry seems to have imposed sessility (in Porifera, Cnidaria and 700 speciose Crinoidea) or a sedentary mode of life in the remaining 6,300 echinoderm species. The radial symmetry ensures 100% clonality or clonal potency, although the potency is reduced to 1.9% in the structurally complex secondarily radial echinoderms (see Pandian, 2018).

Within radials, the structurally more organized echinoderms are less speciose than the structurally simpler Porifera and Cnidaria. On four counts, the echinoderms simulate the primitive radials but deviate from the bilaterals. (i) As in radials, the gonads of echinoderms ranges from no discrete organ/system in crinoids to filamentous tubules in holothurians (see Pandian, 2018). (ii) Adoption of the intraovarian pattern by the 900 speciose Echinoidea necessiates both oogenesis and vitellogenesis to take place within the ovary itself. In contrast, the extraovarian pattern is adopted by most bilaterals, in which oogenesis occurs in the ovary and vitellogenesis elsewhere. (iii) As in other radials, echinoderms do not have an intromittent organ. Consequently, most radials release their gametes into water and ensure the costlier external fertilization, within which either male gametes alone are released or both male and female gametes are released. The former saves investment on more valuable eggs. The recalculated values for these two groups from Pandian (2021b) listed below show that the mode of gamete release of echinoderms

Phylum	Release of (%)	
	sperms and eggs	sperms only
Porifera	33.2	66.8
Cnidaria	35.4	64.6
Acnidaria	0.0	100.0
Echinodermata	99.2	0.8
Mollusca	6.2	93.8

simulates more of the primitive radials rather than that of the bilaterals like the molluscs. (iv) It is known that spatial distribution within a single habitat limits species diversity. Like the radiates, the distribution of echinoderms is also limited solely to marine habitat. Interestingly, in the more speciose radials, Porifera (8,553 species) and Cnidaria (10,831 species), 3% sponges and 0.2% cnidarians have successfully colonized freshwater habitats, whereas the less speciose Acnidaria (166 species) and Echinodermata (7,000 species) are entirely limited to marine habitat alone (Pandian, 2021b).

Despite unicellularity, the manifestation of bilaterality has enriched diversity to 32,950 species in Protozoa. Irrespective of multicellularity and bilaterality, the acoelomate Platyhelminthes (27,700 species) and the coelomate, segmented Annelida (16,911 species) are less speciose than Protozoa. This calls for an explanation. The reason for the limited diversity may be traced to the existence of 99.84% hermaphroditism in Platyhelminthes. Unlike possessing protective shell(s) by their close relatives like the 118,451 speciose Mollusca, almost none of the annelid has a protective shell to escape predation. Being protected by a shell, the Gastropoda have diversified into 90,000 species. In its absence, the diversity is limited to 800 species, despite being highly motiles in Cephalopoda. Clearly, *the diversity depends more on protection against predation than motility.* Among Protozoa, the most speciose Rhizopoda (11,550 species), known for its slowest motility of 2–3 μm/s, Foraminifera (4,500 species) and Arcenellida (2,000 species) are protected by the shell. Some dinoflagellates are also enveloped and protected by a coat of gelatinous or tectinous material. Not surprisingly, protozoans are more speciose than the annelids.

In metazoans, sessility not only imposes radial symmetry but also coloniality. According to Blackstone and Jasker (2003), a colony is an association, in which the constituent members are intimately and organically connected together. In a rotifer colony, the members are not intimately connected, as they are in bryozoans (Wallace and Snell, 2010). At best, the rotifers may form only a pseudocolony. Estimate for the true colonies reveals that 29,647 species are colonials among the sessiles. Of 525,013 sessile metazoan species (see Table 15.7), 5.6% are colonials or 1.1% are eukaryotes.

15.7 Clonality

To gain new gene combinations – the raw material for evolution and speciation, organisms depend on (i) random mutation, (ii) segregation during meiosis and (iii) recombination at fertilization, that brings together the whole sets of haploid chromosomes from two individuals. As clonals agametically arise from additive/totipotent stem cells, they depend on random mutations alone. While the advantageous cloning saves time and resources on progeny production, its probability of gaining new gene combination is limited. Consequently, clonality may reduce species diversity.

15.7.1 Regeneration and Clonality

Regeneration and clonality are generally considered as intermingled processes. The former regains the missing parts from parental cells, tissues and/or organs. But the latter agametically multiplies individuals arising from parental additive/totipotent stem cells. For example, the liver, kidney and others can be regenerated in vertebrates, which, however, can not clonally multiply. Regeneration includes two independent processes namely wound healing and regaining the missing body parts. For information on wound healing in algae and angiosperms, Kumar et al. (2019) and Ivanov (2004) may be consulted. The unicellular Chlorophyta namely Bacillariophyceae (8,397 species), Dinophyceae (2,270) and Rhodophyta (12) undergo binary fission on attaining maximum size. Following fragmentation, they regenerate the missing parts. When injured, plants can heal the injured parts but cannot regenerate the missing parts. In a teared leaf of banana, the wound is healed but the leaf remains teared apart. In leeches too, injury is healed but they cannot regain the missing parts. However, *the unicellular algae are able to regenerate all the missing parts but this regenerative potency is limited to wound healing alone in multicellular plants; even the healing is lost in fruits and seeds of angiosperms*. Animals have retained regenerative potency for many tissues and organs but not tissues/organs like the nerve, sphincter muscle, heart and brain, as in vertebrates. Enormous efforts are being made to reprogram the differentiation process of these cells to restore regeneration. An injury, limited to the superficial layer of the skin/body, is healed and completely regenerated. But when the injury is too deep, regeneration is limited to wound healing and the process leaves a scar. An injury on the human palm, however deep it may be, is entirely regenerated – a reason why the thumb impression is considered as the most reliable mark of identification.

An overview on the incidence of regeneration in eukaryotes suggests that motile animals including Protozoa and Metazoa have retained the potency for regeneration, albeit at different levels in different taxa. Motility holds greater probability for injury due to the encounter with objects and others. Sessile animals like sponges have a limited regeneration potency. Being highly motiles, vertebrates have retained perhaps the greatest regeneration potency. In fungi and plants, *sessility has reduced the regenerative potential only to wound healing*.

15.7.2 Quantitative Estimates

A survey on the cumulative number of clonal eukaryotic species reveals that of 2,056,907 eukaryote species, *clonality decelerates species diversity to 307,914 species or 15% of eukaryotes but sexual reproduction accelerates it to 1,748,634 species (or 85% eukarytoes)* (Table 15.10).

TABLE 15.10

Estimation on the number of eukaryotic species that reproduce clonally and/or sexually. Data for Plantae and Protista are drawn from Pandian (2022, 2023), for Fungi from Table 7.3 and for Metazoa from Table 15.11; however, it is 10,917 (see p 103), **which reproduce sexually alone.

Kingdom	Total species (no.)	Clonal Species		Sexual species**	
		(no.)	(%)	(no.)	(%)
Protista	32,950	32,950	100	0	0
Fungi	106,761	95,485	89	10,917	10
Plantae	374,000	147,631	39.4	226,369	61
Metazoa	1,543,196	31,848	2	1,511,348	98
Total	2,056,907	307,914	15	1,748,634	85

15.7.3 Budders and Fragmenters

Clonal multiplication takes place either in the form of fragmentation (inclusive of fission, autotomy and others) or budding. Fragmenters share the parental body mass. But budders develop new tissues/organs using the parental reserved resources. Hence, fragmentation is costlier than budding. Whereas budders accelerate species diversity, the fragmenters decelerate it.

Quantitative estimates for the number of clonal species are available for Fungi (Table 6.1), Plantae (Pandian, 2022) and Protozoa (Pandian, 2023). However, they have to be assembled from different sources for Metazoa (Table 15.11). *These estimates for Metazoa reveal the existence of 2,206 species or 7% solitary motile fragmenters and 29,642 species or 93% sessile colonial budders.*

TABLE 15.11

Estimate on the number of solitary motile fragmenting and sessile colonial budding metazoans (compiled from Pandian, 2021b).

Phylum/ Class	Solitary fragmenters	Colonial budders	Phylum/ Class	Solitary fragmenters	Colonial budders
Porifera	–	8,553	Nemertea	4	–
Cnidaria	–	10,856	Bdelloids	461	–
Acnidaria	166	–	Sipuncula	1	–
Turbellaria	45	–	Entoprocta	–	200
Annelida	190	–	Phoronida	23	–
Echinodermata	341	–	Bryozoa	–	5,700
Placozoa	3	0	Pterobranchs	5	20
Mesozoa	–	150	Urochordata	967	1,963
Myxozoa	–	2,200	Total	2206 (6.9%)	29,642 (93.1%)

For the first time, Pandian (2021b) made estimates for the numbers of species that are fragmenters and budders in Metazoa. Interestingly, he reported that almost all fragmenters (e.g., planarians) are motile solitaries but all the budders are sessile colonials. In Protozoa also, this holds true, except for the sessile ciliates, which can be solitaries or colonials (Pandian, 2023). The cloners within the kingdoms of Protozoa, Fungi and Metazoa are amenable to this classification (see Tables 6.1, 15.12) but those of plants are not. The chlorophytes like *Ulva* fragment or multiply by clonal zoospores; so are the phaeophytes. The clonal bryophytes may either fragment or mutiply through gemma. So are lycopodiales, gymnosperms but bulbils instead of gemmae. Among flowering plants, at least 22 clonal forms were recognized by Klimes et al. (1997). Of them, (i) adventitious buds of *Alliaria petiolata*, *Rumex acetocella*, (ii) root tubers of *Ranunculus fiearia* and (iii) leaves of *Bryophyllum* alone are shown to be budders. The runners are initiated as budders but terminated as fragmenters. Incidentally, there is a need to distinguish totipotent clonals like the mint herb *Mentha spicata*, shrub *Ananas cosmosus* and drumstick tree *Moringa oleifera*, which buds, shoots and roots from (i) the stem, (ii) the leaf *Bryophyllum* or (iii) the root, as in *Tabernaemontana divaricata* and the Indian cork tree *Millingtonia hortensis*. The other pluripotent clonals like the curry plant *Murraya koenigii*, whose stem can bud to produce shoots but not roots. Notably, in all these terrestrial plants, cloning is achieved by budding. Hence, all the terrestrial clonal plants namely bryophytes (21,925 species), tracheophytes (11,850), gymnospemrs (632) and angiosperms (69,434) are considered as budders. But the 44,000 speciose aquatic algae are fragmenters.

With these assemblages, it is possible to quantify budding and fragmenting eukaryotic clonal species. Table 15.12 shows that *of 307,059 eukaryotic clonal species, 81,990 species or 27% are fragmenters. But 225,069 species or 73% are budders. Clearly, the costlier fragmenters decelerate species diversity but the cheaper budders accelerate it.*

TABLE 15.12

Estimation of fragmenting and budding clonal eukaryotic species.

Kingdom	Fragmenting species		Budding species		Source
	(no.)	(%)	(no.)	(%)	
Protista	32,950	100	–	–	Pandian (2023)
Fungi	2,834	3	91,586	97	Table 6.1
Plantae	44,000	30	103,841	70	Pandian (2022)
Metazoa	2,206	7	29,642	93	Table 15.11
Total	81,990	27	225,069	73	

In animals and plants, some species are clonals, while their respective closely related taxa may not be. The need for an explanation is obvious for the existence and lack of clonality in closely related species. One explanation is that of Pandian's hypothesis. In clonals, each stem cell plays an additive role, and a critical mass of stem cells and the resultant 'mass effect' alone achieve successful cloning. Accordingly, the critical mass of stem cells is present in all clonals but it is missing at a different level in those, which are unable to clone (Pandian, 2021b). It is not yet clear whether the hypothesis can hold true *in toto* for plants. In them too, meristem size and hence the number of stem cells govern clonality. Still, two aspects may have to be considered. In animals, stem cells are irrevocably differentiated with the onset of ontogeny. But in plants, the differentiated cells can be reversed to the stem cells. Furthermore, the unipotent stem cell number can be regained from the existing 5% pluripotent stem cells. Many hormones play an inhibitory or activation role on the meristem, while it is not yet known whether hormones have any role on the animal stem cells (see Pandian, 2022).

Aside from it, the following represents how clonality is influenced by plant hormones in plants. Leaves are considered as the terminal end in the ontogenetic developmental lineage. Unlike roots and shoots, they lack apical meristem. The terminal plant parts do not have basal meristem and when present, it may only be diffused intercalarily. But the meristem is concentrated in succulent leaves. For a list of it, Gorelick (2015) may be consulted. So much so, leaves of *Bryophyllum* and *Kalanchoe* can clonally produce plantlets along the leaf margin. It is possible to locate the meristem concentration, as plantlet primordium in the sunken area along the leaf margin. Yet, the leaves excised with a piece of stem do not produce plantlets. Experimental exposure of the excised leaf to varying concentrations of plant hormones like Indole-3-Acetic Acid (IAA), *Indole*-3-Butyric Acid (IBA) or α-Naphthalene Acetic Acid (NAA) had no significant effect on the growth of the plantlet. But cytokinin inhibited the growth. Hence, cytokinin from the stem inhibits the clonal development of plantlet (see Kulka, 2006).

Though the available information is piecemeal and not reported in a single format and unit, the undermentioned hints show that a minimal supporting somatic tissue(s) is obligately required for the development of clonals, irrespective of whether they are derived from the above- or below-ground stem; incidentally, it confirms the Pandian's hypothesis of mass effect. For example, a minimum 6 g supporting tuber tissue is required in a potato for successful clonal propagation of a single plantlet (see below for more examples). Clearly, the inhibiting stem role of leaf-derived clonals and the supporting role of stem or stem-derived tissue in clonal development display a contrasting picture.

➢ Each potato tuber *Solanum tuberosum* has 2–10 eyes or spirally arranged buds on the surface. As the largest potato size is 60 g. Hence, each bud/eye must have a supporting tissue of 6 g (Padmanabhan et al., 2016).

➢ In vines, the minimum length of ~ 40 cm runner is required for successful propagation of a single clone (*agritech@tnau.ac.in, aces.nmsu.edu*).

➢ In the betel vine, a minimum length of ~ 40 cm is required for the successful propagation of a single clone.

➢ A minimum of 10 cm and 20 cm stem cutting is required for successful propagation of a single softwood and hardwood plant, respectively (*Wikipedia*).

15.7.4 Tissue Types and Clonality

Almost all unicellular protozoan species are clonals. In them, their clonal multiplication is rarely interrupted by sexual reproduction. The ~ 7 speciose *Entamoeba* alone may not be clonals. Approximately, 3,300 species or 10.0% protozoans have secondarily lost sex (Pandian, 2023). For propagation, these anamorphic protozoans solely depend on clonal multiplication. The structurally simpler tissue-grade organization in the fungi does not allow increase in tissue types to more than seven. In them, only the agaricale complex (10,911 species, see p 103) and a few saccharomycetes (see Table 7.3) do not clonally multiply, i.e., of 106,761 species, 10,917 species or 10.2% fungi are not clonals. In them, another 16,889 species or 15.8% fungi (Table 7.1) are anamorphics and clonals.

In plants, the number of tissue types increases from 6–9 in algae to 14 in bryophytes, 30–31 in tracheophytes, ~ 41 in gymnosperms and 56–57 in angiosperms (for more details, see Pandian, 2022). For metazoans, it increases from < 6–7 in diploblastic sponges and cnidarians to 14 in triploblastic planarians (see Pandian, 2021b), 60 in polychaete worm *Nereis* (*lanwebslander.edu*) and to > 200 in mammals.

From an innovative analysis, Pandian (2021b, 2022) found an inverse relation between clonality and number of tissue types. Clonality is reduced from 100% in unicellular protozoans to 89% in the seven tissue typed fungi. In plants, the reduction is gradual from 100% in < 9 tissue typed algae to 59% in 14 tissue typed bryophytes and 41 tissue typed gymnosperms, and to 24% in 60 tissue typed angiosperms (see Fig. 6.1). But, it is drastic in metazoans and decreases from ~ 100% in 7–9 tissue typed sponges and cnidarians to < 1% for 14 tissue typed acoelomate planarians and 60 tissue typed coelomate worms, and to 0% in > 200 tissue typed vertebrates. The reasons for this difference can be traced to the fact that metazoans undergo irreversible differentiation at fertilization or at an early embryonic stage and in gonads and some somatic organs. But in plants, the differentiation is reversible, expressed at a later stage of life and is limited to reproductive organs alone. Despite the ancestoral roots of four eukaryotic kingdoms are aquatic, species diversity increases but clonality decreases with increasing structural complexity (Fig. 15.9). While protozoans have remained aquatic, the others have gone for colonization of land.

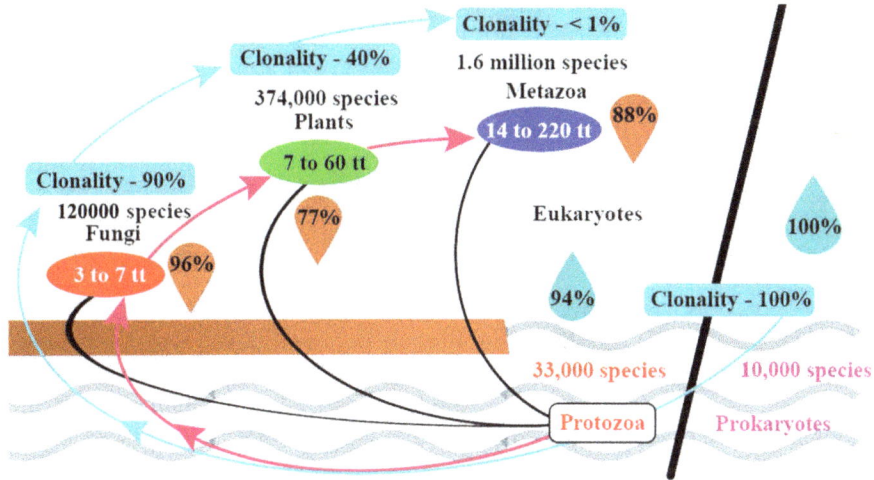

FIGURE 15.9

Origin and evolution of eukaryotic kingdoms. Though originated from waters, increasing structural complexity increases species diversity but decreases clonality. tt = tissue type.

15.8 Sexuality

Sexual reproduction is central to eukaryotic evolution by its ability to accelerate genetic diversity and purge deleterious mutations. It introduces new gene combinations during meiotic gametogenesis and recombination at fertilization. Consequently, it accelerates the processes of evolution and speciation by increasing heterozygosity at the population level. Sex was discovered by organisms ~ 2 BYA and was manifested at different times during the geological past. For example, it was manifested 1.0 BYA in foraminifers but only 100 MYA in volvocids (see Pandian, 2023). It is established as (i) monoecy and dioecy in Protozoa and (ii) in lower plants, (iii) homothallism vs heterothallism in fungi and algae, (iv) monoecy/hermaphroditism and dioecy/gonochorism in higher plants and metazoans. Sexes are phenotypic expression of sexuality.

In a way, mating types represent the 'gametes'. If not in fungi, the existence of sex determining mechanism is known for algae like chlamydomonads and volvocales. For example, the M+ and M– mating types in these algae are now identified as equivalents to female and male gametes. The ulvales and bryophytes engage the U and V sex chromosomes to determine female and male sexes, respectively (Coelho et al., 2019, Umen and Coelho, 2019, Okada et al., 2001, Renner et al., 2017). Notably, mitotic gametogenesis takes place in many protozoans, algae and bryophytes. In unicellular protozoans, true metazoan-like meriosis is not manifested. Even after its manifestation in unicellular protozoans (10%) and tissue grade fungi (16%), sex could not be maintained and is secondarily lost during different geological times (see Pandian, 2023, Table 7.1). Meiotic gametogenesis is characteristic of metazoans and higher plants (Pandian, 2021a, 2022). In Metazoa, motility seems to have accelerated evolution and speciation.

15.8.1 Monoecy vs Dioecy

In Protozoa, only a fraction of 15,142 species are sexualized. Of them, 7,706 and 7,438 species or 23.4 and 22.6% are monoecious and dioecious, respectively (Tables 15.14). *In them, the restriction of sexualization and gametogenesis to < 50% and dioecy (gonochorism) to 22.6% may have limited the protozoan diversity to 32,950 species.* Within 106,761 speciose Fungi, 89,913 species or 84.2% fungi are sexualized and the remaining 15.8% are anamorphs (Table 7.1). Of the sexualized fungi, 26,926 species or 25% fungi are homothallics and the remaining 62,987 species or 59% fungi heterothallics (Table 15.14).

TABLE 15.13

Estimatation on species number of monoecious and dioecious plants. Values arrived from Fig. 1.4, 1.5B, C and Table 10.2, Pandian (2022); *compromised values.

Phylum/Class	Species (no.)			Reference
	Total	Monoecy	Dioecy	
Cyanophyta	3,300	clonals only		
Chlorophyta	32,777	0	32,777*	
Phaeophyta	1,792	896*	896*	
Rhodophyta	6,131	6,131	0	
Bryophyta	21,925	11,615	10,310	Villarreal and Renner (2013)
Tracheophyta	11,850	10,915	935	Renner (2014)
Gymnosperms	1,079	412	667	Walas et al. (2018)
Angiosperms	295,146	280,526	14,620	Renner and Ricklefs (1995)
Total (no.)	374,000	310,495	60,205	
%		83.0	16.1	

For sexuality, data are available for species number in Protozoa, Fungi and Metazoa. But, the quantification of monoecious and dioecious plants posed problems, especially in the 32,777 speciose Chlorophyta. Though the 531 speciose Ulvophyceae generate zoosporic gametes, which morphologically do not resemble the flat sporophyte, they do resemble their clonal zoospores (Fig. 1.4 of Pandian, 2022). Notably, not many chlorophytes have distinct 'gonad(s)'. Despite possessing a sexual tissue, the sorus, sexuality in Phaeophyta remains so plastic, that it is difficult to consider them as sexualized plants. But Rhodophyta are certainly sexualized, as they have a carpogonium and spermatium to generate immotile eggs and male gametes and are likely to be homothallics. *Of 370,700 plant species (excluding the 3,300 speciose clonal Cyanophyta and 34,569 algal species, for which sexuality had to be compromised), 310,495 species or 83% are monoecious but 60,205 species or 16% are dioecious* (Table 15.13). Notably, > 690 species are obligately self-pollinating cleistogamic angiosperms (Culley and Klooster, 2007). Despite the constraints encountered in quantification of sexuality in chlorophyta and phaeophyta, it has been possible to quantify monoecy/homothallism/hermaphroditism and dioecy/heterothallism/gonochorism in red algae. Table 15.14 shows that monoecy is limited to ~ 25% in structurally simpler Protozoa and Fungi. With the beginning of structural complexity, the sessile plants and motile metazoans have opted for dichotomic pathways, i.e., ~ 83% plants have opted for monoecy but 95% metazoans for dioecy.

In 2,056,907 speciose eukaryotes, ~ 2,018,960 species (or 98.2%) are sexualized and gametogenics. Notable is the number of species that are not sexualized due to (i) clonality, (ii) secondary loss of sex and (iii) using vegetative gamete. In this category, (i) the 3,300 speciose Cyanophyta remain as clonals since their origin, when sex was not yet discovered, (ii) the structurally simpler Protozoa (3,296 species) and Fungi (16,889), 20,185 species are unable to maintain the manifested sex and have secondarily lost it; incidentally, a few hymenophyllid ferns are also reported to have lost sex in Britain (see Fusco and Minelli, 2019); however, they may not have lost it elsewhere, and (iii) approximately, 8,128 speciose protozoans are somatogamics, i.e., generate vegetative mostly mitotic (meiotic, e.g., heliozoan protozoa *Actinophyrs*) gametes. The Zygomycota and Saccharomycotina use somatogamic but sexualized gametes (see Table 15.18). Notably, *the high proportion of monoecy in plants is associated with their sessility. However, that of dioecy in metazoans is accompanied by motility. Of 2,018,960 sexualized eukaryotes, only 421,051 species or 20.5% are monoecious, but 77.7% are diecious.*

TABLE 15.14

Estimation on the number of monoecious and dioecious eukaryotic species.

Kingdom	Total (no.)	Monoecious species		Dioecious species		Source
		(no.)	(%)	(no.)	(%)	
Protozoa	32,950	7,706	23.4	7,438	22.6	Pandian (2023)
Fungi	106,761	26,926	25.2	62,987	59.0	Table 7.2*
Plantae	374,000	310,495	83.0	60,205	16.1	Table 15.13
Metazoa	1,543,196	75,924	4.9	1,467,279	95.1	Pandian (2021b)
Total	2,056,907	421,051	20.5	1,597,909	77.7	

*corrected values.

15.8.2 Selfing and its Minimization

To minimize selfing, the 5–83% monoecy, eukaryotes have developed different strategies, some of which are described hereunder:

Protozoa and Fungi: To minimize selfing in ~ 25% monoecy, *the structurally simpler Protozoa and Fungi have devised one or two strategies limited to sexual systems*. The second most speciose Ciliophora in Protozoa have devised a unique mating system that almost eliminates selfing. Firstly, members of the same clone are not eligible to conjugate. Secondly, each ciliate species consists of several *mating types*, each of which, in its turn, consists of several *mating groups*. Conjugation can occur only between members of *mating groups* but not between the *mating types* (Jennings, 1939, Sonneborn, 1939). In the chlamydomonad flagellates, anisogamic conjugation minimizes selfing by complementary conjugants appearing from two individuals (e.g., *Chlamydomonas brauni*, *C. ooganum*). Notably, their nuclei fuse to form a zygote. But in the colonial flagellate *Gonium pectorale*, the conjugant nuclei remain intact adjoiningly and separate following meiosis, as shown below:

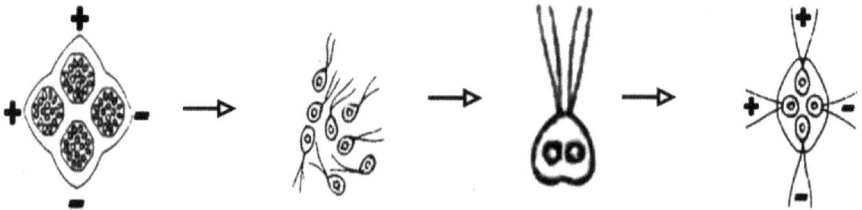

Fungi: Within the basidiomycotes, tetrapolar mating system has been devised to reduce selfing to 25% from that of 50% in dipolar mating system.

Plants: To minimize selfing in the 83% monoecy, plants have developed different strategies at structural and chemical systems. The former involves sexual and floral systems, while the latter four chemical systems (Fig. 15.10C), of which the gametophytic self-incompatibility is most prevalent. It is characterized by a single locus gene. When a haploid pollen and ovule of the same genotype approach each other, the incompatibility eliminates their fusion. Angiosperms have developed seven sexual systems (Fig. 15.10A) and many floral systems like herkogamy, enantiostyly, stylar dimorphism, heterostyly and so on (Fig. 15.10B). However, the chemical self-incompatiblity is the cheapest and is prevalent among 60% angiosperms. The structural systems are costlier and are found in 17% sexual systems and 13% floral systems. In sexual systems, the share for elimination of selfing is reduced in the following order: monoecy (12%) > gynomonoecy (3%) > andromonoecy (1%) > trimonoecy (rare). In the floral system, the long styled distyly is the chepeast, as it is formed in more number of potentially selfing plants.

From a survey, Pandian (2022) has estimated the number of plants species that are self- or cross-fertilizers, for which reliable and updated information could be assembled. Of them, 87,225 species or 23.5% plants still remain as selfers, whereas 283,475 species or 76.5% have escaped from selfing, i.e., *of 310,495 monoecious plant species (Table 15.14), 283,475 species have escaped from selfing due to the structural and chemical strategies developed by mostly angiosperms*. Interestingly, the scope for selfing in 83% monoecious plants is reduced to 24% by the foregone described strategies developed especially by angiosperms. Being the very first attempt, these values may change but the approximate shares between ~ 25% selfers and ~ 75% cross fertilizers may remain valid.

TABLE 15.15

Estimation for the number of selfing and cross fertilizing plant species (modified from Table 6.1 of Pandian, 2022).

Phylum/Class	Self-fertilizers (species no.)	Cross-fertilizers (species no.)
Algae	7,027	33,673
Bryophytes	8,476	13,449
Tracheophytes	10,805	1,045
Gymnosperms	412	667
Angiosperms	60,505	234,641
Total	87,225 (23.5%)	283,475 (76.5%)

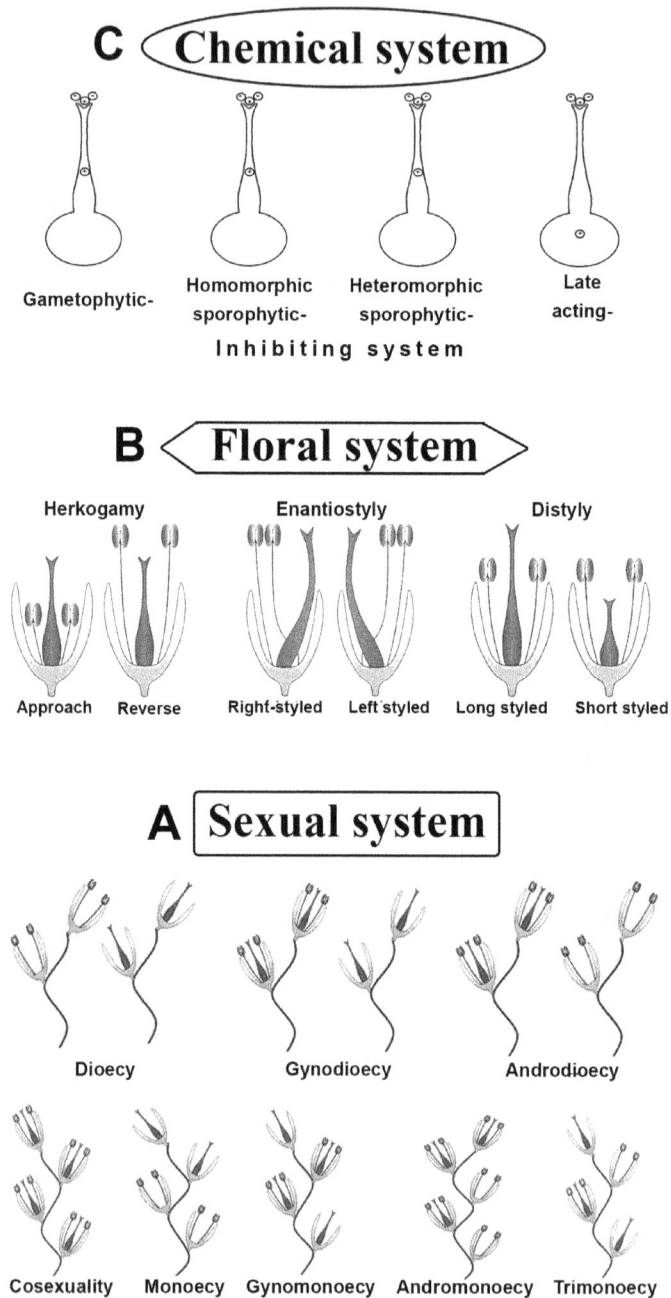

FIGURE 15.10

(A, B) Structural strategies in sexual and floral systems avoid selfing in 17 and 13% flowering plants, respectively. (C) Chemical systems inhibit selfing during post-pollination phase in 60% flowering plants.

Metazoa: In Metazoa, nine major phyla and 26 minor phyla are accommodated. The latter comprise aberrant clades that are not usually considered in the mainstream of evolution. In the phylogenetic tree, most of them terminate as blind offshoots. Furthermore, with an average of 1,795 species/phylum, the minor phyla are not as speciose (157,066 species/phylum) as those of major phyla. A major reason for the wide difference between them is that evolution seems to have proceeded from a 'wrong combination' of gonochorism and structurally simpler slow motility in lower minor phyla (e.g., acoelomate Nemertea 1,300 species, pseudocoelomate Nematomorpha 360 species) to that of structurally complex sessile hermaphrditism (e.g., eucoelomate: Bryozoa 5,700 species, Urochordata 3,000 species) (Fig. 15.11). Contrastingly, major phyletics commenced with the 'right combination' of hermaphroditism and sessility in Porifera and proceeded toward increasing structurally complex gonochorism and motility in higher major phyla like Arthropoda (1,242,040 species) and Vertebrata (64,882 species).

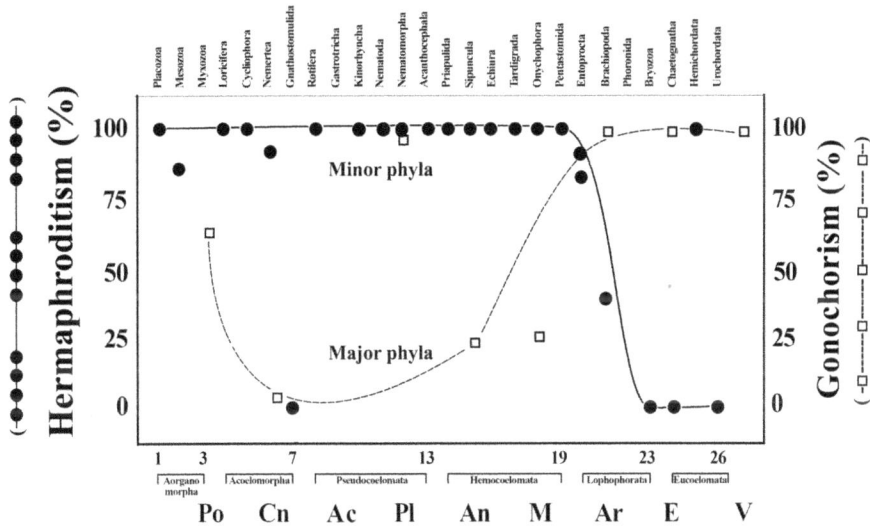

FIGURE 15.11

Contrasting trends showing proportion of gonochorism and hermaphroditism in minor and major phyla. Po = Porifera, Cn = Cnidaria, Ac = Acnidaria, Pl = Platyhelminthes, An = Annelida, M = Mollusca, Ar = Arthropoda, E = Echinodermata, V = Vertebrata (corrected from Pandian, 2021a, b).

To minimize or eliminate selfing within 5% hermaphrodites, Metazoa have also devised two level strategies. 1. The costlier and less preferred sequential hermaphroditism (SQH) involving an array of structural changes in reproductive and associated morphological systems and 2. The cheaper and more preferred behavioral change to act as gonochores.

An example is the reciprocal Simultaneous Hermaphroditism (SH) in molluscs. In *Planorbis planorbis,* one of the suitors acts as male, while the other as female (Pandian, 2017). In serranid fish, the initiator acts as male and subsequently as female (Pandian, 2011a). With enormous efforts of going through a large number of publications and computer searches, it has been possible to approximately assess the number as 60,533 species, which act as behavioural gonochores within SH. Among SQH, there are the costly protandrics (1,432 species), costlier protogynics (279) and costliest serials (29 species). During the life time, the protandrics change sex from male to female once, the protogynics do it from female to male once but the serials do it many time during their lifetime.

Within the 4.91% or 75,924 hermaphroditic metazoan species, 60,533 species or 3.92% simultaneous hermaphrodites (SH) act as behavioral gonochorics, 1,740 species or 0.11% sequential hermaphrodites (SQH); they eliminate selfing and leave only 13,647 species or 0.88% as selfing SH (Table 15.16). Apart from these, it was recently shown that hermaphroditism and parthenogenesis (incidence limited to 0.6% in Metazoa) mutually eliminate from each other. Similarly, parthenogenesis and clonality (incidence limited to < 1% in Metazoa) also mutually eliminate from each other. These show that Metazoa do not simultaneously tolerate potential inbreeding (selfing) depression arising from parthenogenesis and hermaphroditism as well as parthenogenesis and clonality (Pandian, 2021b).

TABLE 15.16

Species number of gonochores, hermaphrodites, behavioral gonochoric simultaneous hermaphrodites (SH) and sequential hermaphrodites (from Pandian, 2021b).

Particular	Species (no.)	Species (%)
Animal species	1,543,196	100.00
Structural gonochores	1,467,279	95.08
Behavioral gonochorics SH	60,533	3.92
Sequential hermaphrodites	1,740	0.11
Selfing hermaphrodites	13,647	0.88
Hermaphrodites	75,924	4.91

15.8.3 Gametes – Transfer – Fusion

Life originated from sea water. Subsequently, it diverged into freshwater and terrestrial habitats. Of 2,056,907 eukaryotic species, 402,262 species or < 20% have remained as aquatics, while the remaining > 80% have successfully colonized terrestrial habitats, as listed below. However, two points should be noted: 1. Some 72,489 species or 3.9% of them have returned to aquatic habitats (see Table 15.3), albeit their gametic pattern and mode of their gamete transfer

Clade	Total species (no.)	Aquatic species	
		(no.)	(%)
Protozoa	32,950	> 1,630	~ 6.0
Fungi	106,761	< 4,000	< 4.0
Plantae	374,000	45,065	7.2
Metazoa	1,543,196	353,567	22.9
Total	2,056,907	402,262	19.7

are retained. 2. The 33,775 speciose bryophytes and tracheophytes are terrestrials, though relegated to moist zones alone. In them, fertilization is achieved by flagellated spermatozoa swimming through aquatic *milieu*. In sexuals, especially dioecics, gametes must be transferred from one individual to another through water, air and/or vectors. As water is 800 times denser than air, this classification of aquatic and terrestrial gametes is important.

Depending on their motility, medium/mode of transfer, fertilization site, i.e., external or internal, the gametic patterns can be classified. Incidentally, vegetative somatogamic gametes may also have to be considered separately. This classification is a critically important aspect in the context of resource allocation for reproduction and associated functions, as many eukaryotes generate millions of gametes to successfully recruit a few progenies. This immense but important task was undertaken. 1. In this task, relevant data assembled in the earlier volumes on Evolution and Speciation in Animals, Plants and Protozoa were handy. However, some values reported in these books had to be corrected and updated. 2. For a few taxonomic groups, for example, the 14,714 speciose Lecanoromycetes, the life cycles remain to be described. 3. Of course, a few compromises had to be made for want of adequate information, or when available, it remains confusing or even contradictory at a few times. There are definite hints for the existence of monoecy and dioecy in the 4,500 speciose foraminifers; as a compromise 2,250 species are assigned to each of them. 4. Approximately, 10 and 16% protozoan and fungal species have secondarily lost sex and their respective numbers had to subtracted and so on. With these constraints, a beginning must be made and it has been made here. As usual, the values assembled in the forthcoming tables may not be exhaustive and precise. However, their proportions shall remain valid.

Protozoa: In unicellular Protozoa, sexuality is limited to monoecy and dioecy, as distinguished by isogamic motile gametes in the former and anisogamic gametes, of which the male gametes alone are motile and achieve oogamic fusion. Approximately, 15,144 species or 46% are sexualized, of which 7,706 (51%) and 7,438 (49%) species are monoecious and dioecious, respectively (Table 15.14). Notably, gametic fusion occurs externally in aquatic *milieu* in 61% (Table 15.17). However, external fusion by ♂ gametes occurs in parasitic protozoans within the host in 3,150 species but in open water in 2,800 species.

TABLE 15.17

Estimation on the number of species releasing all motile gametes or 'male' motile gametes only in Protozoa (compiled from Pandian, 2023).

Clade	Species (no.)		
	Total	External ♂ & ♀ gametes	External ♂ gametes only
Volvocids	400	156	244
Choanoflagellates	125	0	125
Filosians	300	300	0
Foraminifers	4,500	2,250	2,250
Radiolarians	4,200	4,200	0
Proteomyxids	75	0	75
Piroplasmeans	175	175	0
Sporozoa	5,950	2,450	3,500
Total	15,725	9,531 (60.6%)	6,194 (39.4%)

Fungi: For the following reasons, the relevant information could be assembled for 61,143 species (or 57% fungi) alone (Table 15.18). I. Description is wanted for the life cycle of 14,799 speciose Lecanoromycetes. II. The number has to be subtracted for the 16,889 anamorphic species, for the 10,212 speciose Coelomycetes, Hyphomycetes and Mycelia sterilia, which multiply clonally alone. 1. In Fungi, fertilization between immotile spores takes place predominantly in the air. However, in 2,341 species or 2.7% fungi, it occurs between motile gametes of chytrids, oomycotes and myxomycotes in aquatic *milieu*. Terrestriality seems to have diversified the gametic patterns into ~ five groups. 2. Somatogamic non-motile 'gametes' are known from unicellular Saccharomycotina (1,309 species) and hyphal 'gametangia' growing toward each other in Zygomycota (1,065 species). 3. In the agaricomycetes (13,113 species), the sexualized mating hyphae grow toward each other. 4. In the remaining basidiomycotes, mating occurs between the yeasts or spores. Notably, the sexual spores are fused externally in the non-pucciniomycotines. 5. Internal fertilization occurs between compatible spores of pezizomycotines but their fertilized ascal spores are dispersed by air. In the majority of them (29,243 species), the 'male' spore grows and penetrates the oogonium – the process is known as spermatization. Contrastingly, the trichogyne, the female gametangium curves to receive the 'male' spore in Eurotiomycetes (2,871 species). In terrestrial fungi, evolution has proceeded from external to internal fertilization. However, their fertilized spores are dispersed by air. *In all, external and internal fertilization take place in 47 and 53% fungi, respectively. However, the generation of isogamic gametes is limited to 5%, in comparison to 95% anisogamic gametes.*

TABLE 15.18

Estimation on the number of fungal species characterized by different gametic patterns. Gametic fusion is internal in pezizomycotines. In all others, it is external.

Clade	Species (no.)		
	Total	Isogamic mating types	Anisogamic mating types
1. External aquatic gametes			
Chytridiomycota	1,340	605	605
Oomycota	500	500	0
Myxomycota	500	500	0
Coelomycetes	1	0	1
Subtotal	2,341	1,605 (71.4%)[†]	606 (28.6%)[†]
2. External somatogamic vegetative gametes			
Saccharomycotina[‡]	1,309	1,309	0
Zygomycota	1,065	0	1,065
Subtotal	2,374	1,309 (55.0%)	1,065 (45.0%)
3. External mating types[‡]			
Agaricales hyphal[‡]	13,113	0	13,113
Spores/yeasts of non-agaricales[‡]	10,862	0	10,658*
Taphrinomycotina yeasts[‡]	140	0	133
Subtotal	24,115	0	23,904 (99.1%)*
4. Internal: Spermatization: ♂ gamete growing into oogonium			
Pezizomycetes/Orbiliomycetes	1,972	0	1,407*
Dothideomycetes/Arthoniomycetes	20,500	0	17,402*
Sordariomycetes	10,000	0	7,948*
Laboulbeniomycetes	2,486	0	2,486
Subtotal	34,958	0	29,243 (83.6%)
5. Internal: Trichogyne curves toward ♂ spore			
Eurotiomycetes	3,000		2,871 (95.7%)*
Total	66,778	2,914	57,685
% for 60,599 species: **Isogamics = 2,914 (4.8%); Anisogamics = 57,685 (95.2%)**			
% for 61,273 species **External = 29,159 (47.6%), Internal = 32,114 (52.4%)**			

*after substracting the number of anamorphs, †fusion achieved by either both motile isogamic gametes or motile male gametes, ‡mating achieved by growing hyphal conjugants, ‡fusion achieved through dispersal in air..

Plantae: In plants, the gametic pattern is simpler and recognizable under the basic three patterns. Both female and male gametes are motile in phaeophytes and in many chlorophytes but immotile in rhodophytes. The quantification in chlorophytes encounters problems. However, it was surmounted by subtracting the known number of chlorophytes (11,877 species), which generate motile isogamic gametes from the total number (32,777 species) to assess those, in which only the motile 'male' gametes are generated to achieve oogamic fertilization (Table 15.19). In all these algae, external fertilization occurs in open water. As mentioned earlier, although bryophytes and tracheophytes are terrestrials, they accomplish internal oogamic fertilization by flagellated spermatozoa swimming through aquatic *milieu*. Hence, these early land colonizers have not only saved motility cost but also the safety of their eggs. The 296,225 speciose sessile terrestrial spermatophytes (consisting of gymnosperms and angiosperms) share a common conserved pollination mode of transferring immotile pollens to the immotile ova held safely within the ovary of flowers. They profitably engage animals (in 251,687 species or 85%), air (44,415 species or < 15%) or water (123 species or < 0.1%) to transfer pollen from anthers to the pistil of style in flowers. Of course, animal pollinators are suitably rewarded by the offer of honey and/or pollen food. This involves cost. Contrastingly, the transfer by wind involves no cost, but requires the development of pollen capturing, adaptive structures like bracts in sepals. However, they need not develop costly showy petals and nectar. As mentioned earlier, water is 800 times denser than air, which has demanded the transfer of enlarged pollens and rarely the male flower itself (*Vallisneria*). Indeed, this transfer of costliest hydrophilous pollen limits the species number to 123 only. Within anemophily and zoophily, the latter has accelerated species diversity to the highest level in animal kingdom. *In all, male and female motile gametes are limited to 5.3%. But about 80% plants produce*

TABLE 15.19

Estimation on the number of species releasing externally ♂ and ♀, or ♂ gametes only as well as internally or externally fertilizing gametes of Plantae (condensed from Pandian, 2022). Aq Int = Aquatic internal, Aq Ext = Aquatic external.

Clade	Species (no.)				
	Total	♀ and ♂	♂	Int	Ext
Aquatic gametes (Aq)				Aq Fertilization site	
Chlorophyta	32,777	11,877	20,900	20,900	11,877
Phaeophyta	1,792	1,792	0	0	1,792
Rhodophyta	6,131	6,131	0	6,131	0
Subtotal (no.)	40,700	19,800	20,900	27,031	13,669
%		48.6	51.3	66.4	33.6

Table 15.19 contd. ...

...Table 15.19 contd.

Clade	Species (no.)				
	Total	♀ and ♂	♂	Int	Ext
Aquatic gametes in terrestrial plants (T)				Aq Fertilization site	
Bryophyta	21,925	0		21,925	0
Tracheophyta	11,850	0		11,850	0
Subtotal (no.)	33,775	0		33,775	0
%		0		100.0	0
Terrestrial (T) isogamic immotile gametes				T Fertilization site	
Spermatophyta	296,225	296,225	0	296,225	
Pollination by					
i. Animals: 251,687, 85%					
ii. Air: 44,415, < 15%					
iii. Water: 123, < 0.1%					
Total (no.)	370,700	316,025	54,675	357,031	13,669
% for 370,700 species		85.2	14.8	96.3	3.7
Isogamic: motiles: 19,800 (5.3%)		**Anisogamic motiles: 54,675 (14.8%)**			
Isogamic immotiles: 296,225 (79.9%);					
Aq ext: 13,669 (3.7%)		**Aq + T internal: 357,031 (96.3%)**			

immotile isogamic gametes and the remaining 14.8% plants - limited to aquatic habitats - generate motile male gametes (Table 15.19). Internal fertilization occurs in 96% plants, mostly in terrestrial plants.

Metazoa: Viviparity is unknown among protozoans and fungi. Sporadic incidence of viviparity is known in a very few plants (e.g,. *Rhizophora*). However, viviparity is reported from > 17,822 species and occurs across almost all taxonomic groups from Porifera to Mammalia. Hence, they are considered as oogamic internals. However, brooding is also known from polychaetes to urochordates. In the former, development (from zygote to newly born progeny) is dependent on the female. Only protection is provided in brooding by either females or males (e.g., mouthbrooding tilapias). In them, following embryonic development, larvae may be released, as in *Artemia* or the completely developed progeny, the neonates, as in isopods. In vivipares, the embryo is nourished but not in brooders. It must also be noted that oogamic fertilization in vivipares is obligately internal. The oogamy may be accomplished internally by fertilizing the immotile eggs by motile sperm introduced by the intromitant organ, as in reptiles or merely by cloacal juxtapposition, as in birds. Nevertheless, it may also occur externally, as in fish and amphibians, which spawn both immotile eggs and motile spermatozoa. Therefore, it shall be a Himalayan task to classify and subclassify the diversed

TABLE 15.20

Estimation on the number of species releasing externally ♂ and ♀ or ♂ gametes only as well as internally fertilizing gametes of Metazoa (condensed from Tables 2.1, 11.1, 16.14, 16.5 and 20.1 of Pandian, 2021b). *indicates the inclusion of spermatophore transfer, Ext = External, Int = Internal.

Clade	Species (no.)			
	Total	Ext ♀ and ♂	Int ♂	Viviparity
Aquatic gametes				
Porifera	8,553	2,840	5,680	3
Cnidaria	10,856	3,845	7,005	4
Acnidaria	166	0	166	0
Monogenea + Turbellaria	7,950	4,500	3,450	0
Polychaetes + clitellates	15,060	7,254	7,787*	19
Mollusca	118,451	7,404	110,987*	60
Crustacea	54,384	3,600	50,713*	71
Echinodermata	7,000	6,941	0	59
Teleost fish	33,790	31,933	0	1,857
Elasmobranch (Awruch, 2018)	869	0	365	504
Amphibia	5,228	5,156	0	72
Reptilia	55	0	0	55
Mammalia	130	0	0	130
Minor phyla	43,798	2,672	40,018*	1,108
Subtotal (no.)	306,290	76,145	226,171	3,942
% for 306,073 species	19.5	24.9	73.9	1.2
Terrestrial gametes				
Other Polycheates	19,750	0	19,725	25
Other clitellates	1,851	0	1,848	3
Insects	1,020,096	0	1,013,621	6,475*
Myriopods	11,800	0	11,800	0
Chelicerans	155,760	0	154,007	1,753
Minor phyla	444	0	358	86
Reptiles	9,490	0	9,335	155
Birds	1,038	0	1,038	0
Mammals	5,383	0	0	5,383
Subtotal (no.)	1,225,612	0	1,211,732	13,880
% for 1,225,797 species	80.5	0	98.9	1.1
Total (no.)	1,531,902	76,145	1,437,903	17,822
% for 1,531,870		4.9	93.9	1.2
External: 76,145 (5.0%); Internal: 1,455,725 (95.0%)				

gametic patterns. To keep uniformity, the classification for Metazoa is also brought into three categories.

Of 1,543,196 metazoan species, information is assembled for 1,531,902 species or 99.3% (Table 15.20). In them, *the externally released isogamic male and female gametes of 76,145 metazoan species fuse in the aquatic milieu.* In the remaining 1,437,903 species (93.9%), anisogamic gametes accomplish internal fertilization in aquatic and terrestrial media. Notably, *viviparity occurs in 1.2 and 1.1% metazoans in aquatic and terrestrial media.* For more information on the biology of marine viviparous mammals, Vivekanandan and Jeyabaskaran (2012) may be consulted. A look at Table 15.20 suggests that almost every metazoan phylum has proceeded from aquatic external to terrestrial internal fertilization. In terrestrial habitats, 99% metazoans have opted for internal fertilization and 1% internal fertilization is supplemented by viviparity. *In the remaining 1,437,903 species or 94% metazoans, internal fertilization occurs through oogamic fertilization in aquatic and terrestrial media. In them, only 5% metazoans achieve external fertilization.* Internal fertilization in the remaining 95% metazoans allows safety and saves motility cost of eggs.

Eukaryotes: Irrespective of diverse gametic types, their modes of fusion and fertilization site, relevant data are readily available for classification in the four kingdoms of eukaryotoes. Of 2,056,907 eukaryotic species, 1,979,504 species (or 96%) are sexualized (Table 15.21). The 8,128 species protozoans engage in the somatogamic vegetative gametes. In the sexualized eukaryotes, the gametes can be morphologically distinguishable as eggs and sperms or at molecular level in mating types. There are motile (spermatozoa) and immotile (eggs) gametes. Their mode of fusion can be isogamic appearing from the same individual in monoecy and anisogamic from two individuals in dioecy. Incidentally, the immotile male gametes of some metazoans are packed in a structure called spermatophore, which are deposited by a male in and around female's gonopore. This mode of fusion is considered as external oogamic (anisogamic) fertilization. Their fertilization site can be external in water or air. Within a single phylum like the Vertebrata, both gametes are released into water and the immotile eggs are fused by motile spermatozoa in teleost fish and amphibians or the eggs are retained within the female and motile spermatozoa are introduced through an intromitant organ (e.g., reptiles, mammals) or by cloacal juxapposition (e.g., birds).

TABLE 15.21

Estimation on the number of species releasing external ♂ and ♀ or ♂ gametes only as well as internally fertilizing gametes of eukaryotes (compiled from Table 15.17 to 15.20). Ext = External.

Kingdom	Species (no.)			
	Total	External	Internal	Viviparity
Protozoa	15,725	15,725	0	0
Fungi	61,273	29,159	32,114	0
Plantae	370,700	13,669	357,031	?
Animalia	1,531,902	76,145	1,437,903	17,822
Total	1,979,600	134,698	1,827,048	17,822
% for 1,979,600		6.8	92.3	0.9

The species numbers are summarized for the fertilization sites and viviparity in Table 15.21. For 2,056,907 eukaryotic species, relevant values could be assembled for 1,979,600 species or 96% eukaryotes. Understandably, neither internal fertilization nor viviparity is known from the unicellular Protozoa, in which fusion mode of motile gametes is limited to external sites in aquatic milieu. Notably, the external is limited to 29,159 species or 47.6% of aquatic and terrestrial fungi. Further, yeast or hyphal isogametic external fusion takes place in 2,914 fungal species of Saccharomycotina and Zygomycota and anisogamic external fusion in 24,115 basidiomycotes and taphrinomycotines. In pezizomycotines, internal fusion between anisogamic gametes is followed by dispersal of ascal spores by air. Plants share the conserved common mode of isogametic external fusion in water and anisogametic internal fusion in terrestrial bryophytes and tracheophytes. Only internal fertilization occurs in flowering plants. On land, metazoans display diverse strategies. In their 1,531,902 species, external fertilization occurs in water for 4.9% and internal fertilization in the remaining 95%, which includes ~ 1% vivipares. Considering 1,974,600 eukaryotic species, external and internal fertilizations take place in 6.3% and 91.5%, respectively. Viviparity, which includes internal fertilization both in aquatic and terrestrial habitats, involves < 1% eukaryotes. It was earlier shown (p 236), that of 2,056,907 eukaryotes, 402,262 species (or 19.7%) are aquatic. As values for (i) anamorphs and (ii) somatogamics were subtracted, the net values for aquatic sexualized species are limited to 365,056 (18.4% eukaryotes) and 1,617,685 (81.6% eukaryotes) species for aquatic and terrestrial species, respectively.

15.9 Modality

According to the Webster's New Collegiate Dictionary, modality means a model attribute. Being a biological attribute, modality is taken here to refer to modality of life style, i.e., free-living, symbiotic and parasitic modes of life. The values for symbiotic species could readily be drawn from the available sources (Table 15.22). Not surprisingly, the unusual combination of sessility and carnivory in cnidarians has required the supplementary source of nutrients engaging symbiotic dinoflagellates. The incidence of symbiotic algae is notable in the acoelate *Convoluta roscoffensis* and among molluscs in the opisthobranch *Elysia viridis* (Pandian, 1975) and chemoautotrophic bacteria in the 142 speciose vasicomycids from the thermal vents (see Pandian, 2017). Whereas symbiotic protozoans, fungi and algae are endosymbionts that between plants (gymnosperms + angiosperms), and animal pollinators represent exosymbiosism. Only a few pollinators are specialists, whose pollination is obligately associated with a single host plant. Although most pollinators are generalists, they cannot survive in the absence of host plants (e.g., bees, butterflies, moths). Hence, they are considered as symbionts. Surprisingly, *the share for symbionts among eukaryotes progressively increases from 3.3% in Protozoa to 72% in plants* (Fig. 15.12).

TABLE 15.22

Estimation on the species number of symbiotic eukaryotes.

Kingdom		Symbionts	Total	%	Source
Protozoa		1,095	32,950	3.3	
i. flagellates	343				Pandian (2023)
ii. rhizopods	263				
iii. ciliates	489				
Fungi		15,345	106,761	14.4	Table 3.5
Plantae		270,471	374,000	72.3	
Algae	2,270				
Bryophytes	10,086*				Table 11.1*
Tracheophytes	6,162*				
Gymnosperms	1,079*				
Angiosperms	250,874*				
Metazoa		360,372	1,543,196	23.4	
Cnidaria	10,856				
Desmospongiae	1				Pandian
Acoela	1				(1975, 2017, 2021b)
Opisthobranchs	1				
Vesicomyids	142*?				
Pollinators & others	349,371				
	Total	647,283	2,056,907	31.5	

? *in 15 genera (Krylova and Sahling, 2010), however, WoRMS report 137 species only.

For parasitism too, the relevant data could readily be assembled from the available sources (Table 15.23). In contrast to symbionts, the share for parasites decreases from ~ 37% in the structurally simpler Protozoa and Fungi, to 6.4% in Metazoa. Across the length and breadth, the metazoans serve as hosts. But the structurally complex eucoelomatic annelids and vertebrates do not allow manifestation of parasitism except in echinoderms with relatively more open water vascular system. *With the increasing structural complexity among the heterotrophic Protozoa and Fungi parasitism decreases in Metazoa. Strikingly, the value is reduced to 1.2% parasitism in autotrophic plants.* Parasitism is a rarity among lower plants, for example, the parasitic *Sahlingia subintegra* on the red alga *Hynea museiformis*. But there are 2,728 and 1,734 root and stem parasites among higher plants. *Autotrophism does not allow the manifestation of parasitism, especially in the lower plants.*

TABLE 15.23

Estimation on the species number of parasitic eukaryotes.

Clade	Total	Parasites	%	Source	
Heterotrophs					
Protozoa	32,950	11,150	33.8	Table 4.6, Pandian (2023)	
Fungi	106,761	42,733	40.0	Table 3.9	
Metazoa	1,543,196	98,478	6.4	Tables 22.3, 22.4, Pandian (2021b)	
Autotrophs					
Plantae	374,000	4,462	1.2	Table 3.1, Pandian (2022)	
Total	2,056,907	156,438	7.6		
Symbionts	2,056,907	647,283	31.5		
Grand Total		803,721	39.1		
Clade	Total	Symbionts	Parasites	Free-living	
Protozoa	32,950	1,095	11,150	20,705	62.8%
Fungi	106,761	15,345	42,733	48,683	45.6%
Plantae	374,000	270,471	4,462	99,067	26.5%
Metazoa	1,543,196	360,372	98,478	1,084,731	70.3%
Total	2,056,907	647,283	156,823	1,253,186	60.9%

On plotting their share values on one hand and eukaryotic kingdoms on the other, contrasting trends became apparent between symbionts and parasites. It also holds good between symbionts and free-living eukaryotes –

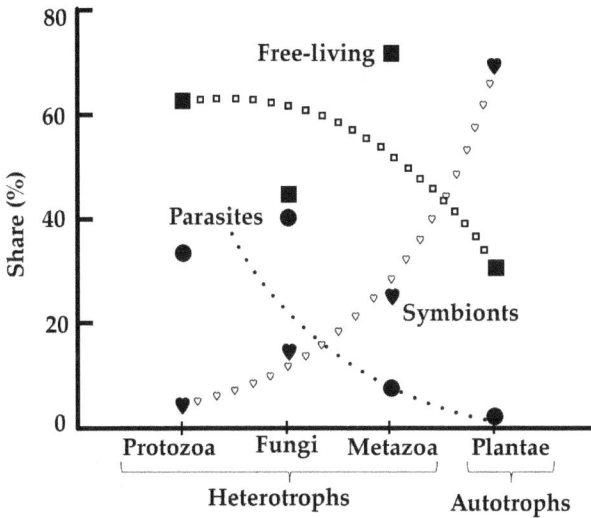

FIGURE 15.12

The shares for free-living, symbionts and parasites as a function of eukaryotic kingdoms (values are drawn from Tables 15.22, 15.23).

the feature brought to light for the first time. The values for free-living eukaryotic species were assessed by deducting the cumulative values for symbionts and parasites from the respective total number (Table 15.23). The cumulative values for free-living, symbiotic and parasitic eukaryotes are scattered. Hence, the trends shown in Fig. 15.12 are more of suggestive than actuals. Free-living species are more common among metazoans than in the unicellular Protozoa and tissue grade Fungi.

15.10 Motility

Motility is the spatial displacement of animals, which creep, crawl, swim, jump, wriggle, walk, run or fly. Unlike symmetry, clonality and sexuality, it is not a unified characteristic but varies widely with regard to many factors (see also Table 15.9), although *motility is more associated with bilaterality.* (i) *Body size* significantly influences the motility speed, which ranges from 2–3 µm/s in Rhizopoda to 400 km/h in Vertebrata and thereby introduces differences in units of measurement. Within a taxonomic group, such as, ciliates, the reported values range from 400 to 2,000 µm/s. (ii) *Motility modes*: Many animals have alternate modes of motility. The fish can swim or leap, tigers can walk or run and thereby render it difficult to reach at a specific value for motility. (iii) *Ontogeny*: Among lepidopterans, caterpillars wriggle but adults fly. Tadpole larvae swim but the adult frogs jump. (iv) *Motility duration and distance*: Not many polychaete worms and frogs can continue motility for longer than a few minutes. On the other hand, many migratory animals like the penaeid shrimp *Penaeus indicus* continue to swim at a stretch covering a distance of 380 km at the speed of 8 km/d (see Pandian, 2016). The anguillids undertake catadromous migration covering a distance of 4,000 km (see Pandian, 2021b). The migratory Monarch butterfly *Danaus plexippus* can fly over a distance of 3,000 km (Miller et al., 2012). So are migratory birds; some of them fly over a distance of 30,000 km (e.g., the Arctic tern, *nationalgeography.org*). (v) *Motility strategy*: Animals engage one or other organelles/appendages to spatially disperse. Scyphozoans (200 species) and cephalopods (800 species) eject water by forcibly pumping water to push them forward – a strategy not employed by many animals. But the ctenodrillid polychaetes alter specific gravity to undertake the low cost vertical migration. Use of swim bladder by fish in water and air-sacs by birds in air is well known.

These wide variations question the comparability of values listed in Table 15.9 and consideration of motility as a factor in regulation of species diversity. As mentioned earlier, the possession of a shell by the slowly creeping 90,000 speciose gastropods to escape predation is a more important factor to ensure species diversity rather than the fast moving 800 speciose cephalopods. Regarding motility speed, *the more stable tetrapodalism among reptiles and mammals and hexapodalism in insects have accelerated diversity but the less stable bipedolism (birds, chiropterans, apes and man) decelerated it.* Nevertheless, motility as a means to escape predation and survive, as well as search for an appropriate mate and ensure sexual reproduction, it becomes a factor in controlling species diversity. Remarkably, the species number for motile metazoans are far higher than those of sessile fungi and plants. Aside from it, there are hierarchially higher taxonomic groups, in which motility has significantly enriched species diversity. For example, the species diversity remains (~ 250 species, *brittanica.com*) less for the non-flying ratite birds but that of flying birds has increased to 9,788 species. Importantly, species diversity is controlled more by a combination of these factors rather than a single factor. Considering the identified factors, *species diversity seems to be conrolled in the following descending order: cellularity > symmetry > clonality > sexuality > modality > motility.*

Summary: Eukaryotes comprise (i) Protista (32,950 species), (ii) Fungi (106,761), (iii) Plantae (374,000), (iv) Metazoa (1,543,196) and consist of ~ 2.1 million species. Notably, with increasing structural complexity the species number increases in these four kingdoms. Currently, 6,200 eukaryotic species are described per year. Hence, the predicted increase in species number may not exceed 3 million. To describe the remaining ~ 1 million species at this rate during the next 160 years may require 40,400 taxonomists and ~ 48.5 billion US$. Despite covering 71% of the Earth's surface, the stable aquatic habitats decelerate species diversity but the labile land with 29% coverage accelerates diversity in fungi, plants and animals. The combination of unicellularity and heterotrophy does not allow colonization of land by protozoans. In contrast, the costlier mode of fungal spore dispersal and rapid solubility of externally secreted digestive enzymes have driven 96% fungi to terrestrial habitats. For, water is 800-times denser medium than air and is a universal solvent. On land, species diversity is determined by temperature and precipitation. It is limited to 0.0045 species/km^2 in deserts and 0.00018 species/km^2 in polar zones, in comparison to 5.4–65.5 species/km^2 in tropical evergreen forests. It is accelerated by higher temperature and heavier precipitation in the tropics but is progressively decelerated toward the cooler and less rainy temperate and cold and least rainy polar zones. Unicellularity decisively decelerates diversity to 2.3% eukaryotes including protozoans, and a few fungi and algae. To reduce the cost of motility in multicellular organisms, evolution seems to have proceeded from sessility to motility. *Sessility imposes radial symmetry; motility is associated with bilaterality.* About 25% eukaryotes are sessile, which imposes radial symmetry. Depending on random mutations alone to gain a new gene combination, clonality decelerates species diversity. In clonals, increasing number of tissue types is shown to reduce clonality. Accordingly, clonality decreases from 100% in unicellular protozoans to 90% in seven tissue typed fungi, 24% in plants with not more than 60 tissue types and < 1% in 14 to 60 tissue typed worms and echinoderms and to 0% in 200 tissue typed vertebrates. Clonal multiplication by fragmentation is costlier and decelerates species diversity, as against the cheaper budding clonality. Gametes are expressed as mating types in most fungi and a few protozoans. Sexuality is expressed as monoecy and dioecy in 46% protozoa and 84% fungi. It is expressed as monoecy in 83% and dioecy in 16% plants, respectively. In contrast, their equivalents are 5% hermaphrodites and 95% gonochores in metazoans. Adopting structural, biochemical and/or behavioral strategies, selfing is minimized to 26% in plants and < 1% in metazoans. Whereas the share for symbionts progressively increases from Protozoa to Plantae, that for parasites decreases from Protozoa to Metazoa. Thereby, they present contrasting trends – a feature brought to light for the first time.

16

References

Aber, J.D., Melillo, J.M. and McClaugherty, C.A. 1990. Predicting long-term patterns of mass loss, nitrogen dynamics, and soil organic matter formation from initial fine litter chemistry in temperate forest ecosystems. Can J Bot, 68: 2201–2208.

Adarsh, K. and Ram, C. 2020. Lignolytic enzymes and its mechanisms for degradation of lignocellulosic waste in environment. Heliyon, 6: e03170.

Admassu, A., Teshome, S., Ensermu, K. et al. 2016. Floristic composition and plant community types of Agama Forest, an Afromontane Forest in Southwest Ethiopia. J Ecol Nat Env, 8: 55–69.

Agrawal, D.C. 2013. Average annual rainfall over the globe. Physics Teacher, 51: 540–541.

Aime, M.C., Bell, C.D. and Wilson, A.W. 2018. Deconstructing the evolutionary complexity between rust fungi (Pucciniales) and their plant hosts. Stud Mycol, 89: 143–152.

Aime, M.C., Toome, M. and McLaughlin, D.J. 2014. Pucciniomycotina. In: Systematics and Evolution: The Mycota (eds). McLaughlin, D.J. and Spatafora, J.W., Springer-Verlag, Berlin, pp 271–295.

Akapo, O.O., Padayachee, T., Chen, W. et al. 2019. Distribution and diversity of cytochrome p450 monooxygenases in the fungal class Tremellomycetes. Int J Mol Sci, 20: 2889, DOI: 10.3390/ijms20122889.

Alexopoulos, C.J. and Mims, C.W. 1979. Introductory Mycology. Wiley, New York, p 664.

Almeida, F., Rodrigues, M.L. and Coelho, C. 2019. The still underestimated problem of fungal diseases worldwide. Front Microbiol, 10: 214, DOI: 10.3389/fmicb.2019.00214.

Alpert, P. 2000. The discovery, scope, and puzzle of desiccation tolerance in plants. Plant Ecol, 151: 5–17.

Ameen, F., Nadhari, S.A., Yassin, M.A. et al. 2021a. Desert fungi isolated from Saudi Arabia: Cultivable fungal community and biochemical production. Saudi J Biol Sci, DOI: 10.1016/j.sjbs.2021.12.011.

Ameen, F., Stephenson, S.L., Nadhari, S.A. and Yassin, M.A. 2021b. A review of fungi associated with Arabian desert soils. Nova Hedwigia, 112: 173–195.

Amselem, J., Cuomo, C., van Kan, J.A.L. and Viaud, M. 2011. Genomic analysis of the necrotrophic fungal pathogens Sclerotinia sclerotiorum and Botrytis cinerea. PLoS Genetics, 7: e1002230.

Anderson, J. and Ullrich, R.C. 1982. Diploids of Armillaria mellea: Synthesis, stability, and mating behaviour. Can J Bot, 60: 432–439.

Andreo-Jimenez, B., Ruyter-Spira, C., Bouwmeester, H.J. and Lopez-Raez, J.A. 2015. Ecological relevance of strigolactones in nutrient uptake and other abiotic stresses, and in plant-microbe interactions below-ground. Plant Soil, 394: 1–19.

Anonymous. 2017a. Stop neglecting fungi. Nat Microbiol, 2: 17120, DOI: 10.1038/nmicrobiol.2017.120.

Anonymous. 2017b. The Burden of Fungal Diseases, on LIFE. http://go.nature.com/2sMKpuN.

Aptroot, A. 1998. New lichens and lichen records from Papua New Guinea, with the description of Crustospathula, a new genus in the Bacidiaceae. Trop Bryol, 14: 25–39.

Arie, T., Kaneko, I., Yoshida, T. et al. 2000. Mating-type genes from asexual phytopathogenic ascomycetes Fusarium oxysporum and Alternaria alternata. Mol Plant Microbe Interact, 13: 1330–1339.

Armstrong, R. and Bradwell, T. 2010. Growth of crustose lichens: A review. Geogr Ann, 92: 3–17.

Arnau, J., Yaver, D. and Hjort, C.M. 2020. Strategies and challenges for the development of industrial enzymes using fungal cell factories. In: *Grand Challenges in Fungal Biotechnology* (eds). Navelainen, H. et al., Springer, Cham, pp 179–210.

Arora, D.R. and Arora, B.B. 2020. *Medical Mycology*. CBS Publications, New Delhi, p 240.

Arumugam, G.K., Srinivasan, S.K., Joshi, G. et al. 2014. Production and characterization of bioactive metabolites from piezotolerant deep sea fungus *Nigrospora* sp. In submerged fermentation. J Appl Microbiol, 118: 99–111.

Asante-Owusu, R.N., Banham, A.H., Bohnert, H.U. et al. 1996. Heterodimerization between two classes of homeodomain proteins in the mushroom *Coprinus cinereus* brings together potential DNA-binding and activation domains. Gene, 172: 25–31.

Astell, C.R., Ahlstrom-Jonasson, L. and Smith, M. 1981. The sequence of the DNAs coding for the mating-type loci of *Saccharomyces cerevisiae*. Cell, 27: 15–23.

Awasthi, D.D. 2013. *A Hand Book of Lichens*. Bishen Singh Mahendra Pal Singh, India, p 157.

Awruch, C.A. 2018. Chondrichthyes (sharks, rays, skates and chimaeras). In: *Encyclopedia of Reproduction* (ed). Skinner, M.K., Academic Press, USA, Vol 6, pp 554–559.

Badouin, H., Hood, M.E., Gouzy, J. et al. 2015. Chaos of rearrangements in the mating-type chromosomes of the anther-smut fungus *Microbotryum lychnidis-dioicae*. Genetics, 200: 1275–1284.

Bailey, R.H. 1966. Studies on dispersal of lichen soredia. J Linn Soc Bot, 59: 479–490.

Bakkeren, G. and Kronstad, J.W. 1994. Linkage of mating-type loci distinguishes bipolar from tetrapolar mating in basidiomycetous smut fungi. Proc Natl Acad Sci USA, 91: 7085–7089.

Baniya, C.B., Solhoy, T., Gauslaa, Y. and Palmer, M.W. 2010. The elevation gradient of lichen species richness in Nepal. Lichenologist, 42: 83–96.

Barnes, C.W., Kinkel, L.L. and Groth, J.V. 2005. Spatial and temporal dynamics of *Puccinia andropogonis* on *Comandra umbellata* and *Andropogon gerardii* in a native prairie. Can J Bot, 83: 1159–1173.

Bar-On, Y.M., Phillips, R. and Milo, R. 2018. The biomass distribution on Earth. Proc Natl Acad Sci USA, 115: 6506–6511.

Barr, D.J.S. 2001. Chytridiomycota. In: *The Mycota: Systematics and Evolution* (eds). McLaughlin, D.J., McLaughlin, E.G. and Lemke, P.A., Springer-Verlag, Berlin, Volume 7, Part A, pp 93–113.

Barr, M.E. 1990. Prodromus to nonlichenized, pyrenomycetous members of Class Hymenoascomycetes. Mycotaxon, 39: 43–184.

Beck, A., Kasalicky, T. and Rambold, G. 2002. Mycophotobiontal selection in a Mediterranean cryptogam community with *Fulgensia fulgida*. New Phytol, 153: 317–326.

Beckett, R.P., Minibayeva, F., Solhaug, K.A. and Roach, T. 2021. Photoprotection in lichens: Adaptations of photobionts to high light. Lichenologist, 53: 21–33.

Begerow, D., Schafer, A.M., Kellner, R. et al. 2014. Ustilaginomycotina. In: *The Mycota: Systematics and Evolution* (eds). McLaughlin, D.J. and Spatafora, J.W. Springer-Verlag, Berlin, Volume 7, Part A, pp 295–329.

Bekker, A., Holland, H.D., Wang, P.-L. et al. 2004. Dating the rise of atmospheric oxygen. Nature, 427: 117–120.

Benedict, K., Jackson, B.R., Chiller, T. and Beer, K.D. 2018. Estimation of direct healthcare costs of fungal diseases in the United States. Clin Infect Dis, 17: 1791–1797.

Bennett, A.E. and Classen, A.T. 2020. Climate change influences mycorrhizal fungi-plant interactions, but conclusions are limited by geographical study bias. Ecology, 101: e02978.

Benny, G.L., Humber, R.A. and Morton, J.B. 2001. Zygomycota: Zygomycetes. In: *The Mycota: Systematics and Evolution* (eds). McLaughlin, D.J., McLaughlin, E.G. and Lempke, P.A., Springer-Verlag, Berlin, Volume 7, Part A, pp 113–146.

Bentham, G. and Hooker, J.D. 1862–1883. *Genera Plantarum*, Spottiswoods, London.

Berg, B., Hannus, K., Popoff, T. and Theander, O. 1982. Changes in organic chemical components of needle litter during decomposition: Long-term decomposition in a Scots pine forest: I. Can J Bot, 60: 1310–1319.

Berner, R.A. 1997. The rise of plants and their effect on weathering and atmosphere CO_2. Science, 276: 544–546.

Bisognin, D.A. 2011. Breeding vegetatively propagated horticultural crops. Crop Breed Appl Biotechnol, 1S: 35–43.

Black, R. 2013. New disease reports. http://www.ndrs.org.uk/.

Blackstone, N.W. and Jasker, B.D. 2003. Phylogenetic consideration of clonality, coloniality and modes of germline development in animals. J Exp Zool, 297B: 35–47.

Blackwell, M. 2011. The fungi: 1, 2, 3 … 5.1 million species? Am J Bot, 98: 426–438.

Blackwell, M., Haelewaters, D. and Pfister, D.H. 2020. Laboulbeniomycetes: Evolution, natural history, and Thaxter's final word. Mycol Soc Am, 1–13, DOI: 10.1080/00275514.2020.1718442.

Bobrowicz, P., Pawlack, R., Correa, A. et al. 2002. The *Neurospora crassa* pheromone precursor genes are regulated by the mating type locus and the circadian clock. Mol Microbiol, 45: 795–804.

Boddy, L. 2016. Interactions with human and other animals. In: *The Fungi* (eds). Watkinson, S.C., Boddy, L. and Money, N.P., Academic Press, Amsterdam, pp 293–336.

Bolker, M., Urban, M. and Kahmann, R. 1992. The *a* mating type locus of *U. maydis* specifies cell signaling components. Cell, 88: 441–450.

Bonfante, P. and Genre, A. 2010. Mechanisms underlying beneficial plant-fungus interactions in mycorrhizal symbiosis. Nat Commu, 1: 48, DOI: 10.1038/ncomms1046.

Bourke, A. 1991. Potato late blight in Europe in 1845: The scientific controversy. In: *Phytophthora*. (eds). Lucas, J.A. et al. Cambridge University Press, New York, pp 12–24.

Bowden, J., Gregory, P.H. and Johnson, C.G. 1971. Possible wind transport of coffee leaf rust across the Atlantic Ocean. Nature, 229: 500–501.

Bowler, P.A. and Rundel, P.W. 1975. Reproductive strategies in lichens. J Linn Soc Bot, 70: 325–340.

Brady, N.C. and Weil, R.R. 2002. *The Nature and Properties of Soils*. Prentice Hall, New Jersey, p 1104.

Branco, S. 2011. Fungal diversity – an overview. In: *The Dynamical Processes of Biodiversity – Case studies of Evolution and Spatial Distribution* (eds). Grillo, O. and Venora, G., InTech Publishers, UK, pp 211–226.

Branco, S., Bi, K., Liao, H.-L. et al. 2017. Continental-level population differentiation and environmental adaptation in the mushroom *Suillus brevipes*. Mol Ecol, 26: 2063–2076.

Brefort, T., Muller, P. and Kahmann, R. 2005. The high-mobility-group domain transcription factor Rop1 is the direct regulator of the prf1 in *Ustilago maydis*. Eukaryot Cell, 4: 379–391.

Brodo, I.M., Sharnoof, S.D. and Sharnoof, S. 2001. *Lichens of North America*. Yale University Press, New Haven, USA, p 427.

Brown, J. 1997. Airborne inoculum. In: *Plant Pathogens and Diseases* (eds). Brown, J.F. and Ogle, H.J., Rockvale Publications, Australia, pp 207–218.

Brown, J.K.M. and Hovmoller, M.S. 2002. Aerial dispersal of pathogens on the global and continental scales and its impact on plant disease. Science, 297: 537–540.

Brunialti, G., Giordani, P., Ravera, S. and Frati, L. 2021. The reproductive strategy as an important trait for the distribution of lower-trunk epiphytic lichens in old-growth vs. non-old growth forests. Forests, 12: 27, DOI: 10.3390/f1010027.

Bruns, T.D., Peay, K.G., Boynton, P.J. et al. 2009. Inoculum potential of *Rhizopogon* spores increases with time over the first 4 yr of a 99-yr spore burial experiment. New Phytol, 181: 463–470.

Buller, A.H.R. 1931. *Researches on Fungi*. Volume 4, Longman, London, p 329.

Burnett, J. 1976. *Fundamentals of Mycology*. 2nd Edition, Edward Arnold Publishers, UK, p 673.

Butlin, R. 2002. Evolution of sex: The costs and benefits of sex: new insights from old asexual lineages. Nat Rev Genet, 3: 311–317.

Cairns, T.C., Nai, C. and Meyer, V. 2018. How a fungus shapes biotechnology: 100 years of *Aspergillus niger* research. Fungal Biol Biotechnol, 5: 13, DOI: 10.11876/s40694-018-0054-5.

Calvin, M. 1964. The path of carbon in photosynthesis, VI. J Chem Edu, 26(12), DOI; 10.1021/ed026p639.

Carvalho, A.B. 2003. The advantages of recombination. Nat Genet, 32: 128–129.

Casadevall, A. 2017. Don't forget the fungi when considering global catastrophic biorisks. Health Secur, 15: 341–342.

Chellappan, S., Jasmin, C., Basheer, S.M. et al. 2006. Production, purification and partial characterization of a novel protease from marine *Engyodontium album* BTMFS10 under solid state fermentation. Process Biochem, 41: 956–961.

Chomicki, G. and Renner, S.S. 2015. Phylogenetics and molecular clocks reveal the repeated evolution of ant-plants after the late Miocene in Africa and the early Miocene in Australasia and the Neotropics. New Phytol, 207: 411–424.

Christenhusz, M. and Byng, J.W. 2016. The number of known plant species in the world and its annual increase. Phytotaxa, 261: 201–217.

Chrzanowska, J., Kolaczkowska, M., Dryjanski, M. et al. 1995. Aspartic proteinase from *Penicillium camemberti*: Purification, properties, and substrate specificity. Enzym Microb Technol, 17: 719–724.

Clemmensen, K.E., Finlay, R.D., Dahlberg, A. et al. 2015. Carbon sequestration is related to mycorrhizal fungal community shifts during long-term succession in boreal forests. New Phytol, 205: 1525–1536.

Coelho, S.M., Mignerot, L. and Cock, J.M. 2019. Origin and evolution of sex-determination systems in the brown algae. New Phytol, 222: 1751–1756.

Cole, G.T. and Baron, S. 1996. Basic biology of fungi. In: *Medical Microbiology* (ed). Baron, S., University of Texas Medical Branch at Galveston.

Collinge, A.J. and Trinci, A.P. 1974. Hyphal tips of wild-type and spreading colonial mutants of *Neurospora crassa*. Arch Microbiol, 99: 353–368.

Cooper, R.D. and Sweeney, A.W. 1986. Laboratory studies on the recycling potential of the mosquito pathogenic fungus *Culicinomyces clavisporus*. J Invertebr Pathol, 48: 152–158.

Costello, M.J. and Chaudhary, C. 2017. Marine biodiversity, biogeography, deep-sea gradient and conservation. Curr Biol, 27: 511–527.

Couch, J.N. and Romney, S.V. 1973. Sexual reproduction in *Lagenidium giganteum*. Mycologia, 65: 250–252.

Coutinho, T.A., Wingfield, M.J., Alfenas, A.C. and Crous, P.W. 1998. *Eucalyptus* rust: A disease with the potential for serious international implications. Plant Dis, 82: 819–825.

Coxson, D.S. and Nadkarni, N.M. 1995. Ecological roles of epiphytes in nutrient cycles of forest ecosystems. In: *Forest Canopies* (eds). Lowman, M.D. and Nadkarni, N.M. Academic Press, San Diego, pp 495–543.

Crawford, R.L. 1981. *Lignin Biodegradation and Transformation*. John Wiley, New York, p 170.

Culley, T.M. and Klooster, M.R. 2007. The cleistogamous breeding system: A review of its genetic transformation. Planta, 243: 847–887.

Currie, J.N. 1917. The citric acid fermentation of *Aspergillus niger*. J Biol Chem, 31: 15–37.

Da Silva, M., Labas, V., Nys, Y. and Rehault-Godbert, S. 2017. Investigating proteins and proteases composing amniotic and allantoic fluids during chicken embryonic development. Poultry Sci, 96: 2931–2941.

Dal Grande, F., Widmer, I., Wagner, H.H. and Sheidegger, C. 2012. Vertical and horizontal photobiont transmission within populations of a lichen symbiosis. Mol Ecol, 21: 3159–3172.

Davison, I.R. 1991. Environmental effects on algal photosynthesis: Temperature. J Phycol, 27: 2–8.

de Bekker, C., Bruning, O., Jonker, M.J. et al. 2011. Single cell transcriptomics of neighboring hyphae of *Aspergillus niger*. Genome Biol, 12: R71.

de Meeus, T., Prugnolle, F. and Agnew, P. 2007. Asexual reproduction: genetics and evolutionary aspects. Cell Mol Life Sci, 64: 1355–1372.

De Menezes, G.C.A., Porto, B.A., Simoes, J.C. et al. 2019. Fungi in snow and glacial ice of Antarctica. In: *Fungi of Antarctica: Diversity, Ecology and Biotechnological Applications* (ed). Rosa, L.H., Springer, Cham, Switzerland, pp 127–146.

de Souza, P.M. and de Oliveira-Magalhaes, P. 2010. Application of microbial α-amylase in industry—A review. Brazil J Microbiol, 41: 850–861.

de Souza, P.M., Bittencourt, M.L.d.A., Caprara, C.C. et al. 2015. A biotechnology perspective of fungal proteases. Braz J Microbiol, 46: 337–346.

Deacon, J. 2005. Fungal spores, spore dormancy, and spore dispersal. In: *Fungal Biology*. Blackwell Publishing, USA, pp 184–212.

Dean, R., Van Kan, J.A.L., Pretorius, Z.A. et al. 2012. The top 10 fungal pathogens in molecular plant pathology. Mol Plant Pathol, 13: 414–430.

Delavaux, C.S., Smith-Ramesh, L.M. and Kuebbing, S.E. 2017. Beyond nutrients: A meta-analysis of the diverse effects of arbuscular mycorrhizal fungi on plants and soils. Ecology, 98: 2111–2119.

DeMaggio, A.E. and Greene, C. 1980. Biochemistry of fern spore germination. Plant Physiol, 66: 922–924.

Deveautour, C., Chieppa, J., Nielsen, U.N. et al. 2019. Biogeography of arbuscular mycorrhizal fungi spore traits along an aridity gradient, and responses to experimental rainfall manipulation. Fungal Ecol, DOI: 10.1016/j.funeco.2019.100899.

Dighton, J. 2009. Mycorrhizae. In: *Encyclopedia of Microbiology* (ed). Schaechter, M., Elsevier, pp 153–162.

Dodds, W.K. 2002. Animals. In: *Freshwater Ecology: Concepts and Environmental Applications*. Academic Press, pp 152–180.

Dong, W., Wang, B. and Hyde, K.D. 2020. Freshwater dothideomycetes. Fungal Div, 105: 319–575.

Drew, E.A. and Smith, D.C. 1967. Studies in the physiology of lichens. VIII. Movement of glucose from alga to fungus during photosynthesis in thallus of *Peltigera polydactyla*. New Phytol, 66: 389–400.

Drew, E.A., Murray, R.S., Smith, S.E. et al. 2003. Beyond the rhizosphere: Growth and function of arbuscular mycorrhizal external hyphae in sands of varying pore sizes. Plant Soil, 251: 105–114.

Dujon, B.A. and Louis, E.J. 2017. Genome diversity and evolution in the budding yeasts (Saccharomycotina). Genetics, 206: 717–750.

Dyer, P.S. and O'Gorman, C.M. 2011. A fungal sexual revolution: *Aspergillus* and *Penicillium* show the way. Curr Opin Microbiol, 14: 649–654.

Early, R. 2009. Pathogen control in primary production: Crop foods. In: *Foodborne Pathogens: Hazards, Risk Analysis and Control* (eds). Blackburn, C.d.W. and McClure, P.J. Woodhead Publishing, Cambridge, UK, pp 205–279.

Eggleton, P. 2000. Global patterns of termite diversity. In: *Termites: Evolution, Sociality, Symbiosis, Ecology* (eds). Abe, T., Bignell, D.E. and Higashi, M., Kluwer Academic Publishers, Dordrecht, p 469.

Elbert, W., Taylor, P.E., Andreae, M.O. and Poschl, U. 2007. Contribution of fungi to primary biogenic aerosols in the atmosphere: Wet and dry discharged spores, carbohydrates, and inorganic ions. Atmos Chem Phys, 7: 4569–4588

El-Elimat, T., Raja, H.A., Figueroa, M. et al. 2021. Freshwater fungi as a source of chemical diversity: A review. J Nat Prod, 84: 898–916.

Elix, J.A. and Stocker-Worgotter, E. 2008. Biochemistry and secondary metabolites. In: *Lichen Biology* (ed). Nash, T.H. III. Cambridge University Press, UK, pp 104–133.

Eriksson, K.E., Blanchette, R.A. and Ander, P. 1990. *Microbial and Enzymatic Degradation of Wood and Wood Components*. Springer, Berlin, p 407.

Eriksson, O.E. 2005. Outline of Ascomycota. Myconet, 11: 1–113.

Eriksson, O.E. and Winka, K. 1997. Supraordinal taxa of Ascomycota. Myconet, 1: 1–16.

Estlander, S., Kahilainen, K.K., Horppila, J. et al. 2017. Latitudinal variation in sexual dimorphism in life-history traits of a freshwater fish. Ecol Evol, 7: 665–673.

Evert, R.F. and Eichhorn, S.E. 2013. *Raven Biology of Plants*. W.H. Freeman, New York, p 919.

Farlow, A., Long, H., Arnoux, S. et al. 2015. The spontaneous mutation rate in the fission yeasts *Schizosaccharomyces pombe*. Genetics, 201: 737–744.

Fay, P. 1992. Oxygen relations of nitrogen fixation in cyanobacteria. Microbiol Rev, 56: 340–373.

Fehrer, J., Reblova, M., Bambasova, V. and Vohnik, M. 2018. The root-symbiotic *Rhizoscyphus ericae* aggregate and *Hyaloscypha* (Leotiomycetes) are congeneric: Phylogenetic and experimental evidence. Stud Mycol, 92: 195–225.

Feofilova, E.P., Ivashechkin, A.A., Alekhin, A.I. and Sergeeva, Ya.E. 2012. Fungal spores: Dormancy, germination, chemical composition and role in biotechnology. Appl Biochem Microbiol, 48: 1–11.

Fernandes, E.G., Valerio, H.M., Feltrin, T. and Sand, S.T.V.D. 2012. Variability in the production of extracellular enzymes by entomopathogenic fungi grown on different substrates. Brazil J Microbiol, 2012: 827–833.

Fiedler, M.R.M., Nitsche, B.M., Franziska, W. et al. 2013. *Aspergillus*: A cell factory with unlimited prospects. In: *Applications of Microbial Engineering* (eds). Gupta, V.K., Schmoll, M. and Maki, M. CRC Press, Boca Raton, pp 1–51.

Field, C.B., Behrenfeld, M.J., Randerson, J.T. and Falkowski, P. 1998. Primary production of the biosphere: integrating terrestrial and oceanic components. Science, 281: 237–240.

Figueroa, M., Hammond-Kosack, K.E. and Solomon, P.S. 2018. A review of wheat diseases—A field perspective. Mol Plant Pathol, 19: 1523–1536.

Finlay, B.J. and Esteban, G.F. 2018. Protozoa. In: *Reference Module in Life Sciences* (ed). Roitberg, B.D., Elsevier, pp 1–12.

Fisher, M.C., Gurr, S.J. and Cuomo, C.A. 2020. Threats posed by the fungal kingdom to humans, wildlife, and agriculture. Am Soc Microbiol, 11: e00449-20.

Fisher, P.J., Stradling, D.J. and Pegler, D.N. 1994. Leaf cutting ants, their fungus gardens and the formation of basidiomata of *Leucoagaricus gongylophorus*. Mycologist, 8: 128–131.

Fletcher, J., Luster, D., Bostock, R. et al. 2010. Emerging infectious plant diseases. In: *Emerging Infections* (eds). Scheld, W.M. et al. ASM Press, Washington, DC, pp 337–366.

Forche, A., Alby, K., Schaefer, D. et al. 2008. The parasexual cycle in *Candida albicans* provides an alternative pathway to meiosis for the formation of recombinant strains. PLoS Biol, 6: e110.

Fortin, J.A., Plenchette, C. and Piche, Y. 2009. *Mycorrhizas: The New Green Revolution*. Quebec, Canada, p 140.

Friedl, T. and Budel, B. 2008. Photobionts. In: *Lichen Biology*. Cambridge University Press, 2nd Edition, pp 9–26.

Frohlich-Nowoisky, J., Pickersgill, D.A., Despres, V.R. and Poschl, U. 2009. High diversity of fungi in air particulate matter. Proc Natl Acad Sci USA, 106: 12814–12819.

Fry, W.E., Goodwin, S.B., Dyer, A.T. et al. 1993. Historical and recent migrations of *Phytophthora infestans*: chronology, pathways and implications. Plant Dis, 77: 653–661.

Fusco, G. and Minelli, A. 2019. Reproduction: A taxonomic survey. In: *The Biology of Reproduction*. Cambridge University Press, UK, pp 342–403.

Ganesh Kumar, A., Balamurugan, K., Vijaya Raghavan, R. et al. 2019. Studies on the antifungal and serotonin receptor agonist activities of the secondary metabolites from piezotolerant deep-sea fungus *Ascotricha* sp. Mycology, 10: 92–108.

Ganesh Kumar, A., Manisha, D., Sujitha, K. et al. 2021. Genome sequence analysis of deep sea *Aspergillus sydowii* BOBA1 and effect of high pressure on biodegradation of spent engine oil. Sci Rep, 11: 9347, DOI: 10.1038/s41958-021-88525-9.

Ganugi, P., Masoni, A., Pietramellara, G. and Benedettelli, S. 2019. A review of studies from the last twenty years on plant-arbuscular mycorrhizal fungi associations and their uses for wheat crops. Agronomy, 9: 840, DOI: 10.3390/agronomy9120840.

256   *Evolution and Speciation in Fungi and Eukaryotic Biodiversity*

Garcia-Rubio, R., de Oliveira, H.C., Rivera, J. and Trevijano-Contador, N. 2020. The fungal cell wall: *Candida, Cryptococcus,* and *Aspergillus* species. Front Microbiol, 10, DOI: 10.3389/fmicb.2019.02993.

Garrido-Benevent, I. and Perez-Ortega, S. 2017. Past, present and future research in bipolar lichen-forming fungi and their photobionts. Bot Soc Am, 104: 1–15.

Gassmann, A. and Ott, S. 2000. Growth-strategy and the gradual symbiotic interactions of the lichen *Ochrolechia frigida.* Plant Biol, 2: 368–378.

Gasulla, F., del Campo, E.M., Casano, L.M. and Guera, A. 2021. Advances in understanding of desiccation tolerance of lichens and lichen-forming algae. Plants, 10: 807, DOI: 10.3390/plants10040807.

Geagea, L.W., Huber, L. and Sache, I. 1997. Removal of urediniospores of brown (*Puccinia recondite* f. sp. *tritici*) and yellow (*Puccinia striiformis*) rusts of wheat from infected leaves submitted to a mechanical stress. Eur J Plant Pathol, 103: 785–793.

Geiser, D.M., LoBuglio, K.F. and Gueidan, C. 2015. Pezizomycotina: Eurotiomycetes. In: *Systematics and Evolution. The Mycota (A Comprehensive Treatise on Fungi as Experimental Systems for Basic and Applied Research)* (eds). McLaughlin, D. and Spatafora, J., Springer, Berlin, pp 121–141.

Ghany, T.M.A. and El-Sheikh, H.H. 2016. *Mycology.* OMICS Group, USA, p 114.

Gianinazzi, S., Gollotte, A., Binet, M.N. et al. 2010. Agroecology: The key role of arbuscular mycorrhizae in ecosystem services. Mycorrhiza, 30: 519–530.

Gillman, L.N., Wright, S.D., Cusens, J. et al. 2015. Latitude, productivity and species richness. Global Ecol Biogeogr, 24: 107–117.

Glare, T.R. 1987. Effect of host species and light conditions on production of conidia by *Nomuraea rileyi.* J Invertebr Pathol, 50: 67–69.

Glare, T.R. and Milner, R.J. 1991. Ecology of entomopathogenic fungi. In: *Handbook on Applied Mycology: Humans, Animals and Insects* (eds). Arora, D.K., Mukeriji, K.G. and Pugh, J.G.F., Dekker, New York, Vol 2, pp 547–612.

Glare, T.R., Milner, R.J. and Chilvers, G.A. 1986. The effect of environmental factors on the production, discharge and germination of primary conidia of *Zoophthora phalloides* Batko. J Invertebr Pathol, 48: 275–283.

Glass, N.L., Grotelueschen, J. and Metzenberg, R.L. 1990. *Neurospora crassa* A mating-type region. Proc Natl Acad Sci USA, 87: 4912–4916.

Gleason, F.H., Lilje, O., Marano, A.V. et al. 2014. Ecological functions of zoosporic hyperparasites. Front Microbiol, 5: 1–10.

Glick, B.S. and Malhotra, V. 1998. The curious status of the Golgi apparatus. Cell, 95: 883–889.

Goettel, M.S., Eilenberg, J. and Glare, T. 2005. Entomopathogenic fungi and their role in regulation of insect populations. Comp Mol Insect Sci, 6: 361–405.

Goncalves, V.N., Cantrell, C.L., Wedge, D.E. et al. 2016. Fungi associated with rocks of the Atacama Desert: Taxonomy, distribution, diversity, ecology and bioprospection for bioactive compounds. Env Microbiol, 18: 232–245.

Gorelick, R. 2015. Why vegetative propagation of leaf cutting is possible in succulent and semi-succulent plants. Haseltonia, 20: 51–57.

Gorniak, A., Wiecko, A. and Cudowski, A. 2013. Fungi biomass in lowland rivers of north-eastern Poland: Effects of habitat conditions and nutrient concentrations. Pol J Ecol, 61: 749–758.

Gould, A.B. 2009. Fungi: Plant pathogenic. In: *Encyclopedia of Microbiology.* Elsevier, pp 457–477.

Gow, N.A.R., Latge, J.-P. and Munro, C.A. 2017. The fungal cell wall: structure, biosynthesis, and function. Microbiol Spectrum, 5: FUNK-0035-2016.

Graham, D.E., Wallenstein, M.D., Vishnivetskaya, T.A. et al. 2012. Microbes in thawing permafrost: The unknown variable in the climate change equation. ISME J, 6: 709–711.

Graham, J.H., Raz, S., Hel-Or, H. and Nevo, E. 2010. Fluctuating asymmetry: Methods, theory and applications. Symmetry, 2: 466–540.

Gregory, P.H. 1961. *The Microbiology of the Atmosphere.* Interscience Publishers, UK, p 252.

Grossart, H.-P., Hassan, E.A., Masigol, H. et al. 2021. Inland water fungi in the Anthropocene: Current and future perspectives. In: *Encyclopedia of Inland Waters* (ed). Likens, G.E., Academic Press, USA, 2nd Edition, pp 667–684.

Gruninger, R.J., Puniya, A.K., Callaghan, T.M. et al. 2014. Anaerobic fungi (phylum Neocallimastigomycota): Advances in understanding their taxonomy, life cycle, ecology, role and biotechnological potential. FEMS Microbiol Ecol, 90: 1–17.

Gueidan, C., Hill, D.J., Miadlikowska, J. and Lutzoni, F. 2015. Pezizomycotina: Lecanoromycetes. In: *Systematics and Evolution. The Mycota (A Comprehensive Treatise on Fungi as Experimental Systems for Basic and Applied Research).* (eds). McLaughlin, D. and Spatafora, J., Springer, Berlin, pp 89–120.

Gueidan, C., Ruibal, C., de Hoog, G.S. and Schneider, H. 2011. Rock-inhabiting fungi originated during period of dry climate in the late Devonian and middle Triassic. Fungal Biol, 115: 987–996.

Hajek, A.E., Wheeler, M.M., Eastburn, C.C. and Bauer, L.S. 2001. Storage of resting spores of the gypsy moth fungal pathogen, *Entomophaga maimaiga*. Biocont Sci Tech, 11: 637–647.

Hale, M.E. 1955. Phytosociology of corticolous cryptogams in the upland forests of southern Wisconsin. Ecology, 55: 45–63.

Hale, M.E. 1974. *The Biology of Lichens.* Edward Arnold, London, p 190.

Hanssen, J.F. 1974. Evaluation of hyphal lengths and fungal biomass in soil by a membrane filter technique. OIKOS, 25: 102–107.

Hartel, P.G. 1999. The soil habitat. In: *Principles and Applications of Soil Microbiology* (eds). Sylvia, D.M., Fuhrmann, J.J., Hartel, P.G. and Zuberer, D.A. Prentice Hall, New Jersey, pp 21–43.

Hatakka, A. 2001. Biodegradation of lignin, Germany. In: *Biopolymers* (eds). Hofrichter, M. and Stembuchel, A. Wiley, UK, pp 129–180.

Hawksworth, D.L. and Lucking, R. 2018. Fungal diversity revisited: 2.2 to 3.8 million species. Microbiol Spectr, 5, DOI: 10.1128/microbiolspec.FUNK-0052-2016.

Hawksworth, D.L., Kirk, P.M., Sutton, B.C. and Pegler, D.N. 1995. *Ainsworth & Bisby's Dictionary of Fungi.* CAB International, Wallingford, 8th Edition, UK, p 632.

Hawksworth, D.L., Sutton, B.C. and Ainsworth, D.C. 1983. *Ainsworth & Bisby's Dictionary of Fungi.* Commonwealth Mycological Institute, 7th Edition, UK, p 783.

Helfer, S. 2014. Rust fungi and global change. New Phytol, 201: 770–780.

Helston, R.M., Box, J.A., Tang, W. and Baumann, P. 2010. *Schizosaccharomyces cryophilus* sp. nov., a new species of fission yeast. FEMS Yeast Res, 10: 779–786.

Henskens, F.L., Green, T.G.A. and Wilkins, A. 2012. Cyanolichens can have both cyanobacteria and green algae in a common layer as major contributors to photosynthesis. Ann Bot, 110: 555–563.

Hessler, A.M. 2011. Earth's earliest climate. Nat Edu Knowl, 3: 24.

Hicks, J., Strathern, J.N. and Klar, A. 1979. Transposable mating type genes in *Saccharomyces cerevisiae*. Nature, 282: 478–483.

Hohmann, S. 2016. Nobel yeast research. FEMS Yeast Res, 16, DOI: 10.1093/femsyr/fow094.

Honegger, R. 1993. Tansley review no. 60: Development biology of lichens. New Phytol, 125: 659–677.

Honegger, R. 2008. *Lichen Biology.* Cambridge University Press, 2nd Edition, pp 27–39.

Honegger, R. 2012. The symbiotic phenotype of lichen-forming ascomycetes and their endo- and epibionts. In: *The Mycota: Fungal Association.* (ed) Hock, B. Springer, Berlin, Vol 9, pp 288–339.

Hongsanan, S., Sanchez-Ramirez, S., Crous, P.W. et al. 2016. The evolution of fungal epiphytes. Mycosphere, 7: 1690–1712.

Hood, M.E. 2002. Dimorphic mating-type chromosomes in the fungus *Microbotryum violaceum*. Genetics, 160: 457–461.

Hood, M.E. and Antonovics, J. 1998. Two-celled promycelia and mating-type segregation in *Ustilago violacea* (*Microbotryum violaceum*). Int J Plant Sci, 159: 199–205.

Hull, C.M. and Johnson, A.D. 1999. Identification of a mating type-like locus in the asexual pathogenic yeast *Candida albicans*. Science, 285: 1271–1277

Humphrey, E. 1891. The comparative morphology of the fungi. Am Nat, 1055–1069.

Hyman, L.H. 1940. Protozoa. In: *The Invertebrates: Protozoa through Ctenophora.* McGraw-Hill Book, New York, pp 44–232.

Hywel-Jones, N.L. and Webster, J. 1986. Scanning electron microscope study of external development of *Erynia conica* on *Simulium*. Trans Brit Mycol Soc, 86: 393–399.

Ingold, C.T. 1971. *Fungal Spores: Their Liberation and Dispersal.* Oxford University Press, p 302.

Insarova, I.D. and Blagoveschenskaya, E.Y. 2016. Lichen symbiosis: Search and recognition of partners. Biol Bull, 43: 408–418.

Isard, S.A., Gage, S.H., Comtois, P. and Russo, J.M. 2005. Principle of the atmospheric pathway for invasive species applied to soybean rust. BioScience, 55: 851–861.

Ishida, T.A., Nara, K., Tanaka, M. et al. 2008. Germination and infectivity of ectomycorrhizal fungal spores in relation to their ecological traits during primary succession. New Phytol, 2: 491–500.

Ivanov, V.B. 2004. Meristem as a self-renewing system: Maintenance and cessation of cell proliferation (A review). Russ J Plant Physiol, 51: 834–847.

Jackson, R.B., Lajtha, K., Crow, S.E. et al. 2017. The ecology of soil carbon: Pools, vulnerabilities, and biotic and abiotic controls. Ann Rev Ecol Evol Syst, 48: 419–445.

Jackson, R.S. 2008. Vineyard practice. In: *Wine Science: Principles and Applications.* Academic Press, Burlington, USA, pp 108–238.

Jaklitsch, W.M., Baral, H.O., Lucking, R. and Lumbsch, H.T. 2016. Ascomycota. In: *Syllabus of Plant Families – Adolf Engler's Syllabus der Pflanzenfamilien* (ed). Frey, W., Borntraeger, Stuttgart, p 288.

Jany, J.-L. and Pawloska, T.E. 2010. Multinucleate spores contribute to evolutionary longevity of asexual Glomeromycota. Am Nat, 175: 424–435.

Jaronski, S.T. 2014. Mass production of entomopathogenic fungi: State of the art. In: *Mass Production of Beneficial Organisms* (eds). Morales-Ramos, J.A. et al. Academic Press, pp 357–413.

Jennings, H.S. 1939. Genetics of *Paramecium bursaria*. I. Mating types and groups, their interrelations and distribution; mating behavior and self sterility. Genetics, 24: 202–233.

Jiang, J., Moore, J.A.M., Priyadarshi, A. and Classen, A.T. 2017. Plant-mycorrhizal interactions mediate plant community coexistence by altering resource demand. Ecology, 98: 187–197.

Jin, L., Schults, T.P. and Nicholas, D.D. 1990. Structural characterization of brown-rotted lignin. Holzforschung, 44: 133–138.

Johnson, D.A. and Cummings, T.F. 2000. Cost of fungicides used to manage potato late blight in the Columbia basin: 1996 to 1998. Am Phytopathol Soc, 84: 399–402.

Jones, E.B.G., Sakayaroj, J., Suetrong, S. et al. 2009. Classification of marine Ascomycota, anamorphic taxa and Basidiomycota. Fungal Divers, 35: 1–187.

Joshi, R. 2018. Role of enzymes in seed germination. Int J Creative Res Thoughts, 6: 1481–1484.

Juge, C., Samson, J., Bastien, C. et al. 2002. Breaking dormancy in spores of the arbuscular mycorrhizal fungus *Glomus intraradices*: A critical cold-storage period. Mycorrhiza, 12: 37–42.

Kappen, L. 1973 Response to extreme environments. In: *The Lichens* (eds). Ahmadjian, V. and Hale, M.E., Academic Press, Cambridge, USA, pp 311–380.

Kappen, L. 1988. Ecophysiological relationships in different climatic regions. In: *Handbook of Lichenology* (ed). Galun, M. CRC Press, Boca Raton, pp37–100.

Karling, J. 1936. A new predacious fungus. Mycologia, 28: 307–320.

Kelly, A.E. and Goulden, M.L. 2008. Rapid shifts in plant distribution with recent climate change. Proc Natl Acad Sci USA, 19: 11823–11826.

Kelly, M., Burke, J., Smith, M. et al. 1988. Four mating-type genes control sexual differentiation in the fission yeast. EMBO J, 7: 1537–1547.

Kemboi, D.C., Antonissen, G., Ochieng, P.E. et al. 2020. A review of the impact of mycotoxins on dairy cattle health: Challenges for food safety and dairy production in Sub-Saharan Africa. Toxins, 12: 222, DOI: 10.3390/toxins12040222.

Kenrick, P. and Crane, P.R. 1997. The origin and early evolution of plants on land. Nature, 389: 33–39

Kenrick, P., Wellman, C.H., Schneider, H. and Edgecombe, G.D. 2012. A timeline for terrestrialization: Consequences for the carbon cycle in the Palaezoic. Phil Trans R Soc, 367B: 519–536

Kinloch, B.B. 2003. White pine blister rust in North America: Past and prognosis. Phytopathology, 93: 1044–1047.

Kirk, P.M., Cannon, P.F., Minter, D.W. and Stalpers, J.A. 2008. *Dictionary of the Fungi*. CABI, UK, p 771.

Kiss, L. 2003. A review of fungal antagonists of powdery mildews and their potential as biocontrol agents. Pest Manag Sci, 59: 475–483.

Klimes, L., Klimesova, J., Hendriks, R. and Groenendael, J.V. 1997. Clonal plant architecture: A comparative analysis of form and function. In: *The Ecology and Evolution of Clonal Plants* (eds). de Kroom, H. and van Groenendari, J., Backhuys Publishers, Leiden, pp 1–29.

Kochkina, G., Ivanushkina, N., Ozerskaya, S. et al. 2012. Ancient fungi in Antarctic permafrost environments. FEMS Microbiol, 82: 501–509.

Kohlmeyer, J. and Kohlmeyer, E. 1977. New genera and species of higher fungi from the deep sea (1615–5315 m). Rev Mycol, 41: 189–206.

Kolmer, J., Ordonez, M.E. and Groth, J.V. 2009. The rust fungi. In: *Encyclopedia of Life Sciences*. John Wiley, Chichester, pp 1–8, DOI: 10.1002/9780470015902.a0021264.

Korner, C. 1995. Alpine plant diversity: A global survey and functional interpretations. In: *Arctic and Alphine Biodiversity: Patterns, Causes and Ecosystem Consequences* (eds). Chapin, F.S. and Korner, C., Springer, Berlin, pp 45–62.

Kos, J., Mastilovic, J., Janic Hajnal, E. and Saric, B. 2013. Natural occurrence of aflatoxins in maize harvested in Serbia during 2009–2012. Food Control, 34: 1–34.

Kramer, J.P. 1980. The housefly mycocis caused by *Entomophthora muscae*: Influences of relative humidity on infectivity and conidial germination. J NY Ent Soc, 88: 236–240.

Kranner, I., Cram, W.J., Zorn, M. et al. 2005. Antioxidants and photoprotection in a lichen as compared with its isolated symbiotic partners. Proc Natl Acad Sci USA, 102: 3141–3146.

Kranner, I., Beckett, R., Hochman, A. and Nash III, T.H. 2008. Desiccation-tolerance in lichens: A review. The Bryologist, 111 : 576–593.

Krebs, H.A. 1940. The citric acid cycle. Biochem J, 34: 460–463.

Krishna, M.P. and Mohan, M. 2017. Litter decomposition in forest ecosystems: A review. Energ Ecol Environ, DOI: 10.1007/s40974-017-0064-9.

Kriticos, D.J., Morin, L., Leriche, A. et al. 2013. Combining a climatic niche model of an invasive fungus with its host species distributions to identify risks to natural assets: *Puccinia psidii* sensu lato in Australia. PLoS One, 8: e64479.

Kroken, S. and Taylor, J.W. 2000. Phylogenetic species, reproductive mode, and specificity of the green alga *Trebouxia* forming lichens with the fungal genus *Letharia*. Bryologist, 103: 645–660.

Krylova, E.M. and Sahling, H. 2010. Vesicomyidae (Bivalia): Current taxonomy and distribution. PLoS One, 5: 39957.

Kuck, U., Poggeler, S., Nowrousian, M. et al. 2009. *Sordario macrospora*, a model system for fungal development. In: *The Mycota: Physiology and Genetics* (eds). Anke, T. and Weber, D., Springer-Verlag, Berlin, pp 17–39.

Kues, U., Richardson, W.V., Tymon, A.M. et al. 1992. The combination of dissimilar alleles of the A-alpha and A-beta gene complexes, whose proteins contain homeodomain motifs, determines sexual development in the mushroom *Coprinus cinereus*. Genes Dev, 6: 568–577.

Kulka, R.G. 2006. Cytokinins inhibit epiphyllous plantlet development on leaves of *Bryophyllum* (Kalanchoe) *marnierianum*. J Exp Bot, 57: 4089–4098.

Kumar, A. and Chandra, R. 2020. Ligninolytic enzymes and its mechanisms for degradation of lignocellulosic waste in environment. Heliyon, 6: e03170.

Kumar, S., Marrero-Berrios, I., Kabet, M. and Berthiaume, F. 2019. Recent advances in the use of algal polysaccharides for skin wound healing. Curr Pharm Des, 25: 1236–1248.

Kumar, S., Sharma, N.S., Saharan, M.R. et al. 2005. Extracellular acid protease from *Rhizopus oryzae*: Purification and characterization. Process Biochem, 40: 1701–1705.

Kurtzman, C.P. and Sugiyama, J. 2015. Saccharomycotina and Taphrinomycotina: The yeasts and yeastlike fungi of the Ascomycota. In: *Systematics and Evolution. The Mycota* (eds). McLaughlin, D. and Spatafora, J., Speringer, Berlin, pp 4–34.

Kyaschenko, J., Clemmensen, K.E., Karltun, E. and Lindahl, B.D. 2017. Below-ground organic matter accumulation along a boreal forest fertility gradient relates to guild interaction within fungal communities. Eco Lett, 20: 1546–1555.

Lacey, J. 1996. Spore dispersal – its role in ecology and disease: The British contribution to fungal aerobiology. Mycol Res, 100: 641–660.

Langley, J.A., Chapman, S.K. and Hungate, B.A. 2006. Ectomycorrhizal colonization slows root decomposition: The post-mortem fungal legacy. Eco Lett, 9: 955–959.

Large, E.C. 1940. *The Advance of the Fungi*. Dover Publications, New York, p 488.

Lauckner, G. 1983. Diseases of Mollusca: Bivalia. In: *Diseases of Marine Animals* (ed). Kinne, O. Biologische Anstalt Helgoland, Hamburg, Vol 2, pp 477–962.

LeBrun, E.S., Taylor, D.L., King, R.S. et al. 2018. Rivers may constitute an overlooked avenue of dispersal for terrestrial fungi. Fungal Ecol, 32: 72–79.

Lee, S.C., Ni, M., Li, W. et al. 2010. The evolution of sex: A perspective from the fungal kingdom. Microbiol Mol Biol Rev, 74: 298–340.

Lengeler, K., Fox, D.S., Fraser, J.A. et al. 2002. Mating-type locus of *Cryptococcus neoformans*: A step in the evolution of sex chromosomes. Eukaryot Cell, 1: 704–718.

Lenoir, I., Fontaine, J. and Sahraoui, A.L.H. 2016. Arbuscular mycorrhizal fungi responses to abiotic stresses: A review. Phytochemistry, 1 23: 4–15.

Levin, A.M., de Vries, R.P., Conesa, A. et al. 2007. Spatial differentiation in the vegetative mycelium of *Aspergillus niger*. Eukaryot Cell, 6: 2311–2322.

Levin, S.A. 2001–2013. *Encyclopedia of Biodiversity*. Academic Press, Vol 1–7, p 5484.

Lewis, D.H. 1973. Concepts of fungal nutrition and the origin of parasitism and mutualism. Biol Rev, 48: 261–278.

Li, J. and Zhang, K.-Q. 2014. Independent expansion of zincin metalloproteinases in Onygenales fungi may be associated with their pathogenicity. PLoS One, 9: e90225.

Li, J., Lu, L., Jia, Y. et al. 2016. Characterization of field isolates of *Magnaporthe oryzae* with mating type, DNA fingerprinting, and pathogenicity assays. Plant Dis, 100: 298–303.

Li, X., Fan, T., Zou, P. et al. 2020. Can the anatomy of abnormal flowers elucidate relationships of the androecial members in the ginger (Zingiberaceae)? Evo Devo, 11: 12.

Limon, J.J., Skalski, J.H. and Underhill, D.M. 2017. Commensal fungi in health and disease. Cell Host Microbe, 22: 156–165.

Lin, X., Hull, C.M. and Heitman, J. 2005. Sexual reproduction between partners of the same mating type in *Cryptococcus neoformans*. Nature, 434: 1017–1021.

Linde, K.v.d. and Gohre, V. 2021. How do smut fungi use plant signals to spatiotemporally orientate on and in Planta? J Fungi, 7: 107, https://doi.org/10.3390/jof7020107.

Linnaeus, C. 1753. *Species Planatarum*. Laurentius Salvius, p 1200.

Lisci, M., Monte, M. and Pacini, E. 2002. Lichens and higher plants on stone: A review. Int Biodet Biodegrad, 51: 1–17.

Liu, N.-G., Lin, C.-G., Liu, J.-K. et al. 2018. Lentimurisporaceae, a new pleosporalean family with divergence times estimates. Cryptogamie Mycologie, 39: 259–282.

Lodge, D.J. 1987. Nutrient concentrations, percentage moisture and density of field-collected fungal mycelia. Soil Biol Biochem, 19: 727–733.

Lucking, R. and Nelsen, M.P. 2018. Ediacarans, protolichens, and lichen-derived *Penicillium*: A critical reassessment of the evolution of lichenization in fungi. In: *Transformative Paleobotany* (eds). Krings, M. et al., Academic Press, UK, pp 551–590.

Lucking, R., Hodkinson, B.P. and Leavitt, S.D. 2017. Corrections and amendments to the 2016 classification of lichenized fungi in the Ascomycota and Basidiomycota. The Bryol, 120: 58–69.

Ma, L., Chen, Z., Huang da, W. et al. 2016. Genome analysis of three *Pneumocystis* species reveals adaptation mechanisms to life exclusively in mammalian hosts. Nat Commun, 7: 10740.

Maciel, M.J.M., Castro e Silva, A. and Ribeiro, H.C.T. 2010. Industrial and biotechnological applications of ligninolytic enzymes of the basidiomycota: A review. Electronic J Biotechnol, 1–6.

Magan, N. and Aldred, D. 2007. Environmental fluxes and fungal interactions: Maintaining a competitive edge. In: *Stress in Yeasts and Filamentous Fungi* (ed). van West, P., Avery, S. and Amsterdam, S.M., Elsevier, Holland, pp 19–35.

Magwene, P.M., Kayikci, O., Granek, J.A. et al. 2011. Outcrossing, mitotic recombination, and life-history trade-offs shape genome evolution in *Saccharomyces cerevisiae*. Proc Natl Acad Sci USA, 108: 1987–1992.

Magyar, D., Shoemaker, R.A., Bobvos, J. et al. 2011. *Pyrigemmula*: A novel hyphomycete genus on grapevine and tree bark. Mycol Prog, 10: 307–314.

Magyar, D., Vass, M. and Li, D.-W. 2016. Dispersal strategies of microfungi. In: *Biology of Microfungi* (ed). Li, D.-W., Springer, Switzerland, pp 315–371.

Maharachchikubura, S.S.N., Hyde, K.D., Jones, E.B.G. et al. 2016. Families of Sordariomycetes. Fungal Diversity, 79: 1–317.

Mallon, D.P. 2011. Global hotspots in the Arabian Peninsula. Zool Mid East, 43: 13–20.

Mandels, M. and Sternberg, D. 1976. Recent advances in cellulase technology. Ferment Technol, 54: 267–286.

Manter, D.K., Reeser, P.W. and Stone, J.K. 2005. A climate-based model for predicting geographic variation in Swiss needle cast severity in the Oregon coast range. Phytopathology, 95: 1256–1265.

Margulis, L., Chapman, M.J. and Hole, W. 2009. Kingdom Fungi. In: *Kingdoms and Domains: An Illustrated Guide to the Phyla of Life on Earth*. Academic Press, London, pp 379–409.

Mathavan, S. and Pandian, T.J. 1977. Patterns of emergence, import of egg energy and energy export via emerging dragonfly populations in a tropical pond. Hydrobiologia, 54: 257–272.

McConnaughey, M. 2014. Physical chemical properties of fungi. In: *Reference Module in Biomedical Research*. Elsevier, pp 1–3, http://dx.doi.org/10.1016/B978-0-12-801238-3.05231-4.

McCook, S. 2006. Global rust belt: *Hemileia vastatrix* and the ecological integration of world coffee production since 1850. J Global Hist, 1: 177–195.

McLaughlin, D. and Spatafora, J.W. 2014. *The Mycota: Systematics and Evolution*. Springer, Heidelberg, p 461.

Medina, A., Rodriguez, A. and Magan, N. 2015. Climate change and mycotoxigenic fungi: Impacts on mycotoxin production. Curr Opinion Food Sci, 5: 99–104.

Mehta, A., Bodh, U. and Gupta, R. 2017. Fungal lipases: A review. J Biotech Res, 8: 58–77.

Meletiadis, J., Meis, J.F.G.M., Mouton, J.W. and Verweij, P.E. 2001. Analysis of growth characteristics of filamentous fungi in different nutrient media. J Clinic Microbiol, 39: 478–484.

Melillo, J.M., Aber, J.D., Linkins, A.E. et al. 1989. Carbon and nitrogen dynamics along the decay continuum plant liter to soil organic matter. Plant Soil, 115: 189–198.

Merinero, S., Mendez, M., Argon, G. and Martinez, I. 2017. Variation in the reproductive strategy of a lichenized fungus along a climatic gradient. Ann Bot, 120: 63–70.

Meyer, V., Basenko, E.Y. and Benz, J.p. 2020. Growing a circular economy with fungal biotechnology: A white paper. Fungal Biol Biotechnol, 7: 5, DOI: 10.1186/s40694-020-00095-z.

Miadlikowska, J., Kauff, F., Hofstetter, V. et al. 2006. New insights into classification and evolution of the Lecanoromycetes (Pezizomycotina, Ascomycota) from phylogenetic analyses of three ribosomal RNA- and two protein-coding genes. Mycologia, 98: 1088–1103.

Michod, R.E., Bernstein, H. and Nedelcu, A.M. 2008. Adaptive value of sex in microbial pathogens. Infect Genet Evol, 8: 267–285.

Miller, G.H. and Andrews, J.T. 1972. Quaternary history of northern Cumberland peninsula, east Baffin Island, North West Territory, Canada. VI. Preliminary lichen growth curve. Geol Soc Am Bull, 83: 1133–1138.

Miller, N.G., Wassenaar, L.I., Hobson, K.A. and Norris, D.R. 2012. Migratory connectivity of the monarch butterfly (*Danaus plexippus*): Patterns of spring recolonization in eastern North America. PLoS One, 7: e31891.

Milliken, W., Zappi, D., Sasaki, D. et al. 2010. Amazon vegetation: how much don't we know and how much does it matter? Kew Bull, 65: 691–709.

Mohan, J.E., Cowden, C.C., Baas, P. et al. 2014. Mycorrhizal fungi mediation of terrestrial ecosystem responses to global change: Mini-review. Fungal Ecol, 10: 3–19.

Monastersky, R. 2012. Ancient fungi found in deep-sea mud. Nature, 492: 163–164.

Money, N.P. 2016. Spore production, discharge, and dispersal. In: *The Fungi* (eds). Watkinson, S.C., Boddy, L. and Money, N.P., Academic Press, UK, pp 67–97.

Mora, C., Tittensor, D.P., Adl, S. et al. 2011. How many species are there on earth and in the ocean? PLoS Biol, 9: e1001227.

Mukhin, V.A. and Votintseva, A. 2002. Basidiospore germination and conidial stages in the life cycles of *Fomes fomentarius* and *Fomitopsis pinicola* (Fungi, Polyporales). Polish Bot J, 47: 265–272.

Muntz, K., Belozersky, M.A., Dunaevsky, Y.E. et al. 2001. Stored proteinases and the initiation of storage protein mobilization in seeds during germination and seedling growth. J Exp Bot, 52: 1741–1752.

Murgia, M., Fiamma, M., Barac, A. et al. 2018. Biodiversity of fungi in hot desert sands. Microbiologyopen, 2019: e595.

Murugavel, P. and Pandian, T.J. 2000. Effect of altitude on hydrobiology, productivity and species richness in Kodayar – a tropical peninsular Indian aquatic system. Hydrobiology, 430: 35–57.

Naef, A., Roy, B.A., Kaiser, R. and Honegger, R. 2002. Insect-mediated reproduction of systematic infections by *Puccinia arrhenatheri* on *Berberis vulgaris*. New Phytol, 154: 717–730.

Nagarajan, S. and Singh, D.V. 1990. Long-distance dispersion of rust pathogens. Annu Rev Phytopathol, 28: 139–153.

Nair, K.K.C. and Anger, K. 1979. Experimental studies on the life cycle of *Jassa falcata* (Crustacea, Amphipoda) in laboratory culture. Helgolander Meeresunters, 32: 279–294.

Nannfeldt, J.A. 1932. *Studien uber die Morphologie und Systematic der Nicht-Lichenisierten Inoperculaten Discomyceten*. Soc der Wissenschaften, Uppsala, p 368.

Nara, K. 2009. Spores of ectomycorrhizal fungi: Ecological strategies for germination and dormancy. New Phytol, 181: 245–248.

Nash III, T.H. 2008. *Lichen Biology*. Cambridge University Press, UK, 2nd Edition, p 486.

Nazem-Bokaee, H., Hom, E.F.Y., Warden, A.C. et al. 2021. Towards a system biology approach to understanding the lichen symbiosis: Opportunities and challenges of implementing network modelling. Front Microbiol, 12: 667864.

Newcombe, G., Campbell, J., Griffith, D. et al. 2016. Revisiting the life cycle of dung fungi, including *Sordaria fimicola*. PLoS One, 11: e0147425.

Newton, S.F. and Newton, A.V. 1997. The effect of rainfall and habitat on abundance and diversity of birds in a fenced protected area in the central Saudi Arabian desert. J Arid Environ, 35: 715–735.

Ni, M., Feretzaki, M., Sun, S. et al. 2011. Sex in fungi. Annu Rev Genet, 45: 405–430.

Nieuwenhuis, B.P.S. 2012. *Sexual Selection in Fungi*. Ph.D. Thesis, Wageningen University, The Netherlands, p 157.

Nieuwenhuis, B.P.S. and James, T.Y. 2016. The frequency of sex in fungi. Phil Trans R Soc, 371B: 20150540.

Niranjan, M. and Sarma, V.V. 2018. A check-list of fungi from Andaman and Nicobar Islands, India. Phytotaxa, 347: 101–126.

Nunez, M.A., Horton, T.R. and Simberloff, D. 2009. Lack of belowground mutualisms hinders Pinaceae invasions. Ecology, 90: 2352–2359.

O'Brien, B.L., Parrent, J.L., Jackson, J.A. et al. 2005. Fungal community analysis by large-scale sequencing of environmental samples. Appl Env Microbiol, 71: 5544–5550.

O'Donnell, D., Wang, L., Xu, J. et al. 2001. Enhanced heterologous protein production in *Aspergillus niger* through pH control of extracellular protease activity. Biochem Eng J, 8: 187–193.

O'Shea, S.F., Chaure, P.T., Halsall, J.R. et al. 1998. A large pheromone and receptor gene complex determines multiple B mating type specificities in *Coprinus cinereus*. Genetics, 148: 1081–1090.

Oberwinkler, F. 2014. Dacrymycetes. In: *Systematics and Evolution. The Mycota* (eds). McLaughlin, D. and Spatafora, J. Volume 7A, Springer, Berlin, https://doi.org/10.1007/978-3-642-55318-9_13.

Oehl, F., Sieverding, E., Palenzuela, J. et al. 2011. Advances in Glomeromycota taxonomy and classification. IMA Fungus, 2: 191–199.

Okada, S., Sone, T., Fujisawa, M. et al. 2001. The Y chromosome in the liverwort *Marchantia polymorpha* has accumulated unique repeat sequences harboring a male-specific gene. Proc Natl Acad Sci USA, 98: 9454–9459.

Oliveira, D.M.P., Gomes, F.M., Carvalho, D.B. et al. 2013. Yolk hydrolases in the eggs of *Anticarsia gemmatilis* Hubner (Lepidoptera: Noctuidae): A role for inorganic phosphate towards yolk mobilization. J Insect Physiol, 59: 1242–1249.

Oneto, D.L., Golan, J., Mazzino, A. et al. 2020. Timing of fungal spore release dictates survival during atmospheric transport. Proc Natl Acad Sci USA, 117: 5134–5143.

Padmanabhan, P., Sullivan, J.A. and Paliyath, G. 2016. Potatoes and related crops. Encyclo Food Health, 2016: 446–451.

Palmer, G.E. and Horton, J.S. 2006. Mushrooms by magic: Making connections between signal transduction and fruiting body development in the basidiomycete fungus *Schizophyllum commune*. FEMS Microbiol Lett, 262: 1–8.

Panchapakesan, A. and Shankar, N. 2016. Fungal cellulases: An overview. In: *New and Future Developments in Microbiol Biotechnology and Bioengineering*. Elsevier, Amsterdam, pp 9–18.

Pandian, T.J. 1975. Mechanism of heterotrophy. In: *Marine Ecology* (ed). Kinne, O., John Wiley, London, 3A: 61–249.

Pandian, T.J. 1980. Impact of dam building on marine life. Helgolander Wissens Meeresunters, 33: 415–421.

Pandian, T.J. 1994. Crustacea. In: *Reproductive Biology of Invertebrates* (eds). Adiyodi, K.G. and Adiyodi, R.G. Oxford & IBH Publishing, New Delhi, 6A: 39–166.

Pandian, T.J. 2000. Hydrobiologia. Special Issue, 430: 1–205.

Pandian, T.J. 2002. Biodiversity status and endeavors of India. ANJAC J Sci, 1: 21–32.

Pandian, T.J. 2011a. *Sexuality in Fishes*. Science Publishers. CRC Press, USA, p 208.

Pandian, T.J. 2011b. *Sex Determination in Fish*. Science Publishers. CRC Press, USA, p 270.

Pandian, T.J. 2015. *Environmental Sex Determination in Fish*. CRC Press, USA, p 299.

Pandian, T.J. 2016. *Reproduction and Development in Crustacea*. CRC Press, USA, p 301.

Pandian, T.J. 2017. *Reproduction and Development in Mollusca*. CRC Press, USA, p 299.

Pandian, T.J. 2018. *Reproduction and Development in Echinodermata and Prochordata*. CRC Press, USA, p 270.

Pandian, T.J. 2019. *Reproduction and Development in Annelida*. CRC Press, USA, p 276.

Pandian, T.J. 2020. *Reproduction and Development in Platyhelminthes*. CRC Press, USA, p 303.

Pandian, T.J. 2021a. *Reproduction and Development in Minor Phyla*. CRC Press, USA, p 320.

Pandian, T.J. 2021b. *Evolution and Speciation in Animals*. CRC Press, USA, p 320.

Pandian, T.J. 2022. *Evolution and Speciation in Plants*. CRC Press, USA, p 354.

Pandian, T.J. 2023. *Evolution and Speciation in Protozoa*. CRC Press, USA, p 177.

Pang, K. and Jones, E. 2016. Phylogenetic diversity of fungi in the sea including the Opisthosporidia. In: *Biology of Microfungi* (ed). Li, D-W., Springer, Switzerland, pp 267–284.

Parmesan, C. and Yohe, G. 2003. A globally coherent fingerprint of climate change impacts across natural systems. Nature, 421: 37–42.

Pasanen, A.L., Pasanen, P., Jantunen, M.J. and Kalliokoski, P. 1991. Significance of air humidity and air velocity for fungal spore release into the air. Atmos Environ, 25: 459–462.

Peberdy, J.F. 1980. *Developmental Microbiology (Tertiary Level Biology)*. Springer, New York, p 240.

Pem, D., Jeewon, R., Chethana, K.W.T. et al. 2021. Species concepts of Dothideomycetes: Classification, phylogenetic inconsistencies and taxonomic standardization. Fungal Div, 109, DOI: 10.1007/s13225-021-00485-7.

Petersen, R.H. 1974. The rust fungus life cycle. Bot Rev, 40: 453–513.

Pethybridge, S.J., Hay, F.S., Esker, P.D. et al. 2008. Diseases of Pyrethrum in Tasmania: Challenges and prospects for management. Am Phytopathol Soc, 92: 1260–1272.

Pfister, D.H. 2015. Pezizomycotina: Pezizomycetes, Oribiliomycetes. In: *Systematics and Evolution. The Mycota (A Comprehensive Treatise on Fungi as Experimental Systems for Basic and Applied Research)* (eds). McLaughlin, D. and Spatafora, J., Speringer, Berlin, pp 35–57.

Phillips, M. 2017. *Mycorrhizal Planet*. Chelsea Green Publishing, USA, p 256.

Pillai, J.S. and O'Loughlin, I.H. 1972. *Coelomomyces opifexi* (Pillai and Smith) (Coelomomycetaceae: Blastocladiales). 2. Experiments in sporangial germination. Hydrobiologia, 40: 77–86.

Pipenbring, M. 2015a. Biologische Schemata, gezeichnet und freigegeben von M. Pipenbring from *common.wikimedia.org* accessed in 20 May 2022.

Pipenbring, M. 2015b. *Introduction to Mycology in the Tropics*. American Phytopathological Society, p 366.

Poinar Jr, G. 2016. A mid-Cretaceous ectoparasitic fungus, *Spheciophila adercia* gen et sp. nov. attached to a wasp in Myanmar amber. Fungal Genomics Biol, 6: 2.

Porada, P., Lenton, T.M. and Pohl, A. et al. 2016. High potential for weathering and climate effects of non-vascular vegetation in the late Ordovician. Nat Comm, 7: 2113.

Prasad, R. 2018. *Fungal Nanotechnology: Applications in Agriculture, Industry, and Medicine*. Springer, India, p 295.

Pringle, A. and Taylor, J.W. 2002. The fitness of filamentous fungi. Trends Microbiol, 10: 474–481.

Purdy, L.J., Krupa, S.V. and Dean, J.L. 1985. Introduction of sugarcane rust into the Americas and its spread to Florida. Plant Dis, 69: 689–693.

Putnam, M.L. and Sindermann, A.B. 1994. Eradication of potato wart disease from Maryland. Am Potato J, 71: 743–747.

Radwan, G.L.H.E. 2013. Molecular comparison and DNA fingerprinting of *Sporisorium reilianum* and *Peronosclerospora sorghi* relating to host specificity and host resistance. Ph.D. Thesis, Texas A&M University, College Station, USA.

Raghukumar, C., Raghukumar, S., Sharma, S. and Chandramohan, D. 1992. Endolithic fungi from deep-sea calcareous substrata: Isolation and laboratory studies. In: *Oceanography of the Indian Ocean* (ed). Desai, B.N. Oxford IBH Publication, New Delhi, pp 1–7.

Raghukumar, C., Raghukumar, S., Sheelu, G. et al. 2004. Buried in time: Culturable fungi in a deep-sea sediment core from the Chagos Trench, Indian Ocean. Deep-Sea Res, 51: 1759–1768.

Raghuwanshi, R., Singh, S., Aamir, M. et al. 2016. Dispersal strategies of microfungi. In: *Biology of Microfungi* (ed). Li, D-W., Springer, Switzerland, pp 543–572.

Ramesh, T., Yamunadevi, R., Sundaramanickam, A. et al. 2021. Biodiversity of the fungi in extreme marine environments. In: *Fungi Bio-Prospects in Sustainable Agriculture, Environment and Nanotechnology: Extremophilic Fungi and Myco-mediated Environmental Management*. Academic Press, pp 75–100.

Rana, A., Sahgal, M. and Johri, B.N. 2017. *Fusarium oxysporum*: Genomics, diversity and plant–host interaction. In: *Developments in Fungal Biology and Applied Mycology* (eds). Satyanarayana, T. et al., Springer Nature, Singapore, pp 159–199.

Ranzoni, F.V. 1968. Fungi isolated in culture from soils of the Sonoran Desert. Mycologia, 60: 356, DOI: 10.2307/3757166.

Raper, J. 1966. *Genetics of Sexuality in Higher Fungi*. Ronald Press, New York, p 283.

Raspor, P. and Zupan, J. 2006. Yeasts in extreme environments. In: *Biodiversity and Ecophysiology of Yeasts* (eds). Rosa, C.A. and Peter, G., Springer-Verlag, Berlin, pp 371–418.

Rayner, A.D.M. and Boddy, L. 1988. *Fungal Decomposition of Wood: Its Biology and Ecology*. John Wiley, Chichester, p 587.

Read, N.D. 2017. Fungal cell structure and organization. In: *Oxford Textbook of Medical Mycology* (eds). Kibbler, C.C. et al., Oxford University Press, pp 1–20.

Reddy, M.V. and Venkataiah, B. 1989. Influence of microarthropod abundance and climate factors on weight loss and mineral nutrient contents of *Eucalyptus* leaf litter during decomposition. Biol Fertility Soils, 8: 319–324.

Reinhart, K.O. and Callaway, R.M. 2006. Soil biota and invasive plants. New Phytol, 170: 445–457.

Renner, S.S. 2014. The relative and absolute frequencies of angiosperm sexual systems: Dioecy, monoecy, gynodioecy, and an updated online database. Am J Bot, 101: 1588–1596.

Renner, S.S. and Ricklefs, R.E. 1995. Dioecy and its correlates in the flowering plants. Am J Bot, 82: 596–606.

Renner, S.S., Heinrichs, J. and Sousa, A. 2017. The sex chromosome of bryophytes: Recent insights, open questions, and reinvestigations of *Frullania dilatata* and *Plagiochila asplenioides*. J Syst Evol, 55: 333–339.

Ribeiro, R.V., Machado, E.C. and de Oliveira, R.F. 2006. Temperature response of photosynthesis and its interaction with light intensity in sweet orange leaf discus under non-photorespiratory condition. Cienc Agrotec, 30: 670–678.

Rigling, D. and Prospero, S. 2018. *Cryphonectria parasitica*, the causal agent of chestnut blight: Invasion history, population biology and disease control. Mol Plant Pathol, 19: 7–20.

Rikkinen, J. 2002. Cyanolichens: An evolutionary overview. In: *Cyanobacteria in Symbiosis* (eds). Rai, A.N., Bergman, B. and Rasmussen, U., Kluwer Academic Publishers, Dordrecht, pp 31–72.

Richardson, G.H., Nelson, J.H., Lubnow, R.E. and Schwarberg, R.L. 1967. Rennin-like enzyme from *Mucor pusillus* for cheese manufacture. J Dairy Sci, 50: 1066–1072.

Rodriguez-Saiz, M., Fuente, J.L.d.l. and Barredo, J.L. 2010. *Xanthophyllomyces dendrorhous* for the industrial production. Appl Microbiol Biotechnol, 88: 645–658.

Rodriquez, M.d.C.H., Evans, H.C., de Abreu, L.M. et al. 2021. New species and records of *Trichoderma* isolated as mycoparasites and endophytes from cultivated and wild coffee in Africa. Sci Rep, 11: 5671.

Rodrigues, M.L., Nimrichter, L., Oliveira, D.L. et al. 2008. Vesicular trans-cell wall transport in fungi: A mechanism for the delivery of virulence-associated macromolecules? Lipid Insights, 2008: 27–40.

Rodrigues, M.L., Nosanchuk, J.D., Schrank, A. et al. 2011. Vesicular transport systems in fungi. Future Microbiol, 6: 1371–1381.

Ronnas, C., Werth, S., Ovaskainen, O. et al. 2017. Discovery of long-distance gamete dispersal in lichen-forming ascomycete. New Phytol, 216: 216–226.

Rotem, J. 1994. *The Genus Alternaria*. APS Press, Minnesota, USA, p 326.

Roth Jr. F.J., Orpurt, P.A. and Ahearn, D.G. 1964. Occurrence and distribution of fungi in a subtropical marine environment. Can J Bot, 42: 375–383.

Rousk, J., Baath, E., Brookes, P.C. et al. 2010. Soil bacterial and fungal communities across a pH gradient in an arable soil. ISME J, 4: 1340–1351.

Roy, B.A. 1994. The use and abuse of pollinators by fungi. Trends Ecol Evol, 9: 335–339.

Rozewicz, M., Wyznska, M. and Grabinski, J. 2021. The most important fungal diseases of cereals —Problems and possible solutions. Agronomy, 11: 714, DOI: 10.3390/agronomy11040714.

Ruderfer, D.M., Pratt, S.C., Seidel, H.S. and Kruglyak, L. 2006. Population genomic analysis of outcrossing and recombination in yeast. Nat Genet, 38: 1077–1081.

Rundel, P.W., Dillon, M.O., Palma, B. et al. 1991. The phytogeography and ecology of the coastal Atacama and Peruvian deserts. Aliso, 11: 1–50.

Sager, R. and Granick, S. 1954. Nutritional control of sexuality in *Chlamydomonas reinhardtii*. J Gen Physiol, 37: 729–742.

Samarakoon, M.C., Hyde, K.D., Promputtha, I. et al. 2016. Divergence and ranking of taxa across the kingdoms Animalia, Fungi and Plantae. Mycosphere, 7: 1678–1689.

Samuels, G.J. and Blackwell, M. 2001. Pyrenomycetes-fungi with perithecia. In: *The Mycota* (eds). McLaughlin, D.J., McLaughlin, E.G. and Lemke, P.A., Springer, Berlin pp 221–255.

Sanderman, J. and Amundson, R. 2003. Biogeochemistry of decomposition and detrital processing. In: *Treatise on Geochemistry* (eds). Hollands, H. and Turekian, K.K., Elsevier, Vol 8, pp 249–316.

Santiago, I.F., Goncalves, V.N., Gomez-Silva, B. et al. 2018. Fungal diversity in the Atacama Desert. Antonie van Leeuwenhoek, DOI: 10.1007/s10482-018-1060-6.

Sapir, A., Dillman, A.R., Connon, S.A. et al. 2014. Microsporidia-nematode associations in methane seeps reveal basal fungal parasitism in the deep sea. Front Microbiol, 5, DOI: 10.3389/fmicb.2014.00043.

Saranraj, P. and Stella, D. 2013. Fungal amylase—A review. Int J Microbiol Res, 4: 203–211.

Sarma, V.V. 2019. Marine fungal diversity: Present status and future perspectives. In: *Microbial Diversity in Ecosystem Sustainability and Biotechnological Applications* (eds). Satyanarayana, T. et al. Springer Nature, Singapore, pp 267–291.

Sarma, V.V. and Raghukumar, S. 2013. Manglicolous fungi from Chorao mangroves, Goa, West coast of India: Diversity and frequency of occurrence. Nova Hedwigia, 97: 533–542.

Savitha, S., Sadhasivam, S., Swaminathan, K. et al. 2011. Fungal protease: production, purification and compatibility with laundry detergents and their wash performance. J Taiwan Inst Chem Eng, 42: 298–304.

Scheidegger, C. 1985. Systematische Studien zur Krustenflechte *Anzina carneonivea* (Trapeliaceae, Lecnorales). Nova Hedwigia, 41: 191–218.

Schirawski, J., Heinze, B., Wagenknecht, M. and Kahmann, R. 2005. Mating type loci of *Sporisorium reilianum*: Novel pattern with three *a* and multiple *b* specificities. Eukaryot cell, 4: 1317–1327.

Schisler, D.A., Janisiewicz, W.J., Boekhout, T. and Kurtzman, C.P. 2010. Agriculturally important yeasts: biological control of field and postharvest diseases using yeast antagonists, and yeasts as pathogens of plants. In: *The Yeasts. A Taxonomic Study* (eds). Kurtzman, C.P., Fell, J.W. and Boekhout, T., Elsevier, Vol 1, pp 45–52.

Schoch, C. and Grube, M. 2015. Pezizomycotina: Dothideomycetes and Arthoniomycetes. In: *Systematics and Evolution. The Mycota (A Comprehensive Treatise on Fungi as Experimental Systems for Basic and Applied Research)* (eds). McLaughlin, D. and Spatafora, J., Springer, Berlin, pp 143–176.

Seifert, K.A. and Gams, W. 2001. The taxonomy of anamorphic fungi. In: *The Mycota: Systematics and Evolution* (eds). McLaughlin, D.G. et al., Springer-Verlag, New York, Volume 7, Part A, pp 307–347.

Sephton-Clark, P.C.S. and Voelz, K. 2018. Spore germination of pathogenic filamentous fungi. Adv Appl Micobiol, 102: 117–157.

Sette, L.D. and Santos, R.C.B. 2013. Ligninolytic enzymes from marine-derived fungi: Production and applications. In: *Marine Enzymes for Biocatalysis: Sources, Biocatalytic Characteristics and Bioprocesses of Marine Enzymes.* Woodhead Publishing, UK, pp 403–427.

Sharma, H.S.S. 1989. Economic importance of thermophilous fungi. Appl Microbiol Biotechnol, 31: 1–10.

Sharnoff, S. 2014. *A Field Guide to California Lichens*. Yale University Press, New Haven, p 416.

Shen, W., Bobrowicz, P. and Ebbole, D.J. 1999. Isolation of pheromone precursor genes of *Magnaporthe grisea*. Fungal Genet Biol, 27: 253–263.

Sherwood, R.K. and Bennett, R.J. 2009. Fungal meiosis and parasexual reproduction – lessons from pathogenic yeast. Curr Opin Microbiol, 12: 599–607.

Shi, L.-L., Mortimer, P.E., Silk, J.W.F. et al. 2013. Variation in forest soil fungal diversity along a latitudinal gradient. Fungal Div, 64: 305–315.

Shimazu, M. and Soper, R.S. 1986. Pathogenicity and sporulation of *Entomophaga maimaiga* Humber, Shimazu, Soper and Hajek (Entomophthorales Entomphthoraceae) on larvae of the gypsy moth *Lymantria dispar* L. (Lepidoptera Lymantriidae). Appl Entomol Zool, 21: 589–596.

Shlezinger, N., Irmer, H., Dhingra, S. et al. 2017. Sterilizing immunity in the lung relies on targeting fungal apoptosis-like programmed cell death. Science, 357: 1037–1041.

Silva-Flores, P., Arguelles-Moyao, A., Aguilar-Paredes, A. et al. 2021. Mycorrhizal science outreach: scope of action and available resources in the face of global change. Plants People Planet, 3: 506–522.

Singh, A.K. and Mukhopadhyay, M. 2012. Overview of fungal lipase: A review. Appl Biochem Biotechnol, 166: 486–520.

Sivakumar, D., Gayathri, G., Nishanthi, V. et al. 2014. Role of fungi species in colour removal from textile industry wastewater. Int J ChemTech Res, 6: 4366–4372.

Skaloud, P., Steinova, J., Ridka, T. et al. 2015. Assembling the challenging puzzle of algal biodiversity: species delimitation within the genus *Asterochloris* (Trebouxiophyceae, Chlorophyta). J Phycol, 51: 507–527.

Skinner, D.Z. and Stuteville, D.L. 1995. Host range expansion of the alfalfa rust pathogen. Plant Dis, 79: 456–460.

Skvorova, Z., Cernajova, I., Steinova, J. et al. 2022. Promiscuity in lichens follows clear rules: Partner switching in *Cladonia* is regulated by climatic factors and soil chemistry. Front Microbiol, 12: 781585.

Smith, D.C. 1963. Studies in the physiology of lichens. IV. Carbohydrates in *Peltigera polydactyla* and the utilization of absorbed glucose. New Phytol, 62: 205–216.

Smith, D.C. 1980. Mechanisms of nutrient movement between the lichen symbionts. In: *Cellular Interactions in Symbiosis and Parasitism* (eds). Cook, C.B. et al., Ohio State University Press, USA, pp 197–227.

Solhaug, K.A., Gauslaa, Y., Nyabakken, L. and Bilger, W. 2003. UV-induction of sun-screening pigments in lichens. New Phytol, 158: 91–100.

Sonneborn, T.M. 1939. Paramecium aurelia: Mating types and groups; lethal interactions; determination and inheritance. Am Nat, 73: 390–413.

Sparrow, F.K. 1960. *Aquatic Phycomycetes*. University of Michigan Press, Michigan, 2nd edition, p 1224.

Spribille, T., Resl, P., Stanton, D.E. et al. 2022. Evolutionary biology of lichen symbioses. New Phytol, 234: 1566–1582.

Staben, C. and Yanofsky, C. 1990. *Neurospora crassa* a mating type region. Proc Natl Acad Sci USA, 87: 4917–4921.

Steinberg, G. 2007. Hyphal growth: A tale of motors, lipids, and the Spitzenkorper. Eukaryotic Cell, 6: 351–360.

Steinkraus, D.C., Howard, M.N., Hollingsworth, R.G. and Boys, G.O. 1999. Infection of sentinel cotton aphids (Homoptera: Aphididae) by aerial conidia of *Neozygites fresenii* (Entomophthorales: Neozygitaceae). Biol Cont, 14: 181–185.

Stelzer, C.-P. 2015. Does the avoidance of sexual costs increase fitness in asexual invaders? Proc Natl Acad Sci, USA, 112: 8851–8858.

Stretch, R.C. and Viles, H.A. 2002. The nature and rate of weathering by lichens on lava flows on Lanzarote. Geomorphology, 47: 87–94.

Subramanian, C.V. 1983. *Hyphomycetes, Taxonomic and Biology*. Academic Press, London, p 502.

Svrcek, M. 1954. Revise Velenovske drusni Rodu *Orbilia* (Discomycetes). Revisio critica J velenovskyi specierum generis *Orbilia*. Sbornik Nrodniho Musea v Praze, 10B: 3–23.

Swift, M.J., Heal, O.W. and Anderson, J.M. 1979. *Decomposition in Terrestrial Ecosystems*. University of California Press, USA, p 388.

Taylor, D.L., Herriott, I.C., Stone, K.E. et al. 2010. Structure and resilience of fungal communities in Alaskan boreal forest soils. Can J Res, 40: 1288–1301.

Taylor, J.W., Hann-Soden, C., Branco, S. et al. 2015. Clonal reproduction in fungi. Proc Natl Acad Sci USA, 112: 8901–8909.

Taylor, T.N., Krings, M. and Taylor, E.L. 2015. *Fossil Fungi*. Academic Press, London, p 401.

Tedersoo, L., Hansen, K., Peery, B.A. and Kjoller, R. 2006. Molecular and morphological diversity of pezizalean ectomycorrhiza. New Phytol, 170: 581–596.

Tedersoo, L., May, T.W. and Smith, M.E. 2010. Ectomycorrhizal lifestyle in fungi: Global diversity, distribution, and evolution of phylogenetic lineages. Mycorrhiza, 20: 217–263.

Terminel, E., Ferguson, K. and Tubac, V. 2010. Toxic Species of the Sonoran Deserts: Perception vs. Reality. Publication of University of Arizona, Tucson, p 16.

Tommerup, I.C. 1983. Spore dormancy in vesicular-arbuscular mycorrhizal fungi. Trans Br Mycol Soc, 81: 37–45.

Torres, P. and Honrubia, M. 1994. Basidiospore viability in stored slurries. Mycol Res, 98: 527–530.

Tsai, I.J., Benesasson, D., Burt, a. and Koufopanou, V. 2008. Population genomics of the wild yeast *Saccharomyces paradoxus*: quantifying the life cycle. Proc Natl Acad Sci USA, 105: 4957–4962.

Tucker, K., Stolze, J.L., Kennedy, A.H. and Money, N.P. 2007. Biomechanisms of conidial dispersal in the toxic mole *Stachybotrys chartarum*. Fungal Genet Biol, 44: 641–647.

Umen, J. and Coelho, S. 2019. Algal sex determination and the evolution of anisogamy. Ann Rev Microbiol, 73: 12.1–12.25.

Vacher, C., Vile, D., Helion, E. et al. 2008. Distribution of parasitic fungal species richness: Influence of climate versus host species diversity. Diversity Distrib J, 14: 786–798.

van der Heijden, M.G.A., Martin, F.M., Selosse, M.-A. and Sanders, I.R. 2014. Mycorrhizal ecology and evolution: The past, the present, and the future. New Phytol, 205: 1406–1423.

van Grunsven, R.H.A., Yuwati, T.W., Kowalchuk, G.A. et al. 2014. The northward shifting neophyte *Tragopogon dubius* is just as effective in forming mycorrhizal associations as the native *T. pratensis*. Plant Ecol Divers, 7: 533–539.

Vargas-Gastelum, L. and Riquelme, M. 2020. The mycobiota of the deep sea: What omits can offer. Life, 10: 292, DOI: 10.3390/lif10110292.

Vaughan, M.M., Huffaker, A., Schmelz, E.A. et al. 2014. Effects of elevated CO_2 on maize defense against mycotoxigenic *Fusarium verticillioides*. Plant Cell Environ, 37: 2691–2706.

Vega, F.E., Meyling, N.V., Luangsa-ard, J.J. and Blackwell, M. 2012. Fungal entomopathogens. In: *Insect Pathology* (eds). Vega, F.E. and Kaya, H.K., Elsevier, London, UK, pp 171–220.

Vega, N.W.O. 2007. A review on beneficial effects of rhizosphere bacteria on soil nutrient availability and plant nutrient uptake. Rev Fac Nal Agri Medellin, 60: 3621–3643.

Verma, M., Brar, S.K., Tyagi, R.D. et al. 2007. Antagonistic fungi, *Trichoderma* spp.: panoply of biological control. Biochem Engi J, 37: 1–20.

Veron, G., Patterson, B.D. and Reeves, R. 2008. Global diversity of mammals (Mammalia) in freshwater. Hydrobiologia, 595: 607–617.

Videira, S.I.R., Groenewald, J.Z., Braun, U. et al. 2016. All that glitters is not *Ramularia*. Stud Mycol, 83: 49–163.

Vidhyasekaran, P. 1997. *Fungal Pathogenesis in Plants and Crops: Molecular Biology and Host Defense Mechanism*. CRC Press, Boca Raton, USA, p 568.

Villarreal, J.C. and Renner, S.S. 2013. Correlates of monoicy and dioicy in hornworts, the apparent sister group to vascular plants. BMC Evol Biol, 13: 239.

Vinck, A., Terlou, M., Pestman, W.R. et al. 2005. Hyphal differentiation in the exploring mycelium of *Aspergillus niger*. Mol Microbiol, 58: 693–699.

Vishwanathan, K.S., Rao, A.G.A. and Singh, S.A. 2009. Characterization of acid protease expressed from *Aspergillus oryzae* MTCC 5341. Food Chem, 114: 402–407.

Vivekanandan, E. 2011. Climate change and Indian Marine Fisheries. Central Marine Fisheries Research Institute, Kochi, Special Publication, 105: 1–97.

Vivekanandan, E. and Jeyabaskaran, R. 2012. *Marine Mammal Species of India*. Central Marine Fisheries Research Institute, Kochi, p 228.

Voigt, K., James, T.Y., Kirk, P.M. et al. 2021. Early-diverging fungal phyla: Taxonomy, special concept, ecology, distribution, anthropogenic impact, and novel phylogenetic proposals. Fungal Div, https://doi.org/10.1007/s13225-021-00480-y.

Walas, L., Thomas, P. and Iszkulo, G. 2018. Sexual systems in gymnosperms: A review. Basic Appl Ecol, https://doi.org/10.1016/j.baae.2018.05.009.

Wallace, R.A. 1991. *Biology: The World of Life*. Harper Collin Publishers, USA, p 695.

Wallace, R.L. and Snell, T.W. 2010. Rotifera. In: *Ecology and Classification of North American Freshwater Invertebrates* (eds). Thorp, J.H. and Covich, A.P., Elsevier, pp 173–235.

Wallen, R.M. and Perlin, M.H. 2018. An overview of the function and maintenance of sexual reproduction in dikaryotic fungi. Front Microbiol, 9: 503, DOI: 10.3389/fmicb.2018.00503.

Wang, B. and Qui, Y.-L. 2006. Phylogenetic distribution and evolution of mycorrhizas in land plants. Mycorrhiza, 16: 299–363.

Wang, H.Y., Guo, S.Y., Huang, M.R. et al. 2010. Ascomycota has a faster evolutionary rate and higher species diversity than Basidiomycota. Sci China Life Sci, 53: 1163–1169.

Ward, D. 2009. *Biology of Deserts*. Oxford University Press, UK, p 352.

Watkinson, S.C., Boddy, L. and Money, N.P. 2015. *The Fungi*. Academic Press, London, p 466.

Webster, J. 1959. Experiments with spores of aquatic hyphomycetes. I. Sedimentation and impaction on smooth surfaces. Ann Bot, 23: 595–611.

Webster, J. and Weber, R. 2007. *Introduction to Fungi*. Cambridge University Press, UK, p 867.

Wellings, C.R., McIntosh, R.A. and Walker, J. 1987. *Puccinia striiformis* f. sp. *tritici* in Eastern Australis – possible means of entry and implications for plant quarantine. Plant Pathol, 36: 239–241.

Whisler, H.C., Zebold, S.L. and Shemanchuk, J.A. 1975. Life history of *Coelomomyces psorophorae*. Proc Natl Acad Sci USA, 72: 693–696.

Willig, M.R. and Presley, S.J. 2013. Latitudinal gradients of biodiversity. In: *Encyclopedia of Biodiversity* (ed). Levin, S.A., Academic Press, 612–627.

Witman, J.D., Etter, R.J. and Smith, F. 2004. The relationship between regional and local species diversity in marine benthic communities: A global perspective. Proc Natl Acad Sci USA, 101: 15664–15669.

Wolters, V. 2000. Invertebrate control of soil organic matter stability. Biol Fertility Soils, 31: 1–19.

Womiloju, T.O., Miller, J.D., Mayer, P.M. and Brook, J.R. 2003. Methods to determine the biological composition of particulate matter collected from outdoor air. Atmos Environ, 37: 4335–4344.

Wright, S.F., Franke-Snyder, M., Morton, J.B. and Upadhyaya, A. 1996. Time-course study and partial characterization of a protein on hyphae of arbuscular mycorrhizal fungi during active colonization of roots. Plant Soil, 181: 193–203.

Wurzbacher, C., Kerr, J.L. and Grossart, H.-P. 2011. Aquatic fungi. In: *The Dynamical Processes of Biodiversity – Case studies of Evolution and Spatial Distribution* (eds). Grillo, O. and Venora, G., InTech Publishers, UK, pp 227–258.

Zachariah, S.A. and Varghese, S.K. 2018. The lichen symbiosis: A review. Int J Sci Res Rev, 7: 1160.

Zakhartsev, M. and Reuss, M. 2018. Cell size and morphological properties of yeast *Saccharomyces cerevisiae* in relation to growth temperature. FEMS, 18, DOI: 10.1093/femsyr/foy052.

Zhang, N. and Wang, Z. 2015. Pezizomycotina: Sordariomycetes and Leotiomycetes. In: *Systematics and Evolution. The Mycota (A Comprehensive Treatise on Fungi as Experimental Systems for Basic and Applied Research)* (eds). McLaughlin, D. and Spatafora, J., Springer, Berlin, pp 58–88.

Zhang, N., Luo, J. and Bhattacharya, D. 2017. Advances in fungal phylogenomics and their impact on fungal systematics. Adv Genet, 100: 309–328.

Zhang, T.-T., Qiu, Z.-X., Wang, W.-Y. et al. 2019. The mRNA expression and enzymatic activity of three enzymes during embryonic development of the hard tick *Haemaphysalis longicornis*. Parasites Vectors, 12: 96, DOI: 10.1186/s13071-019-3360-8.

Zuther, K., Kahnt, J., Utermark, J. et al. 2012. Host specificity of *Sporisorium reilianum* is tightly linked to generation of the phytoalexin luteolinidin by *Sorghum bicolor*. Mol Plant Microbe Interact, 25: 1230–1237.

Author Index

Species Index

Subject Index

A

Anaerobic, 4, 19-20, 41, 173, 257
Anamorph, 30, 35, 112
Anisogamy, 117, 268
Ascogonium, 9, 32, 124, 126, 142, 166
Autoecy, 36

B

Ballistospore, 24, 66, 134
Basidium, 7-9, 22, 27, 35, 44-45, 47-48, 109,
    122-123, 126, 131, 135
Biodiversity, 2, 4, 6, 8, 10, 12, 14, 16, 18, 20, 22,
    24, 26, 28, 30, 32, 34, 36, 38, 40, 42, 44, 46,
    48, 50, 52, 54, 56, 58, 60, 62, 64, 66, 68, 70,
    72, 74, 76, 78, 80, 82, 84, 86, 88, 90, 92, 94,
    96, 98, 100, 102, 104, 106, 108, 110, 112,
    114, 116, 118, 120, 122, 124, 126, 128, 130,
    132, 134, 136, 138, 140, 142, 144, 146, 148,
    150, 152, 154, 156, 158, 160, 162, 164, 166,
    168, 170, 172, 174, 176, 178, 180, 182, 184,
    186, 188, 190, 192, 194, 196, 198, 200, 202,
    204, 206-208, 210, 212, 214, 216, 218, 220,
    222, 224, 226, 228, 230, 232, 234, 236, 238,
    240, 242, 244, 246, 248
Biotrophs, 178, 196
Bipolar, 109, 115, 142, 144, 149, 152-153

C

Chytridiomycosis, 19
Clamp connection,
Cleistothecium, 51
Coenocyte, 5, 7, 162
Conidium, 7-9, 27, 46, 53, 88-89, 128-129,
    136-137
Conjugation, 19-20, 28, 37, 39, 41, 43, 45, 47,
    110, 114-115, 118-119, 125, 142, 232

G

Gloiospores, 66, 132, 134, 185

H

Haplophasic, 16, 18, 37
Haustorium, 7, 167
Heteroecy, 36
Heterogony, 35
Hydrothermal vents, 64
Hymenium, 8-9, 28

I

Idiomorph, 143, 146-147, 150
Immune system, 75, 78, 83, 150
Ingold, 132, 135-138
Isogamy,

M

Monopedal,
Mycotoxin, 198

O

Oogamy, 110, 117, 127, 241

P

Perithecium, 9, 30, 53, 55, 163
Pheromone, 108, 144-145, 148-151
Poikilohydry, 152, 182
Poikilothermics, 192, 198, 209-210
Promycelium, 24, 42, 148-149, 151

R

Rhizines, 163, 176, 215, 218
Rhizoids, 3, 18-20, 41, 176, 214, 218

S

Saprotrophs, 20, 23, 25, 29, 178
Sclerotium, 7, 9-10, 35, 37, 51, 116, 131
Somatogamy, 117
Spermatization, 112, 118-119, 130-131, 238-239

Spitzenkorper, 5-6
Sporocarp, 7-8, 21, 35

T

Teleomorph, 25, 35
Tetrapolar, 109-110, 115, 142, 144, 149, 232

X

Xenospores, 66

Z

Zoosporangium, 7-8, 10, 18
Zygospore, 19, 4

Epilogue

It has been an enjoyable learning experience for me to author 15 books in the Series 1. Sexuality, Sex Determination and Differentiation in Fish (5 volumes), 2. Reproduction and Development in Aquatic Invertebrates (6 volumes) and 3. Evolution and Speciation in Eukaryotes (4 volumes). The first series was initiated by Prof. T. Balasubramanian and the second by Prof. R. Gadagkar, to whom I remain grateful. During the initial 5 years, token grants were provided to me by Annamalai University, Chidhambaram, Department of Science and Technology, Government of India, New Delhi, Central Marine Fisheries Research Institute (CMFRI), Kochi, especially by its Director Dr. A. Gopalakrishnan, and Indian National Science Academy, New Delhi – to all of them, my thanks are due. Like a rock, Dr. E. Vivekanandan (Chennai) stood by me to carefully and critically review the manuscripts of all the 15 volumes; thanks are also due to many reviewers, especially Dr. P. Murugesan (Parangipettai). I wish to record my sincere thanks to CRC Press (Boca Raton), especially Shri Raju Primlani (New Delhi), for continuous support. Sheerly by physical weakness at 83+, I regret to announce that Evolution and Speciation in Fungi and Eukaryotic Biodiversity shall be my last contribution to Biology. I earnestly hope that CRC Press shall soon find an expert to author Evolution and Speciation in Prokaryotes.

Thankfully,

T.J. Pandian

September, 2022
Madurai – 625 014

Author's Biography

Recipient of the S.S. Bhatnagar Prize, the highest Indian award for scientists, one of the 10 National Professorships, T.J. Pandian has served as editor/member of editorial boards of many international journals. His books on Animal Energetics (Academic Press) identify him as a prolific but precise writer. His five volumes on Sexuality, Sex Determination and Differentiation in Fish, published by CRC Press, are ranked with five stars. He has authored a multi-volume series on Reproduction and Development of Aquatic Invertebrates, of which the volumes on Crustacea, Mollusca, Echinodermata and Prochordata, Annelida, Platyhelminthes and Minor Phyla have been published. The CRC Press has recently published his new book series on Evolution and Speciation in Animals, Evolution and Speciation in Plants and Evolution and Speciation in Protozoa. The fourth volume on Evolution and Speciation in Fungi and Eukaryotic Biodiversity is in your hands.

For Product Safety Concerns and Information please contact our EU
representative GPSR@taylorandfrancis.com
Taylor & Francis Verlag GmbH, Kaufingerstraße 24, 80331 München, Germany